Wolfgang Linert
Michel Verdaguer (eds.)

Molecular Magnets
Recent Highlights

SpringerWienNewYork

Dr. Wolfgang Linert
Institute of Applied Synthetic Chemistry, Vienna University of Technology
Vienna, Austria

Dr. Michel Verdaguer
"Chimie Inorganique et Matériaux Moléculaires" Laboratory, Pierre et Marie Curie University
Paris, France
Chairman of the "Molecular Magnets" Programme
of the European Science Foundation

Printed in Austria

Cover illustration:
V. Marvaud et al, "High Spin ...", Fig. 2, page 18, this volume
Typesetting: Thomson Press Ltd., Chennai, India
Printing: Manz Crossmedia, A-1051 Wien
Binding: Papyrus, A-1100 Wien
Printed on acid-free and chlorine-free bleached paper

SPIN: 10896440

With 106 Figures and Tables

CIP data applied for

Special Edition of
Monatshefte für Chemie/Chemical Monthly Vol. 134, No.2, 2003

ISBN 3-211-83891-0 Springer-Verlag Wien New York

Editorial: Molecular Magnets

Between magnetochemistry which was measuring magnetism of chemicals with the hope to discover their structures, to molecular magnetism which can be defined as a discipline which conceives, realizes, studies, and uses new molecular materials bearing new but predictable magnetic (and other) physical property, a few decades took place.

Instead of going to the laboratory shelves to study the magnetic properties of already existing chemicals, chemists are now designing new (supra)molecular systems to get low density, transparent, biocompatible magnets, bistability, single molecule magnet behaviour, materials whose properties can be changed by temperature, pressure or light, or still molecular objects combining several physical functions (magnetism and conductivity, magnetism and optics...).

This new field is blossoming in many countries, in North America, in Japan, and in Europe. Different European networks (including a Molecular Magnets programme of the European Science Foundation) were recently very active and produced major results. It is our pleasure to welcome in this volume some of the actors of this scientific saga, which is on the way to transform molecules in useful devices.

Our volume opens by a brief and lively historical overview of the domain by one of the leaders in the field, Prof. *Dante Gatteschi*, a chemist who received in 2002 with four other colleagues, chemists and physicists, the Agilent Technologies Prize of the European Physical Society: this fact alone demonstrates how the field is indeed multidisciplinary. The content of the volume is another illustration of the diversity of the scientists engaged hand in hand in molecular magnetism: quantum chemists and physicists, computing more and more accurately the properties of molecules and solids and opening astonishing prospects in the field of electronic quantum computing; chemists and physicists using large instruments (neutrons and synchrotron radiation sources) to extract unique information unavailable by other techniques and then comparing to theoretical models. Some of the synthetic chemists present here, tailor spin cross-over systems which present a fascinating bistability between two states with different magnetic properties and colours, opening the way to the use in display devices. Others assemble molecules bearing very large spins, which do not exist "naturally" in the elements of the periodic table; anisotropy will confer to these systems the ability to store magnetic information at the molecular level, to reach the highest, ultimate storage density of information; some transform the beautiful C60 fullerene molecule in a molecule-based magnet; some others combine cleverly the power and the flexibilities of organic and inorganic molecular chemistry to get multifunctional materials, magnetic and conducting, magnetic and optically active.... All of them provide the physicists community with wonderful new objects to study and the engineers with new functions to be used tomorrow in useful devices.

The field is not only multidisciplinary but also more and more international: many of the contributions imply various laboratories belonging to different countries, each of them having a unique expertise which is shared when necessary. Active international cooperation is indeed at work here.

We hope that the reader will enjoy the volume as much as we enjoyed ourselves and that the book will help in developing more rapidly the exchange of ideas and experience in this exciting new scientific area, molecular magnetism.

Wolfgang Linert
Michel Verdaguer
Issue Editors

Foreword: Molecular Magnetism, an Interdisciplinary Field

Molecular magnetism is a relatively recent scientific field which originated from the transformation of magnetochemistry in an interdisciplinary area where chemists and physicists started to collaborate very closely with the stated goal of designing, synthesizing, and characterizing the magnetic properties of molecular based materials. The idea of using molecules, rather than the ionic and metallic lattices of typical magnets, stems from the rapid development of functional molecular materials which started in the second half of the last century. The transition from magnetochemistry to molecular magnetism required some time, because in order to develop efficiently a deeper understanding of the magnetic phenomena was needed than that available at that time in chemistry. The interaction with some physicists who were curious to see the developments which might be associated with the exotic objects provided by molecules was the event which officially determined the birth of the area. A testimony of this is given by the book "Structural-Magnetic Correlations in Exchange Coupled Systems" edited in 1983 by *Roger Willett*, the late *Olivier Kahn*, and myself. The title itself gives an indication of the efforts the chemists were making at that time to understand how structural differences gave rise to different magnetic properties. The book reports the proceedings of a NATO ASI which was extremely stimulating. The contributions from chemists were based on simple dinuclear or oligonuclear species, while those of physicists started from infinite arrays, with a bias towards one dimensional materials which were extremely popular at that time. That ASI was the start of attempts of the two communities of using similar languages and it allowed chemists to dare to tackle more complex materials, and convinced the physicists that it was indeed possible to observe new types of magnetic phenomena using molecular based materials.

The development of molecular magnetism has been rapid. I would like to quote some of the milestones in the development of the area. *Kahn* and *Verdaguer* reported the first example of molecular ferrimagnet based on copper(II)-manganese(II) derivatives, while *Miller* and *Epstein* reported the first example of molecular ferromagnet containing organic building blocks like *TCNE*. Perhaps the most important development was the discovery of *Kinoshita* that purely organic matter can indeed order ferromagnetically after the elegant work of *Itoh* and *Iwamura* who showed how extremely strong ferromagnetic coupling can be observed in polycarbenes. The subsequent discoveries of higher critical temperatures in the work of *Rassat*, *Wudl*, *Mihailovic*, *Rawson*, and *Palacio* confirmed the vitality of this area, the last ones describing organic weak ferromagnets ordering at temperatures as high as 35 K. The rush to room temperature was won by *Miller* and *Epstein* again who showed that V($TCNE$)$_2$ is a disordered ferrimagnet above room temperature. *Verdaguer* opened the saga of Prussian blue derivatives, finally reporting a ferrimagnet comprising chromium(III), vanadium(II), and vanadium(III) which orders above room temperature.

Nowadays the attempts to prepare molecular magnets which can be used at room temperature are continuing but at the same time there are many attempts to synthesise molecular magnets which have properties which are difficult to be met in inorganic magnets. *Day* reported an organic superconductor coexisting in the same lattice of a molecular paramagnet, while more recently *Coronado* reported a molecular ferromagnet hosting in the lattice an organic conductor. This is an example of ferromagnetic conductor in which the active electrons, namely conducting and magnetic, are separated. Many efforts are currently done to investigate chiral magnets and to use light to influence the magnetic properties of the materials.

One of the fields where the interdisciplinary approach has produced much success is that of zero-dimensional magnets. By this I mean systems which comprise a finite number of interacting magnetic centres. Chemists are learning how to make either in a rational or in a serendipitous way large clusters with different spin topologies. The fortunate case has been that these systems have proved to be extremely timely, because physicists were looking for small magnetic particles, all identical to each other, in order to test theories that suggested that it was possible to observe quantum effects in magnets. The main difficulty was just that of obtaining assemblies of identical particles, because the quantum effects are expected to scale exponentially with the size of the particles. Further it is mandatory to be able to minimise the magnetic interactions between particles. Several different approaches were used in order to reach this goal, including attempts to use iron loaded ferritin, but no unambiguous evidence was reached. Finally, as *P.C.E. Stamp* wrote in a "News and Views" in Science magazine "Then the chemists came ....". In fact it was discovered that in a cluster comprising twelve manganese ions, characterised by a ground spin $S = 10$ state possessing a huge *Ising* type magnetic anisotropy, the magnetisation relaxes very slowly at low temperature, following a thermally activated behaviour with a barrier of *ca.* 65 K and a pre-exponential factor of 2×10^{-7} s. This means that at 2 K the relaxation time of the magnetisation becomes of the order of months and magnetic hysteresis of molecular origin is observed. This discovery showed that indeed single molecules may behave as tiny magnets and they were called single molecule magnets, SMM. Soon it was realised that these molecules were providing the long looked for magnetic particles which could show quantum effects. In fact molecules are all identical to each other, allowing to measure a large ensemble and collecting the individual response. In 1996 two teams (*Friedman*, *Sarachik*, *Tejada*, and *Ziolo* on one side, *Thomas*, *Lionti*, *Barbara*, *Sessoli*, and *Gatteschi* on the other side), almost at the same time, discovered that the hysteresis curves of oriented polycrystalline powders or single crystals show a stepped behaviour which is the signature of quantum tunnelling effects. Finally, it was possible to test the theories of quantum relaxation in mesoscopic magnets with the experimental properties of real objects! Other clear evidence of quantum size in magnets came later with the observation of temperature independent relaxation time of the magnetisation in a cluster comprising eight iron(III) ions, Fe_8. *Wernsdorfer* and *Sessoli* later showed that quantum interference effects were observed in the same compound in the presence of an applied field along the hard axis. Again this was the evidence that the so-called *Berry* phase could be observed in magnets, as long theoretically predicted.

The importance of this research has been publicly recognised by the European Physical Society by awarding the Agilent Technology Europhysics Prize 2002 to *Bernard Barbara*, *Jonathan Friedman*, *Dante Gatteschi*, *Roberta Sessoli*, and *Wolfgang Wernsdorfer* "for developing the field of quantum dynamics of nanomagnets including the discovery of quantum tunnelling and interference in dynamics of magnetisation". I find extremely significant that the Prize has been awarded to three physicists and two chemists, recognizing the interdisciplinary nature of the field. I think that this is the official acknowledgement of the coming of age of molecular magnetism, showing that indeed the effort of using exotic materials, exploiting large organic moieties, is worth to be done, because it affords new physics. Indeed, the dreams of the pioneers of the field, among whom I always like to indicate the forward dreamer *Olivier Kahn*, have become true.

Dante Gatteschi
Firenze
October 2002

Contents

Invited Reviews

Contributions

Invited Review

Polyfunctional Two- (2D) and Three- (3D) Dimensional Oxalate Bridged Bimetallic Magnets

René Clément[1], **Silvio Decurtins**[2], **Michel Gruselle**[3], and **Cyrille Train**[3,*]

[1] Laboratoire de Chimie Inorganique, UMR-CNRS, 8613, Université Paris Sud, F-91405 Orsay Cedex, France
[2] Departement für Chemie und Biochemie, Universität Bern, CH-3012 Bern, Switzerland
[3] Laboratoire de Chimie Inorganique et Matériaux Moléculaires, CIM[2] UMR-CNRS, 7071, Université Pierre et Marie Curie, F-75252 Paris Cedex 05, France

Received April 11, 2002; accepted May 27, 2002
Published online January 8, 2003

Summary. We report major results concerning polyfunctional two- (2D) and three- (3D) dimensional oxalate bridged bimetallic magnets. As a consequence of their specific organisation they are composed of an anionic sub-lattice and a cationic counter-part. These bimetallic polymers can accommodate various counter-cations possessing specific physical properties in addition to the magnetic ones resulting from the interactions between the metallic ions in the anionic sub-lattice. Thus, molecular magnets possessing paramagnetic, conductive and optical properties are presented in this review.

Keywords. Oxalate; Molecular magnets; NLO; Photochromism; Cotton effect; Fluorescence.

1. Introduction

1.1 Why polyfunctional materials?

The need for new materials that have more diversified and more sophisticated properties is continuously increasing [1]. Opportunities offered by the flexibility of inorganic chemistry led to a blossoming of new research fields in inorganic molecular materials [2]. One of the goals is to obtain materials that possess not only one expected property (mechanical, optical, magnetic, electric...) but also

* Corresponding author. E-mail: train@ccr.jussieu.fr

combine two or more of them in polyfunctional systems. For example, in the field of molecular magnetism [3], superconducting paramagnets [4], photomagnetic ferrimagnets [5] and room temperature magnets [6] were reported.

1.2 Polyfunctional molecular magnets

Polyfunctional molecule based magnets represent an important class of polyfunctional materials. Their magnetic properties (ferro-, ferri-, anti-ferro) depend on the nature of the interacting metal ions [3a]. These magnetic properties can be modulated by the nature of the bridging ligand and also by the whole structure and environment created by the supramolecular arrangement of the building blocks. Additional physical properties can be introduced by different ways: (i) the ligand can be the centre of this phenomenon when bearing a specific property, for example optic in the case of an optically active ligand, magnetic when the ligand is a free radical; (ii) one of the building blocks can possess a specific property like chirality or fluorescence activity; (iii) when the materials comprise two sub-lattices (hybrid materials), one of them can bring out the magnetic properties while the other one brings a different property.

1.3 Oxalate based magnets

In literature, many oxalate based magnets are described. They can occur as polymetallic discrete entities – named 0D – in which transition metal ions interact through the oxalate ligand [7]. These systems are of importance to understand the nature of the orbital exchange at the molecular level. Chains (1D) [8], layered (2D) and three-dimensional structures (3D) [9] are also known. In these structures the oxalate ligand bridging the metal ions can be replace by a dithiooxalate ligand [10], a change that leads to modifications in both the exchange interaction J values (magnetic effect) and the size of the basic structure of the polymer (structural effect). In other cases, 2,2′-bipyridine [11], 4,4′-bipyridine [12], or 2,2′-bipyrimidine [13] were used together with oxalate ligand to synthesise compounds with mixed bridging ligands.

In the scope of this review, we restrict ourselves to the major works published on polyfunctional tris(oxalato)metalate based magnets.

2. Tris(oxalato)metalate Based Magnets

2.1 2D and 3D structures

In the last few years *Okawa* [14], *Decurtins* [15], *Day* [16], *Coronado* [17], and *Ovanesyan* [18] have thoroughly investigated two- and three-dimensional networks (noted 2D and 3D) prepared from the combination of tris(oxalato)metalates, $[M^{III}(ox)_3]^{3-}$ (M^{III} = Cr, Co, Fe, Ru), with other metallic precursors such as alkali cations (Li^+, Na^+) or di-cationic transition metal ions (Mn^{2+}, Ni^{2+}, Fe^{2+} ...), according to the reaction:

$$n[M_1{}^{III}(ox)_3]^{3-} + nM_2 + nA^{x+} = \{A^{x+}, [M_1M_2(ox)_3]^{x-}\}_n$$

These bimetallic networks of general formula $\{A^{x+}, [M_1M_2(ox)_3]^{x-}\}_n$ (noted $[M_1M_2]$) are composed of an anionic sub-lattice $\{[M_1M_2(ox)_3]^{x-}\}_n$ and a cationic counter-part $[A^{x+}]_n$. In these compounds, the charge of each sub-unit of the anionic sub-lattice is one or two according to the oxidation state of each metal centre. For instance, it is (-1) for $[M_1{}^{II}M_2{}^{III}(ox)_3]^{1-}$ and (-2) for $[M_1{}^{I}M_2{}^{III}(ox)_3]^{2-}$.

When two transition metal ions are connected through the oxalate ligands, exchange interaction occurs leading to a magnetic order below the *Curie* temperature T_c. The ferro-, (canted) antiferro- or ferri-magnetic properties depend on the nature of the connected metal ions.

Additionally, such tris(bidentate) complexes display a propeller-like chirality [19]. Therefore each chiral element exists as Δ or Λ enantiomeric forms [20] (noted ΔM or ΛM), as shown in Fig. 1. Furthermore, the relative configuration of the connected hexacoordinated centres (Fig. 2) determines the 2D or 3D architecture of the polymer. A hetero-chiral arrangement $[\Delta M_1{-}\Lambda M_2]$ or $[\Lambda M_1{-}\Delta M_2]$ leads to a 2D network (Fig. 2a). In this situation the anionic sub-lattice displays a honeycomb structure while the cationic moiety, which is in general a tetra-alkyl ammonium ($NR_4{}^+$) [21, 22] or phosphonium ($PR_4{}^+$) [15, 23, 24] ion, is located between the anionic layers. On the other hand, a homo-chiral arrangement $[\Delta M_1{-}\Delta M_2]$ or $[\Lambda M_1{-}\Lambda M_2]$ leads to a helical organisation of the connected metallic ions (Fig. 2b). Therefore a three dimensional structure is obtained giving a 3-connected decagon anionic network with the associated cationic counter-part fitting in the vacancies (Fig. 3) [25, 26].

To built such 2D and 3D networks in an optically active form, two geometrical elements must be controlled:

- The absolute Δ or Λ configuration of each hexacoordinated metal centre.
- The relative configuration of the adjacent metal centres.

The nature of the template cation appears as a determining factor to orient the reaction towards 2D or 3D structures. Three elements have to be taken in consideration: the charge, the size and the symmetry of the cation.

The charge of the template cation has to be equal to the charge of the anionic sub-unit $[M_1M_2(ox)_3]^{x-}$ $(x=1$ or $2)$. Therefore the template cations are

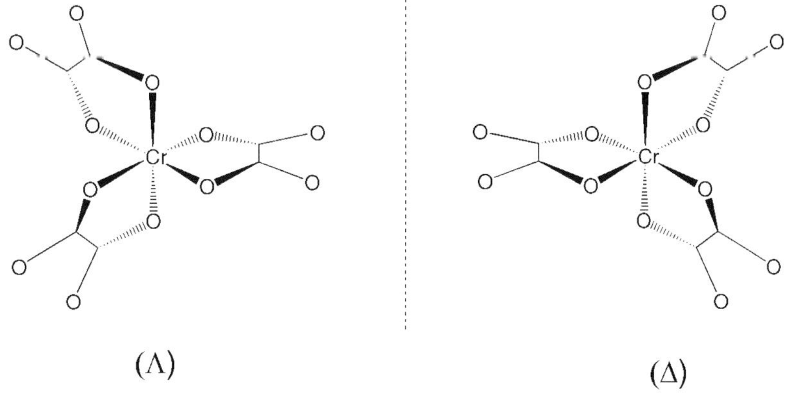

Fig. 1. Δ and Λ enantiomeric forms of $[Cr(ox)_3]^{3-}$

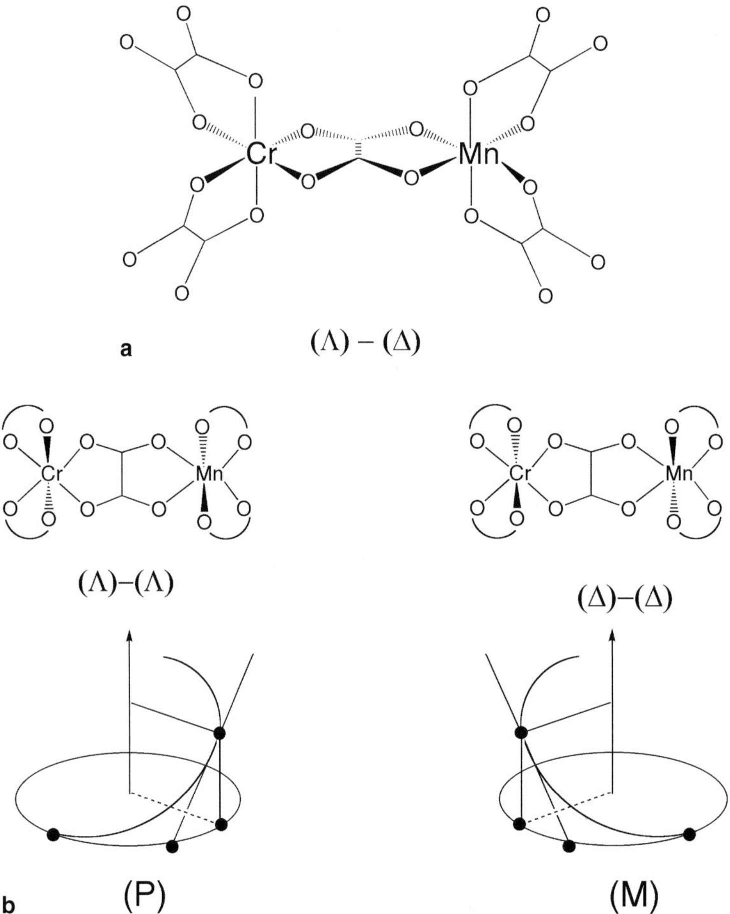

Fig. 2. Relative hetero- (a) or homo-chiral (b) arrangement of oxalate bridged centres

mono- or di-cations (tetra-alkyl Ammonium or Phosphonium, $[Ni(phen)_3]^{2+}$, $[Ru(bpy)_3]^{2+}$,... with *phen* = phenantroline, *bpy* = bipyridine). When the charge of the anionic sub-unit is (– 1), mono-cations belonging to hexacoordinated metal complexes were also used, namely: $[Ru(bpy)_2(ppy)]^+$, $[Ru(bpy)_2(quo)]^+$ with *ppy* = (2-phenylpyridine – H^+), *quo* = 8-Hydroxyquinoleate [27]. In certain cases, a di-cation in tandem with a mono-anion can be used in place of a mono-cation: for example $[Ru(bpy)_3ClO_4]^+$ [17, 27].

The size of the template cation is of primary importance, allowing or not the formation of the anionic framework. In 2D networks, along the series of highly symmetrical tetra-alkyl ammonium salts it was shown that the anionic auto-assembling process is possible from *R* = *n*-propyl to *n*-pentyl. In the case of the $[Fe^{II}Fe^{III}]$ or $[Mn^{II}Fe^{III}]$ compounds, the inter-layer distances were systematically investigated by *Day* [24b], varying from 8.2 to 10.23 Å and 8.18 to 10.15 Å, respectively. However, in the case of cations having a lower symmetry, it was found that cations of various sizes could intercalate into the anionic framework. This is the case for $[(C_6H_5)_3PNP(C_6H_5)_3]^+$ (14.43 Å) [24a] and for a series of ferrocenic ammonium salts $[C_5H_5FeC_5H_4CH_2NR_3]^+$ with *R* = Et, Pr, But (10.0 Å) [28]. Symmetrical

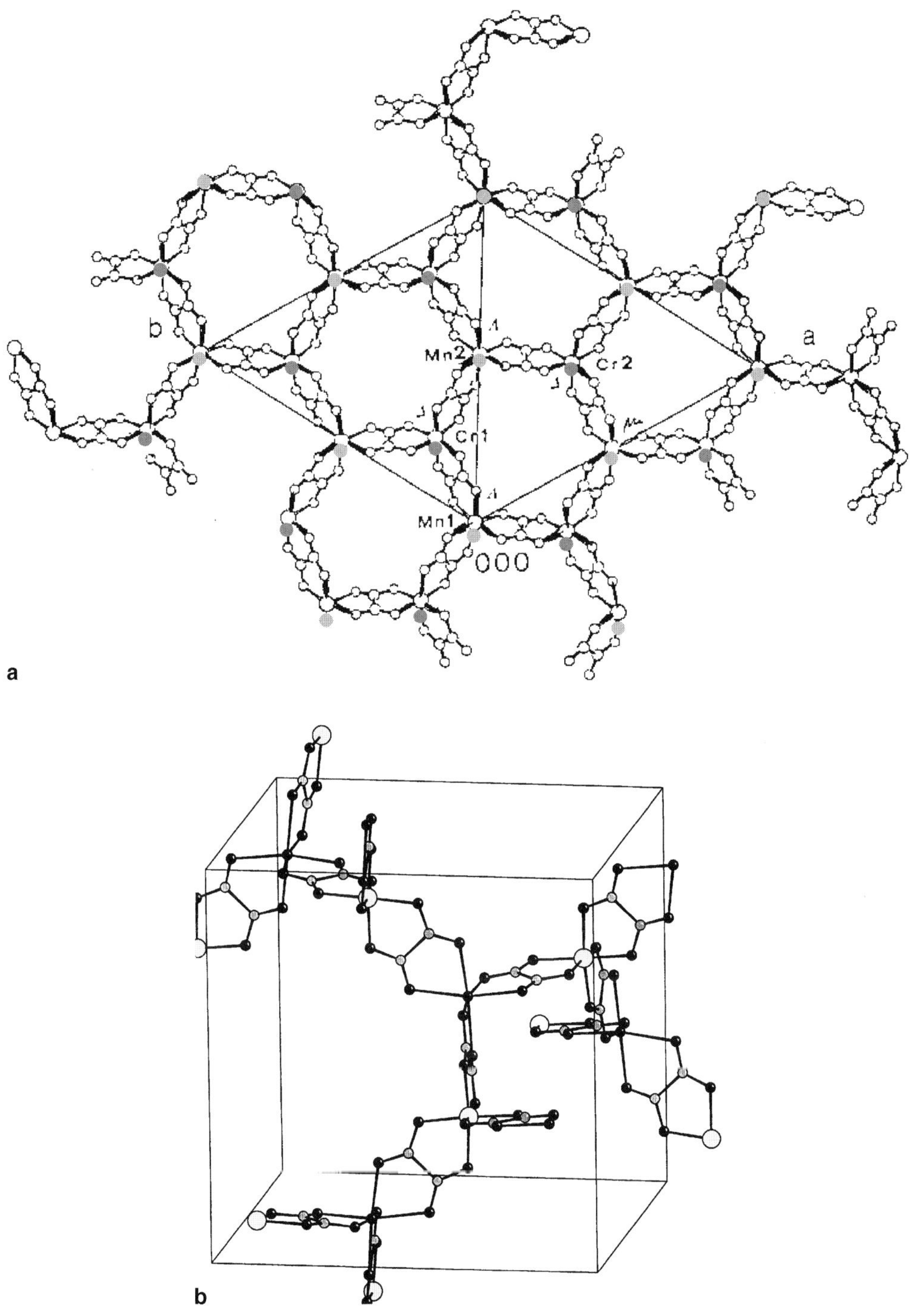

Fig. 3. (a) 2D honeycomb structure (adapted from Ref. [25b]); (b) 3D 3-connected 10-gon anionic network (adapted from Ref. [37])

phosphonium or arsenium cations [24a], are also able to template the formation of 2D anionic framework. Other cations were used as template namely stilbazolium [29], ferricinium and cobalticinium [30]. In contrast, for 3D networks the cavities in the anionic network are such that the number of cations able to play a template role is limited: $[Ni(phen)_3]^{2+}$, $[Ru(bpy)_3]^{2+}$ and $[Ru(bpy)_2(ppy)]^{+}$.

The symmetry of the template cation plays an important role to orient the reaction towards 2D or 3D networks. Chiral or achiral tetra-alkyl ammonium, phosphonium, stilbazolium, ferricinium as well as ferrocenic ammonium cations lead to 2D structures. In contrast, chiral hexacoordinated metal complexes belonging to D_3 or quasi-D_3 symmetry lead to 3D networks.

2.2 Magnetic properties

Most of the magnetic properties of tris(oxalato)metalate based networks have been established on two dimensional compounds starting either from $[Cr(ox)_3]^{3-}$ [14], $[Fe(ox)_3]^{3-}$ [24] or $[Ru(ox)_3]^{3-}$ [22].

In the former case, the exchange coupling in $\{A[M^{II}Cr^{III}(ox)_3]\}_n$ is ferromagnetic (F) leading to ferromagnets with *Curie* temperatures (T_c) ranging from 6 ($M^{II}=Mn$) to 18 K ($M^{II}=Ni$). Starting from iron (III), the exchange coupling is antiferromagnetic (AF) leading to canted antiferromagnets ($M^{II}=Mn-T_c=27$ K) or ferrimagnets ($M^{II}=Fe-T_c=33$–48 K). In some $[Fe^{II}Fe^{III}]$ compounds, *Day et al.* have observed negative magnetisation below the so-called compensation temperature T_{comp} over a wide temperature range [24]. With the ruthenium (III) precursor [22], the exchange coupling in the network is AF for $M^{II}=Fe$, Cu and F for $M^{II}=Mn$. An ordered phase is observed only for $M^{II}=Fe$ ($T_c=13$ K).

The sign and the magnitude of the exchange coupling J in these networks have been compared with those measured on 0D tetra-metallic complexes [7b, 31] and/or deduced from the orbital approach developed by *Kahn* [3a]. The agreement is rather good except from ruthenium (III) based magnets. This disagreement is attributed to the slight increase in the spin-orbit coupling when going from iron (III) to ruthenium (III) [22].

The existence of a long-range magnetic order in 2D compounds raises the question of the nature of the exchange coupling in these compounds. Do we have an Ising, a *XY* or a Heisenberg system, e.g. does the exchange coupling present some kind of anisotropy? The neutron measurement performed on $[Mn^{II}Cr^{III}]$, using NBu_4^{+} as counter-cation, concludes that this system is *Ising*-type with an easy-axis perpendicular to the (a, b) planes [32]. *Mössbauer* studies on $[M^{II}Fe^{III}]$ with $M^{II}=Mn$, Fe assesses that most of these networks are *XY*-planar magnets [18]. In particular, for $\{(NBu_4)[Mn^{II}Fe^{III}]\}$ with 3-fold site symmetry, the unusual magnetic relaxation behavior well below T_c has been attributed to a low in-plane anisotropy constant [18b]. The *Mössbauer* studies lead to the same conclusion for $[Fe^{II}Cr^{III}]$ polymers [23] while the internal magnetic field changes gradually from perpendicular to parallel (to the planes) when x decreases in $\{(NBu_4)[Fe^{II}{}_x Mn^{II}{}_{1-x}Cr^{III}(ox)_3]\}_n$ [33].

Exploiting the versatility of the 2D networks towards the choice of the counter cation A^{+}, a great variety of cations has been introduced in these compounds leading to important variations of the interplanar distances [21, 23–24]. The value of T_c is rather sensitive to such substitution. Nevertheless, no clear correlation appears between the variations of T_c and that of the interplanar distance. A possible interpretation to the sensitiveness of T_c upon cation substitution is the *Ising vs. XY* nature of the exchange coupling. In the former case, there is a low influence of substitution upon T, as observed on the [MnCr] system, because 2D *Ising* systems are able to present a long range magnetic order. On the contrary, in the latter case, additional interaction is needed in order to observe long range magnetic order instead of *Kosterlitz-Thouless* phases [34]. Structural "details" like the interplanar distance or the relative positions of metal ions of two adjacent layers then play a crucial role on the critical temperature as observed in the case of the $[Fe^{II}Fe^{III}]$ system [24].

The antagonist nature of the J value has been exploited by *Bhattacharjee et al.* [35] and *Coronado et al.* [30]. They have studied the competition between ferro- and ferri-magnetic ordering in $\{(NBu_4)[Fe^{II}(Fe^{III}{}_xCr^{III}{}_{1-x})(ox)_3]\}_n$ [35] and $\{A[M^{II}(Fe^{III}{}_xCr^{III}{}_{1-x})(ox)_3]\}_n$ (M^{II} = Mn, Fe, Co, Ni; $A^+ = NBu_4{}^+$, $[FeCp^*{}_2]^+$, $[CoCp^*{}_2]^+$). For intermediate values of x, they have observed spin-glass behaviour in these compounds.

The study of the magnetisation *vs.* field in the ordered phase revealed important variations of the coercive force H_c with the counter cation. It is yet unclear whether these variations should be attributed to intrinsic effect (structural anisotropy due to the cation, effect of the paramagnetism of the cation...) or to extrinsic effect (shape anisotropy depending on the cation). The highest coercive force in those compounds is observed in $\{[CoCp^*{}_2][Fe^{II}(Fe^{III}{}_{0.52\pm0.05}Cr^{III}{}_{0.480\pm0.05})(ox)_3]\}_n$ [30]. It reaches 1.675 T at 2 K exploiting both the effect of the cobalticinium counter cation and the glassy behaviour of the compound.

Magnetic studies on 3D compounds are not so rich. Most of the compounds are actually paramagnets [25c, 27b] or (canted) anti-ferromagnets [36, 37, 26a]. The first studies on ferromagnetic phases came out from the strategy developed by *Coronado et al.* [26b] and *Andrés et al.* [27a, b] that allowed the synthesis of $\{[M_1{}^{II}M_2{}^{III}(ox)_3]_n\}^{n-}$ 3D anionic network (see above). The general trends appear to be a decrease of T_c compared to the parent 2D compounds while the identity of the $[M_1{}^{II}M_2{}^{III}(ox)_3]$ pattern should lead to similar exchange coupling and hence critical temperature. This evolution can be tentatively attributed to the relative stiffness of the network imposed by the use of a template cation which prevent the oxalate bridge from adopting the geometry that maximises the J value. In such a vision, an interesting magneto-structural correlation has been made by *Andrés et al.* [27b] on $\{[Ru(bpy)_3][ClO_4][MnCr(ox)_3]\}_n$ and $\{[Ru(bpy)_2(ppy)][MnCr(ox)_3]\}_n$. In the former case, the oxalate bridge is rather unsymmetrical leading to a low value of T_c. In the later case, the oxalate-bridge is symmetrical leading to a value of T_c close to the one observed in 2D $[Mn^{II}Cr^{III}]$ networks.

3. Polyfunctionnal Tris(oxalato)bimetallic Based Magnets

To the long-range magnetic order arising from the anionic network, it is possible to add another physical property. This property can come from (i) a proper choice of the cationic counter part, (ii) the whole structure itself, and (iii) a combination of the two previous effect.

3.1 Magnetism and magnetism

Among the counter cations introduced in these networks, some are paramagnetic species. The use of ferricinium cation leads to 2D compounds [30] which combines a long-range magnetic order due to the anionic network with the paramagnetic properties of the cationic counterpart. The use of $[M^{II}(bpy)_3]^{2+}$ ($M^{II} = Ni^{II}$, Fe^{II}, Co^{II}) and $[Cr(bpy)_3]^{3+}$ sometimes associated to a monoanion (ClO_4^- or BF_4^-) leads to three dimensional networks [15b, 27b, 17] with an anionic network which is a para-, a (canted) antiferro- or a ferro-magnets. In both 2D and 3D cases, the presented studies do not show any clear interplay between the magnetic properties of the anionic network and those of the cationic counterpart.

An interesting result obtained on 3D compounds has to be mentioned here. *Sieber et al.* [38] has shown that the magnetic properties of the template cation can be modified by a proper choice of the anionic network: the $[Co^{II}(bpy)_3]^{2+}$ ion shows a thermal spin transition from high spin state to low spin state when decreasing the temperature in $\{[Co(bpy)_3][LiCr(ox)_3]\}_n$ while it stays in the high spin state down to 2 K in $\{[Co(bpy)_3][NaCr(ox)_3]\}_n$. Nevertheless, this effect is linked to structural changes in the anionic network and not to the influence of the magnetic properties of this network.

3.2 Magnetism and conduction

Kurmoo et al. [4] and *Coronado et al.* [39] have synthetised compounds including BEDT-TTF (Bis(ethylenedithio)tetrathiafulvalene) cations. *Coronado et al.* has shown that conduction and long range magnetic order coexist in $\{[BEDT\text{-}TTF]_3[MnCr(ox)_3]\}_n$ [39]. The two properties seems to be nearly independent though the appearance of a negative magneto-resistance below 10 K for a magnetic field applied perpendicularly to the layers may be caused by the internal field generated at low temperature by the $\{[MnCr(ox)_3]_n\}^{n-}$ magnetic layers. *Kurmoo et al.* have shown in $[BEDT\text{-}TTF]_4[(H_2O)Fe(ox)_3]$ that it is possible for a paramagnet to present a superconducting phase at low temperature (Fig. 4) despite the long believed antagonism between magnetism and conductivity [4].

3.3 Magnetism and optics

3.3.1 Cotton and Faraday effect

In a provocative article entitled: "Can a magnetic field induce asymmetric synthesis?" *Barron* [40] predicted that, in certain conditions, asymmetric induction might take place in a magnetic field. Recently, the discovery of the magneto-chiral dichroism (MchD) in paramagnetic chiral compounds [41] has stimulated interest in molecular material possessing both *Cotton* and *Faraday* effects. These two properties can be found simultaneously in optically active molecule based magnets. It is noteworthy that chiral magnets obtained in a racemic way were described by *Gatteschi* [42], *Inoue* [43], and *Julve* [44], but the described compounds were not obtained in an optically active form resulting from the whole structure of the network. Recently *Veciana et al.* [45] have proposed the synthesis of a molecular ferromagnet starting from bis(hexafluoroacetylacetonate)manganese(II) coordonated by an optically active nitroxide radical.

3.3.1.i 2D networks Some optically active 2D networks were obtained using as starting materials Δ or Λ $[Cr(ox)_3]^{3-}$ anionic bricks in the presence of Mn^{2+} and

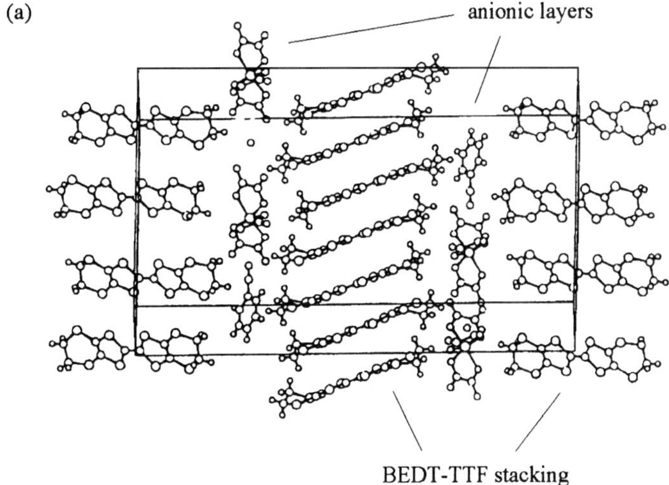

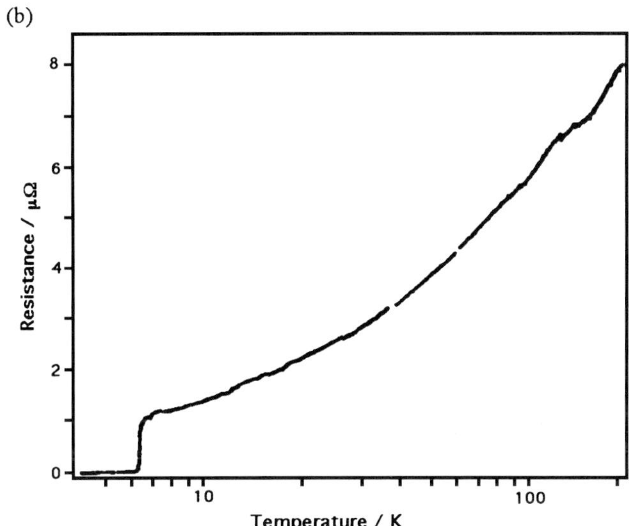

Fig. 4. Structure (a) and resistance *vs.* T (b) for the $\{(BEDT\text{-}TTF)_4[(H_2O)Fe(ox)_3]\cdot C_6H_5CN\}_n$ compound (adapted from Ref. [4])

n-Bu_4N^+ as template cation [37]. These compounds show opposite *Cotton* effect related to the configuration Δ or Λ of the starting material. This effect is attributed to the d–d transition of chromium. Compared to the starting brick, the maxima are shifted by 20 nm. Cristallographic data recorded on powders show that the space group changes from the R3c for *rac*-[MnCr] to $P6_3$ for optically active [ΔMn–ΛCr] or [ΛMn–ΔCr] compounds, indicating a significant change in the organisation of the crystal structure [46].

Others optically active 2D networks were synthesised using as starting materials Δ or Λ $[Cr(ox)_3]^{3-}$ anionic bricks in the presence of Ni^{2+} or Mn^{2+} and of achiral ferrocenic ammonium salts $[C_5H_5FeC_5H_4\text{–}CH_2NR_3]^+$ with $R = n$-ethyl to n-butyl [28, 47]. The Circular dichroism (CD) curves for $R = n$-propyl are shown in Figure 5.

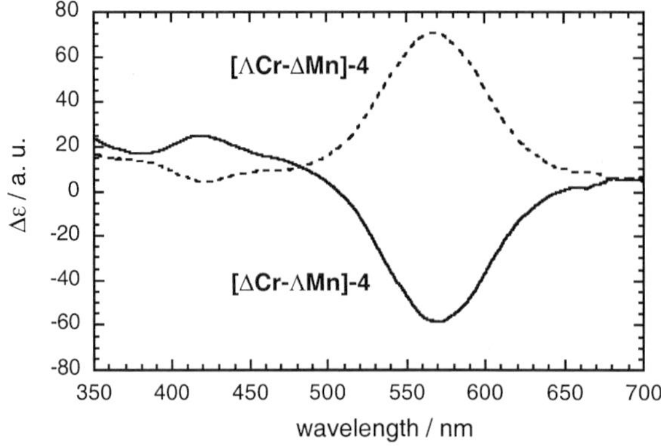

Fig. 5. Circular dichroism spectra of [ΔCr–ΛMn] and [ΛCr–ΔMn]

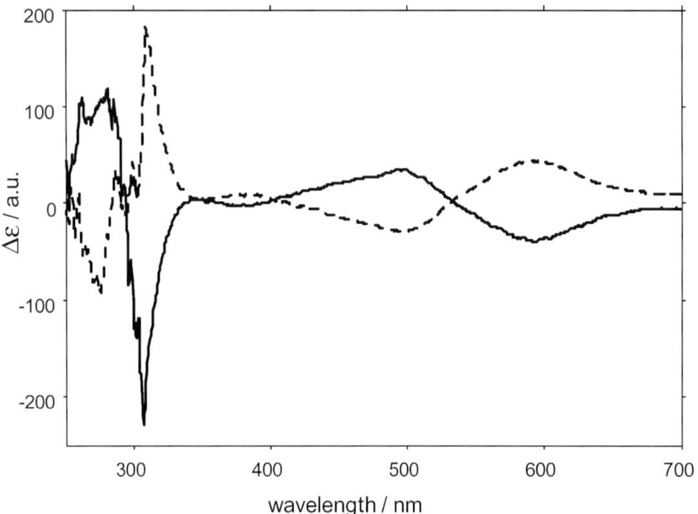

Fig. 6. Circular dichroism spectra of M-$\{[\Delta Ru(bpy)_2ppy][\Delta Cr\Delta Mn(ox)_3]\}_n$ (full line) and P-$\{[\Lambda Ru(bpy)_2ppy][\Lambda Cr\Lambda Mn(ox)_3]\}_n$ (dotted line)

3.3.1.ii 3D networks In the 3D bimetallic networks connected by oxalate ligands, chirality results from the helical auto-assembling. To build optically active materials, it is possible to use either an optically active anionic brick or an optically active cationic template. Compounds having the general formula: $\{C[M_1M_2(ox)_3]\}_n$ with $C = [Ru(bpy)_3]^{2+}$ [27b, 37], $[Ru(bpy)_2(ppy)]^+$ [27c] ($M_1 = M_2 = Mn^{2+}$; $M_1 = Mn^{2+}$, Ni^{2+}, Fe^{2+}, Cu^{2+} and $M_2 = Cr^{3+}$, Co^{3+}, Fe^{3+}) were obtained in homochiral enantiomeric forms. CD curves for the two enantiomers (M)-$\{[\Delta Ru(bpy)_2ppy][\Delta Mn\Delta Cr(ox)_3]\}_n$ and (P)-$\{[\Lambda Ru(bpy)_2ppy][\Lambda Mn\Lambda Cr(ox)_3]\}_n$ are shown in Fig. 6.

The structures of the compounds (P)-$\{[\Lambda Ru(bpy)_2ppy][\Lambda Mn\Lambda Cr(ox)_3]\}_n$ and (M)-$\{[\Delta Ru(bpy)_3][ClO_4][\Delta Mn\Delta Cr(ox)_3]\}_n$, (Fig. 7) determined by X-ray diffraction, support the homochiral arrangement of these 3D networks.

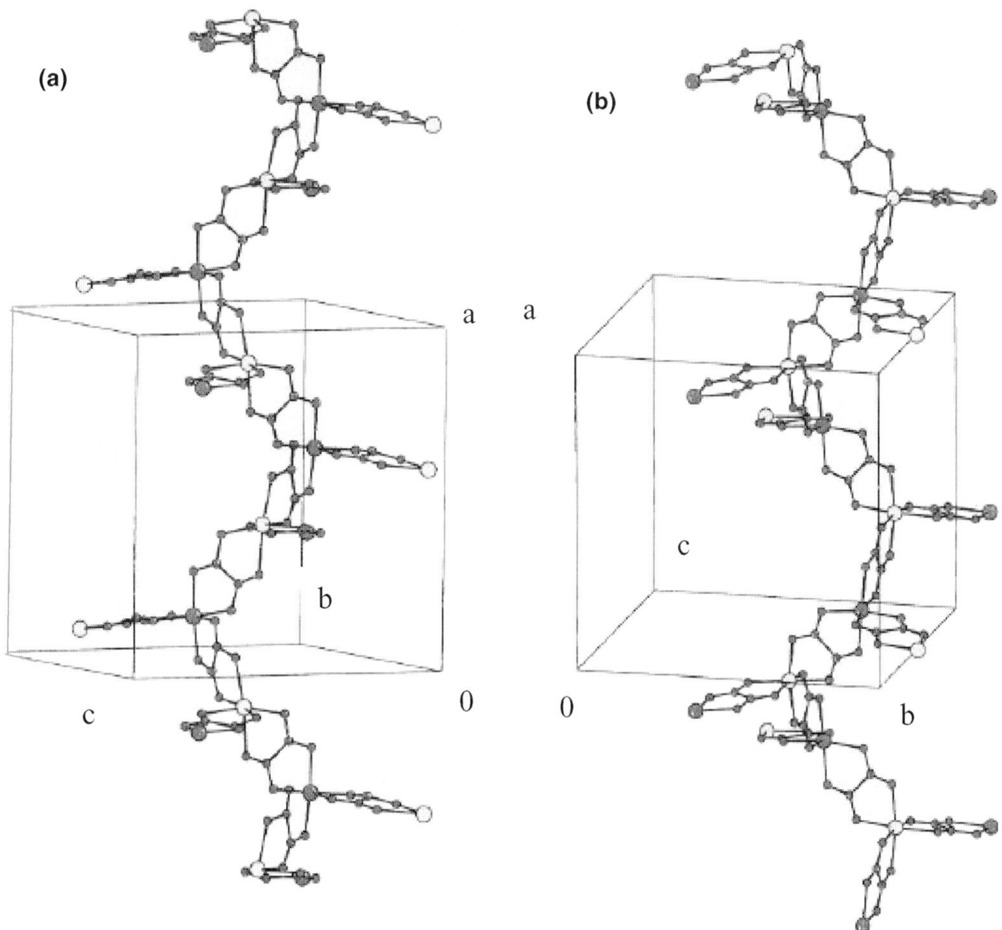

Fig. 7. Helical structures of the compounds M-$\{[\Delta Ru(bpy)_3][ClO_4][\Delta Mn\Delta Cr(ox)_3]\}_n$ (a) and P-$\{[\Lambda Ru(bpy)_2ppy][\Lambda Mn\Lambda Cr(ox)_3]\}_n$ (b)

Together with the measurement of spectroscopic NCD (Natural circular dichroism), the magnetic measurements performed on $\{[\Delta Ru(bpy)_3][ClO_4][\Delta Mn\Delta Cr(ox)_3]\}_n$ and $\{[\Delta Ru(bpy)_2ppy][\Delta Mn\Delta Cr(ox)_3]\}_n$, establish that these compounds are the first optically active ferromagnets [27b]. $\{[\Delta Ru(bpy)_3][ClO_4][\Delta Mn\Delta Cr(ox)_3]\}_n$ has been shown to possess magnetic circular dichroism (MCD) below its *Curie* temperature. This experiment, together with the observed NCD signal, shows the capability of these compounds to exhibit an important cross-effect between NCD and MCD, the so-called magneto-chiral dichroism (MChD). It thus opens a new field of research in polyfunctional molecular materials since the measurement of such a cross effect would be an unambiguous manifestation of the possibility of interaction between two physical properties in polyfunctional molecule-based magnets [48].

3.3.2 Quadratic non-linear optical (NLO) properties

Materials possessing large quadratic (second order) optical non linearities are of interest for their potential use in second harmonic generation (SHG), electro-optic devices and more generally to the emerging field of photonics [49]. Up to a recent period, there were no compounds combining both a spontaneous magnetisation and second order NLO properties. The layered oxalate system provides an opportunity to incorporate hyperpolarisable cationic chromophores in between magnetic layers, which in some cases can exert a constraint strong enough to enforce a non-centrosymmetric packing of the chromophores – a pre-requisite for quadratic non linearity. This strategy has been employed to line up stilbazolium-type chromophores within the $MnPS_3$ layered host lattice [50]. A series of thirty five layered compounds $\{\mathbf{A}[M^{II}Cr^{III}(ox)_3 solvent]\}_n$ has been synthesised with five divalent ions ($M = $ Mn, Fe, Co, Ni, Cu) and the seven hyperpolarisable stilbazolium-shaped [**A**] chromophores shown in Scheme 1 [50, 51].

All compounds order ferromagnetically below Curie temperatures that range from 6 K to 13 K. About two third of these exhibit NLO activity, those containing DAMS or DAZOP chromophores being even particularly efficient (data in Table 1).

Despite extensive efforts, it has been only possible to grow single crystals of two (NLO-inactive) compounds ($M = $ Mn, $A = $ MIPS and DAPS). Both of them crystallise in the centrosymmetric space group $P2_1/c$, a feature consistent with the lack of NLO effect [51]. The dipolar moments of the chromophores lying in a given organic layer are arranged in an antiparallel manner. Recently the structure of the strongly NLO active compound $\{DAZOP[Mn^{II}Cr^{III}(ox)_3]CH_3CN_{0.6}\}_n$ has been solved by *Rietvelt* refining of X-ray diffraction powder data [52]. All the chromophores are

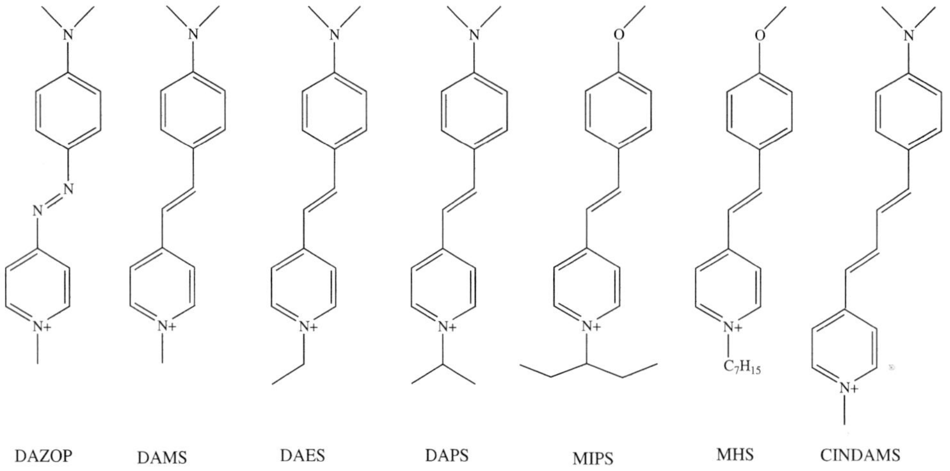

Scheme 1. The seven cationic chromophores introduced within the oxalate lattice

Table 1. Efficiency of second harmonic generation of several oxalate–chromophore hybrid compounds, expressed with respect to urea (powdered samples, incident wavelength 1.9 μm)

	DAZOP	DAMS	DAES	DAPS	MIPS	MHS	CINDAMS
[MnCr]	100	100	0	0	0	30	40
[FeCr]	19	17	5	19	4	21	10
[CoCr]	100	10	0	7	6	0	21
[NiCr]	25	25	0	0	0	0	1
[CuCr]	32	10	0	1	0	0	8

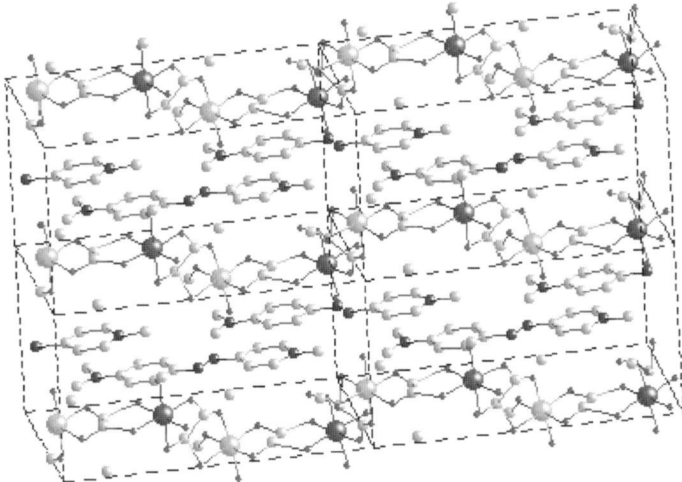

Fig. 8. The ⟨⟨ferroelectric⟩⟩ arrangement of the DAZOP chromophores in the {DAZOP[Mn Cr(ox)$_3$]}$_n$ hybrid compound

lined up in a ⟨⟨ferroelectric⟩⟩ way throughout the whole crystal (Fig. 8). The possible interplay between magnetic and NLO properties in this class of materials has been examined but no clear-cut effect has been evidenced so far [53].

Remarkably, the DAZOP chromophores fill in the interlayer space in an uniform way. The oxalate sheets, which are no longer puckered, come closer to each other. A search for structure-property correlation among the 35 compounds suggests that NLO activity goes along with a shorter interlayer distance. This feature in turn suggests that the gain in coulombic energy due to a closer approach of the anionic and cationic layers may counterbalance the natural tendency of the chromophores to pair up their dipolar moments.

3.3.3 Magnetism and photochromism: towards photosensitive magnetic switches

The incorporation of photosensitive organic species in between the layers of the magnetic oxalate lattice has led recently to the design of a new type of photosensitive magnetic switch. Thus, incorporation of the cationic spiropyran (SP) shown in Scheme 2, yields a hybrid compound {(**SP**)[Mn^{II}Cr^{III}(ox)$_3$]}$_n$, that exhibits an unusual set of optical and magnetic features [54]:

(i) This compound behaves as a photochromic material, that is, the spiropyran (SP closed form) – merocyanin (MC, open form) transformation still comes about in both directions as a result of either UV irradiation (opening) or visible irradiation (closure), see Scheme 2 [55]. Remarkably, both the closed SP form and the photo-induced open MC form are found to be thermally stable, that is, there is no thermal relaxation towards the most stable of the two forms, a feature which contrasts with the usual properties of spiropyrans [56]. Most probably, this feature is due to the effect of the ionic packing forces between the anionic oxalate layers and the cationic organic layers, which increase the potential energy barrier hindering interconversion.

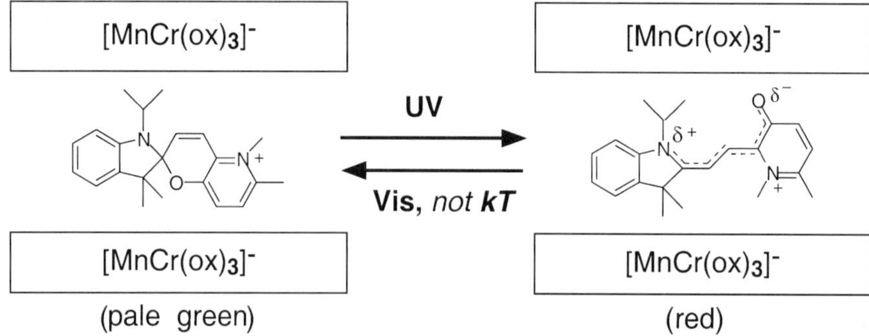

Scheme 2. Spiropyran–merocyanin switching process in between oxalate sheets

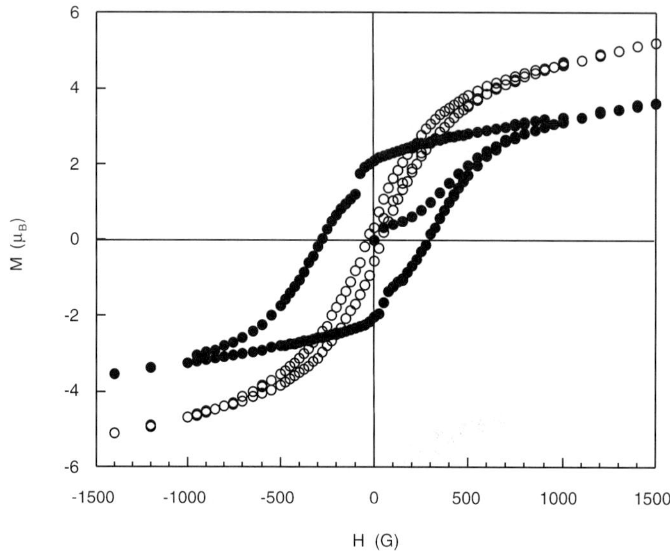

Fig. 9. Hysteresis loop at 2 K of the $\{(SP)[MnCr(ox)_3]\}_n$ compound before (white circles) and after (black circles) UV irradiation (at 365 nm)

(ii) UV irradiation transforms the initially very soft $\{(\mathbf{SP})[Mn^{II}Cr^{III}(ox)_3]\}_n$ magnet into a much *harder* one, as evidenced by the change in the shape of the hysteresis loop (Fig. 9) [55].

This feature appears to be irreversible and its origin is thought to reside in the formation of structural defects and local disorder arising from the strains that must accompany the photostructural changes of the chromophores. Such defects are likely to pin the *Bloch* walls of the magnetic domains and hence decrease their mobility [57].

3.3.4 Fluorescence

In this section, a short comment on observations of excitation energy-transfer processes within the 3D supramolecular host/guest compounds will be presented. Consider that chemical variation and combination of metal ions of different valencies in the oxalate-backbone as well as in the tris-bipyridine cation offer unique opportunities for studying a large variety of photophysical phenomena. Naturally, the sensitizer can be incorporated into the oxalate-backbone or the tris-bipyridine cation, either in low concentration as dopant, at higher concentration in mixed crystals, or fully concentrated in neat compounds. Thereby, depending upon the relative energies of the excited states of the chromophores, energy-transfer is observed either from the guest system with the tris-bipyridine cations as donors to the host system where the oxalate-backbone acts as acceptor sites or *vice versa*. In addition, energy-transfer between identical chromophores occurs within the host as well as within the guest system [36, 58]. Figure 10 illustrates these energy-transfer processes schematically with different host/guest stoichiometries.

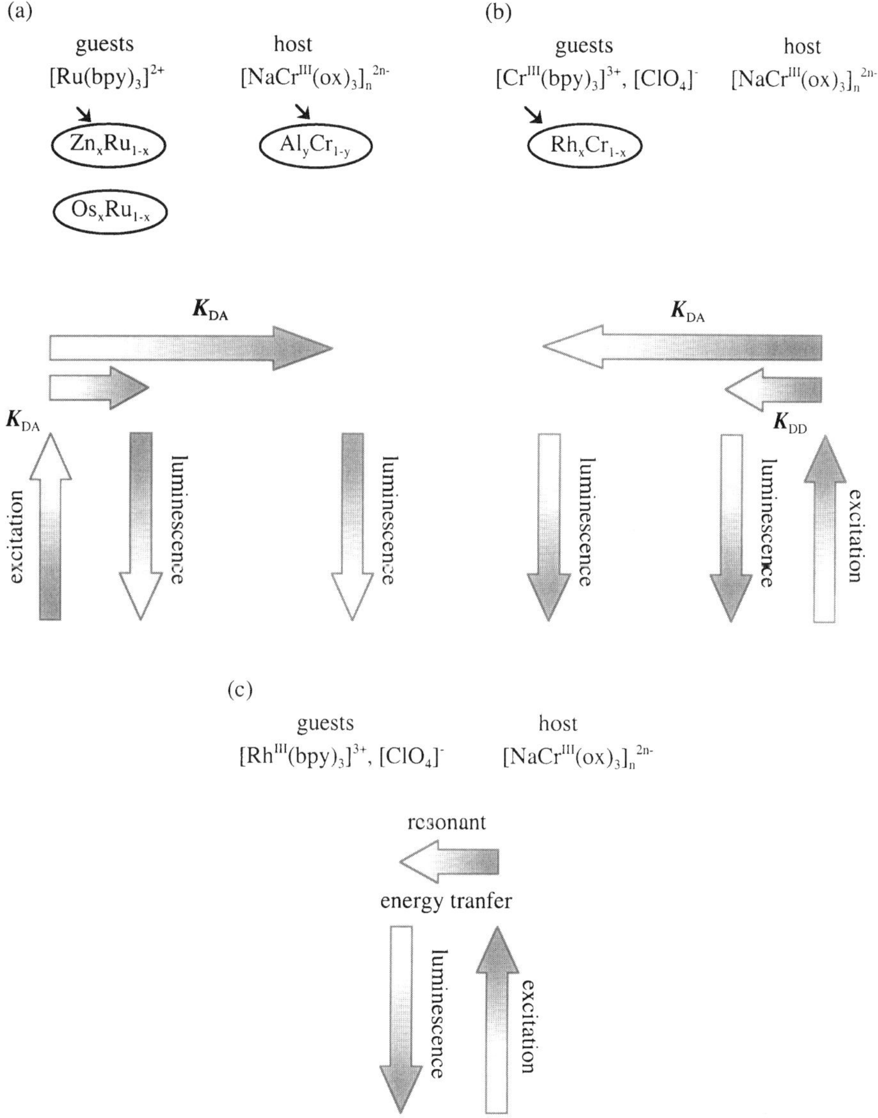

Fig. 10. Schematic representation of energy-transfer processes for different stoichiometries; excitation into the guest system (a); excitation into the host system (b) and (c)

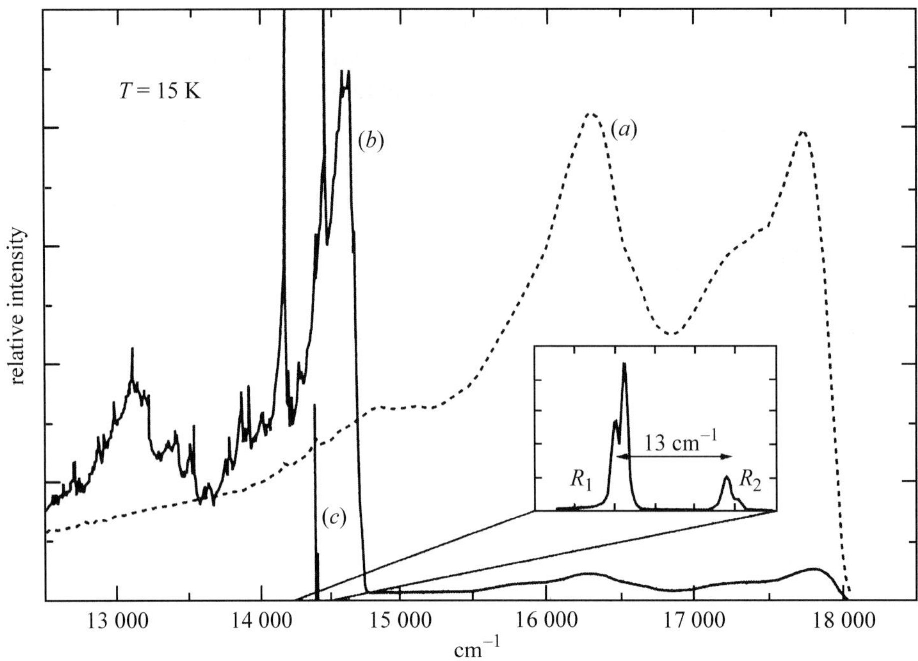

Fig. 11. Luminescence spectra at $T = 15\,\mathrm{K}$ of $\{[Ru(bpy)_3][NaAl(ox)_3]\}_n$ (a), $\{[Ru_{1-x}Os_x(bpy)_3][NaAl(ox)_3]\}_n$, $x = 1\%$ (b), $\{[Ru(bpy)_3][NaCr(ox)_3]\}_n$ (c) ($\lambda = 476\,\mathrm{nm}$)

As an example of the energy-transfer processes, Fig. 11 shows the luminescence spectra of three representative compounds. For instance, if Al^{3+} is replaced by Cr^{3+}, the $[Ru(bpy)_3]^{2+}$ luminescence from the spin-forbidden MLCT transition is completely quenched, and the sharp luminescence bands characteristic for the zero-field components of the $^2E \rightarrow {}^4A_2$ transition of octahedrally coordinated and trigonally distorted Cr^{3+} are observed at $14400\,\mathrm{cm}^{-1}$. This is a clear indication for very efficient energy-transfer from the initially excited $[Ru(bpy)_3]^{2+}$ to $[Cr(ox)_3]^{3-}$. However, not only acceptors on the oxalate-backbone may quench the $[Ru(bpy)_3]^{2+}$ luminescence. Replacing a fraction of the $[Ru(bpy)_3]^{2+}$ by $[Os(bpy)_3]^{2+}$ results in luminescence from $[Os(bpy)_3]^{2+}$ and a quenching of the $[Ru(bpy)_3]^{2+}$ luminescence, too. Indeed, the energy-transfer to $[Os(bpy)_3]^{2+}$ is even more efficient than to $[Cr(ox)_3]^{3-}$.

Alternatively, as schematically drawn in Fig. 10(b), irradiating into the spin-allowed $^4A_2 \rightarrow {}^4T_2$ absorption band of $[Cr(ox)_3]^{3-}$ from the host system results in intense luminescence from the ^2E state of $[Cr(bpy)_3]^{3-}$ within the guest system, again demonstrating a rapid energy-transfer process. Finally, according to Fig. 10(c), the stoichiometry $\{[Rh(bpy)_3][ClO_4][NaCr(ox)_3]\}_n$ allows us to study the energy-transfer within the R_1 line of the ^2E state of Cr^{3+}. In that case, from a fluorescence-line-narrowing experiment, clear evidence for a resonant energy-transfer process could be gained [59, 60].

4. Concluding Remarks

Two- and three-dimensional oxalate bridged bimetallic networks appear as fruitful molecular materials to study various polyfunctionnal magnets with optical properties. Their matrix organisation based on two sub-lattices permit to design a variety of hybrid materials. They offer unique opportunity to study physical fundamental properties in particular to unveil the magneto-optical relationship. Moreover, these materials may bring further developments. For instance, the fact that a specific photochromic chromophore turns out to be bistable once inserted may lead to high density optical memories if nanoparticles of the hybrid materials could be processed. Nanoparticles where all the dipolar moments of the chromophores are aligned could become electrical analogs of the so-called high spin molecules. Such examples show that properties of the whole hybrid may be considerably more than the superposition of the properties of the individual components.

References

[1] a) Traité des matériaux. Presses Polytechniques Universitaires Romandes, Lausanne; b) Schubert U, Hüsing N (2000) Synthesis of Inorganic Materials. Wiley-VCH, Weinheim

[2] Simon J, Bassoul P (2000) Design of Molecular Materials. Wiley, Chichester

[3] a) Kahn O (1993) Molecular Magnetism. VCH, Weinheim; b) Lacroix PG (2001) Chem Mat **13**: 3495

[4] Kurmoo M, Graham AW, Day P, Coles SJ, Hursthouse M, Caulfield JL, Singleton J, Pratt FL, Hayes W, Ducasse L, Guionneau P (1995) J Am Chem Soc **117**: 12209

[5] Sato O, Iyota T, Fujijishima A, Hashimoto K (1996) Science **272**: 704

[6] Ferlay S, Mallah T, Ouahes R, Veillet P, Verdaguer M (1995) Nature **378**: 701

[7] a) Rashid S, Turner Scott S, Day P, Light ME, Hursthouse MB (2000) Inorg Chem **39**: 2426; b) Ohba M, Tamaki H, Matsumoto N, Okawa H (1993) Inorg Chem **32**: 5385; c) Alvarez S, Julve M, Verdaguer M (1990) Inorg Chem **29**: 4500; d) Cortés R, Urtiaga MK, Lezama L, Arriortua MI, Rojo T (1994) Inorg Chem **33**: 829

[8] Verdaguer M, Julve M, Michalowicz A, Khan O (1983) Inorg Chem **22**: 2624

[9] a) Sunderg MR, Kikeväs R, Koskimies JK (1991) J Chem Soc Chem Commun **7**: 526; b) Mörtl KP, Sutter JP, Golhen S, Ouahab L, Kahn O (2000) Inorg Chem **39**: 1626

[10] a) Broderick WE, Thompson JA, Godfrey MR, Sabat M, Hoffman BM (1989) J Am Chem Soc **111**: 7656; b) Kojima N, Aoki W, Itoi M, Ono Y, Seto M, Kobayashi Y, Maeda Yu (2001) Solid State Commun **120**: 165; c) Kojima N, Aoki W, Seto M, Kobayashi Y, Maeda Yu (2001) Synth Met **121**: 1796

[11] a) Rochon FD, Melanson R, Andruh M (1996) Inorg Chem **35**: 6086; b) Andruh A, Melanson R, Stager CV, Rochon FD (1996) Inorg Chim Acta 309; c) Munoz C, Julve M, Lloret F, Faus J, Andruh M (1998) J Chem Soc Dalton Trans 3125

[12] a) Xu-Fang Chen, Peng Cheng, Xin Liu, Bin Zhao, Dai-Zeng Liao, Shi-Ping Yan, Zong-Hui Jiang (2001) Inorg Chem **40**: 2652; b) Lu JY, Lawandy A, Li Jing, Tan Yuen, Lin CL (1999) Inorg Chem **38**: 2695

[13] De Munno G, Armentano D, Julve M, Lloret F, Lescouëzec R, Faus J (1999) Inorg Chem **38**: 2234

[14] a) Tamaki H, Zhong ZJ, Matsumoto N, Kida S, Koikawa M, Achiwa N, Hashimoto Y, Okawa H (1992) J Am Chem Soc **114**: 6974; b) Tamaki H (1992) Chem Lett 1975

[15] a) Decurtins S, Schmalle HW, Oswald HR, Linden A, Ensling J, Gütlich P, Hauser A (1994) Inorg Chim Acta **216**: 65; b) Decurtins S, Schmalle HW, Schneuwly P, Ensling J, Gütlich P, Hauser A (1994) J Am Chem Soc **116**: 9521

[16] Day P (1997) J Chem Soc Dalton Trans 701

[17] Coronado E, Galan-Mascaros JR, Gomez-Garcia CJ, Martinez-Agudo JM (2001) Inorg Chem **40**: 113

[18] a) Ovanesyan NS, Shilov GV, Atovmyan LO, Lyubovskaya RN, Pyalling AA, Morozov YG (1995) Mol Cryst Liq Cryst **273**: 175; b) Ovanesyan NS, Pyalling AA, Sanina NA, Kashuba AB, Bottyan L (2000) Hyperfine Interactions **126**: 149
[19] Dwyer FP, Sargeson AM (1956) J Phys Chem **60**: 1331
[20] Werner A (1912) Ber **45**: 3061
[21] Atovmyan LO, Shilov GV, Lyubovskaya RN, Zhilyaeva EL, Ovanesyan NS, Pirumova SI, Gusa Kovskaya IG, Morozov YG (1993) JETP Lett **58**: 766
[22] Larionova J, Monbelli B, Sanchiz J, Kahn O (1998) Inorg Chem **37**: 679
[23] Shilov GV, Ovanesyan NS, Sanina NA, Pyalling AA, Atovmyan LO (1998) Russian J Coord Chem **24**: 802
[24] a) Carling SG, Mathoniére C, Day P, Abdul Malik KM, Coles SJ, Hursthouse MB (1996) J Chem Soc Dalton Trans 1839; b) Mathoniére C, Nutall CJ, Carling SG, Day P (1996) Inorg Chem **35**: 1201
[25] a) Decurtins S, Schmalle HW, Schneuwly P, Oswald HR (1993) Inorg Chem **32**: 1888; b) Decurtins S, Schmalle HW, Oswald HR, Linden A, Ensling J, Gütlich P, Andreas H (1994) Inorg Chim Acta **216**: 65; c) Decurtins S, Schmalle HW, Pellaux R, Schneuwly P, Andreas H (1996) Inorg Chem **35**: 1451; d) Decurtins S, Schmalle HW, Pellaux R, Huber R, Fischer P, Ouladdiaf B (1996) Adv Mater **8**: 647; e) Pellaux R, Decurtins S, Schmalle HW (1999) Acta Cryst 1075
[26] a) Hernandez-Molina, Lloret F, Ruiz-Perez C, Julve M (1998) Inorg Chem **37**: 4131; b) Coronado E, Galan-Mascaros JR, Gomez-Garcia CJ, Martinez-Agudo JM (2001) Inorg Chem **40**: 113
[27] a) Gruselle M, Andrés R, Malézieux B, Brissard M, Train C, Verdaguer M (2001) Chirality **13**: 712; b) Andrés R, Brissard M, Gruselle M, Train C, Vaissermann J, Malézieux B, Jamet JP, Verdaguer M (2001) Inorg Chem **40**: 4633; c) Brissard M, Gruselle M, Malézieux B, Thouvenot R, Guyard-Duhayon C, Convert O (2001) Eur J Inorg Chem 1745; d) Brissard M, Amouri H, Gruselle M, Thouvenot R (2002) CR Chimie **5**: 53
[28] Malézieux B, Andrés R, Brissard M, Gruselle M, Train C, Herson P, Troitskaya L, Sokolov V, Ovseenko S, Demeschik T, Ovanesyan N, Mamed'yarova I (2001) J Organomet Chem **637–639**: 182
[29] Bénard S (2000) Thèse de l'Université de Paris X1 Orsay, France
[30] Coronado E, Galan-Mascaros JR, Gomez-Garcia CJ, Martinez-Agudo JM (1999) Adv Mater **11**: 558
[31] Pei Y, Journaux Y, Kahn O (1989) Inorg Chem **28**: 100
[32] Pellaux R, Schmalle HW, Huber R, Fischer P, Hauss T, Ouladiaff B, Decurtins S (1997) Inorg Chem **36**: 2301
[33] Iijima S, Koner S, Mizutani F (1999) J Radioanalytical Nuclear Chem **239**: 245
[34] Kosterlitz DM, Thouless JM (1973) J Phys Chem **6**: 1181
[35] Bhattacharjee A, Ijima S, Mizutani F (1996) J Mag Mag Mat **153**: 235
[36] Decurtins S, Schmalle HW, Pellaux R, Schneuwly P, Hauser A (1996) Inorg Chem **35**: 1451
[37] Andrés R, Gruselle M, Malézieux B, Verdaguer M, Vaissermann J (1999) Inorg Chem **38**: 4637
[38] Sieber R, Decurtins S, Stoeckli-Evans H, Wilson C, Yufit D, Howard JAK, Capelli SC, Hauser A (2000) Chem Eur J **6**: 361
[39] Coronado E, Galan-Mascaros JR, Gomez-Garcia CJ, Laukhin V (2000) Nature **408**: 447
[40] Barron LD (1994) Science **266**: 1491
[41] a) Rikken GLJA, Raupach E (1997) Nature **390**: 493; b) Rikken GLJA, Raupach E (2000) Nature **405**: 932; c) Raupach E, Rikken GLJA, Train C, Malézieux B (2000) Chem Phys **261**: 373
[42] Caneschi A, Gatteschi D, Rey P, Sessoli R (1991) Inorg Chem **30**: 3936
[43] Kumaiga H, Inoue K (1999) Angew Chem Inter Ed Engl **38**: 1601
[44] Hernandez-Molina M, Lloret F, Ruiz-Perez C, Julve M (1998) Inorg Chem **37**: 4131

[45] Minguet M, Luneau D, Lhotel E, Villar V, Paulsen C, Amabilino DB, Veciana J (2002) Angew Chem Int Ed **41**: 586
[46] Ovanesyan NS, Shilov GV, Sanina NA, Train C, Gredin P, Gruselle M (2002) The 1th Russian Conference on High-Spin Molecule and Molecular Ferromagnets (Chernogolovka)
[47] Troitskaya LL, Demeshchik TV, Sokolov VI, Mamed'yarova IA, Malézieux B, Gruselle M (2001) Russian Chem Bull Intern Ed **50**: 497
[48] Barron LD, Buckimgham AD (2001) Acc Chem Res **34**: 781
[49] a) Zyss J (1994) Molecular Non Linear Optics. Academic Press, New York; b) Lacroix PG, Clément R, Nakatani K, Zyss J, Ledoux I (1994) Science **263**: 658
[50] Bénard S, Yu Pei, Coradin T, Rivière Z, Nakatani K, Clément R (1997) Adv Mat **9**: 981
[51] Bénard S, Yu P, Audière JP, Clément R, Guilhem J, Tchertaniv L, Nakatani K (2000) J Amer Chem Soc **12**: 9444
[52] Evans JSO, Bénard S, Pei Y, Clément R (2001) Chem Mater **13**: 3813
[53] Lacroix PG, Malfant I, Bénard S, Yu P, Rivière E, Nakatani K (2001) Chem Mater **13**: 441
[54] Bénard S, Yu P (2000) Adv Mater **12**: 48
[55] Bénard S, Rivière E, Yu P, Nakatani K, Delouis JF (2001) Chem Mater **13**: 159
[56] a) Crano JC, Guglielmetti RJ (1999) Organic Photochromic and Thermochromic Compounds. Plenum Press, New York; b) (2000) Chem Rev **100**(5) Special Issue
[57] Nakatani K, Yu P (2001) Adv Mater **13**: 1411
[58] Von Arx M, Burattini E, Hauser A, Van Pieterson L, Pellaux R, Decurtins S (2000) J Phys Chem A **104**: 883
[59] Hauser A, Riesen H, Pellaux R, Decurtins S (1996) Chem Phys Letters **261**: 313
[60] Von Arx M, Hauser A, Riesen H, Pellaux R, Decurtins S (1996) Phys Rev B **54**: 15800

Invited Review

Ferromagnetism in Metallocene-Doped Fullerenes

Dragan Mihailovic

Jozef Stefan Institute, Department of Complex Matter, SI-1000 Ljubljana, Slovenia

Received September 4, 2002; accepted September 6, 2002
Published online January 8, 2003

Summary. Ferromagnetism in fullerene-based systems doped with metallocenes is reviewed. These compounds form a ferromagnetic state by spin-coupling between π electrons on fullerene units, while the metallocene molecules do not contribute to the spin ordering. One of these compounds has the highest critical temperature (19 K) for this class of compound. The magnetic properties of these materials are very strongly dependent on the crystallization conditions.

Keywords. Fullerene; Ferromagnetism; Adducts; Metallocene; Cobaltocene.

1. Introduction

The discovery of high-T_c fullerene magnetism is presenting an exceptional challenge to our understanding of magnetic phenomena the solid state. Indeed p-electron magnetism has been considered unlikely, ever since *Heisenberg* considered it essential to have atoms (or ions) with d-electron orbitals to form a ferromagnetic state. Magnetic behaviour in fullerene compounds was first discovered in 1991 by *Fred Wudl* and collaborators [1]. Their discovery of ferromagnetism in tetrakis-dimethylamino-ethylene-fullerene [60] (*TDAE*-C_{60}) with a *Curie* temperature of 16 K was quite a surprise. The critical temperature signifying the onset of a magnetic state was an order of magnitude higher than the previous record and this brought the research field from the realms of the esoteric into the mainstream.

Inspired by the discovery of *TDAE*-C_{60}, several research groups immediately started to work on synthesis of new fullerene ferromagnetic compounds. A logical step was to replace the donor *TDAE* with other organic or organometallic donors. Unfortunately this approach to our knowledge so far did not result in any new reproducibly synthesised ferromagnetic compounds [2–4]. Charge-transfer salts

* Corresponding author. E-mail: Dragan.Mihailovic@ijs.si

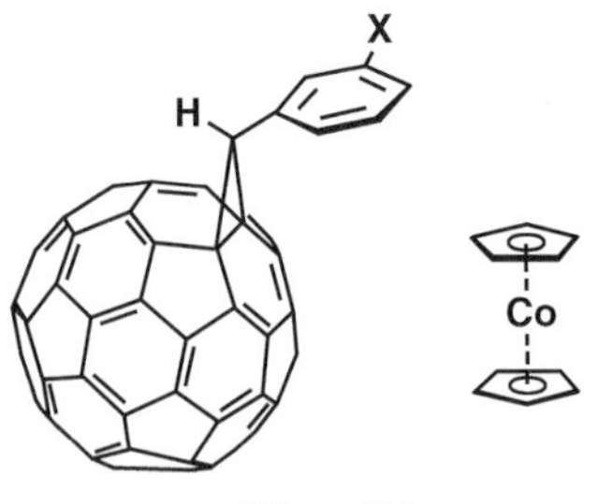

Fig. 1. The building blocks of cobaltocene-doped fullerene-based ferromagnets

of C_{60} with donors other then *TDAE* proved – without exception – to be paramagnetic, as were doped higher fullerenes [5].

Another approach was to functionalise the C_{60} molecule and change its properties slightly and subsequently to dope it with the same set of donors [6]. This direction was more promising and recently, the compound 1-(3-aminophenyl)-1*H*-methanofullerene C [60] doped with cobaltocene (*APHF-CO*) was discovered [7] with the highest reported *Curie* temperature of $T_c = 19\,\mathrm{K}$ (3 K higher then *TDAE*-C_{60}). Within two years this compound could be reproducibly synthesised in a way which would reliably exhibit a low-temperature magnetic phase. A related compound, 1-(3-nitrophenyl)-1*H*-methanofullerene (*NIPHF-CO*), also doped with cobaltocene was also found to exhibit clear signs of ferromagnetic behaviour [8]. The difference between the two compounds is that the aminophenyl adduct on the fullerene is replaced by a nitrophenyl group. The structural units of these compounds are shown schematically in Fig. 1.

In this review we discuss the synthesis, electronic structure and magnetic properties of these materials, emphasizing the connection between synthesis conditions and magnetic properties.

2. Synthesis

The 1-(3-nitrophenyl)-1*H*-methanofullerene-[C_{60}]-cobaltocene (*NIPHF-CO*) and 1-(3-aminophenyl)-1*H*-methanofullerene[C_{60}]-cobaltocene (*APHF-CO*) are synthesised in a two-step procedure. First, the fullerene derivative (*APHF* and *NIPHF*) is synthesised and purified. Subsequently they are doped with organic dopant by similar methods as *TDAE* C_{60}. [7, 8]. The detailed procedure for preparation of these compounds and chemical characterisation data for these compounds are reported by *Mrzel et al.* [6] and *Umek et al.* [8].

It was found that the temperature at which the doping is performed plays a crucial role in determining the magnetic properties of these materials. A detailed study revealed the optimum conditions, particularly the temperature for the synthesis of ferromagnetic material. Only after a prolonged systematic study was the reproducibility of the synthesis sufficiently reliable to be able to seriously examine the magnetic properties of these materials. A number of attempts to determine the crystal structure of the materials were not successful, indicating that the degree of crystallinity of these materials is rather small. The nanocrystallinity of the materials seems to be the cause of a small magnetic moment at saturation, which is always much smaller than would be expected if all the spins were ferromagnetically aligned. Attempts to grow single crystals of these materials were so far not successful.

3. The Electronic Structure

Upon formation of a charge transfer complex, the organic donor such as *TDAE* or cobaltocene transfers one electron to the fullerene. This results in a charge transfer state with a *Jahn-Teller* distorted C_{60} molecule [9]. In the case of the fullerene derivatives, an additional distortion is also caused by the presence of the adduct. The donor itself is also in a charged state, and also changes conformation upon CT. In the case of *TDAE*, this has been studied in some detail [10]. In the case of metallocene-doped fullerenes, it is important to note that the cobaltocene dopant molecule has no net spin (*i.e.* has a closed shell) in the $+1$ charge state and thus does not contribute to the magnetic properties of cobaltocene-doped fullerene ferromagnets. Previous to the discovery of ferromagnetic state in cobaltocene-doped fullerene derivatives, it was not clear how important was the role of the *TDAE* itself (with $S = 1/2$) in the formation of a ferromagnetic state. The fact that cobaltocene has no spin in the case of *APHF-CO* or *NIPHF-CO* provided conclusive evidence that the spin ordering takes place primarily on the fullerenes and not on the dopants.

The main signatures of a charge transfer between the donor and the fullerene acceptor are a) the appearance of a near-infrared absorption due to the new HOMO-LUMO transitions and b) the appearance of shifted vibrational bands of the fullerene in the infrared spectrum. In order to ascertain that the electrons on $CoCp_2$ do not participate in the magnetic interactions, it is important to show that a *complete* charge transfer from the donor to the fullerene acceptor has taken place.

Optical spectra of *APHF-CO* and *NIPHF-CO* show a clear absorption peak near 1.1 eV (1100 nm). In Fig. 2 we compare the optical absorption spectra for *TDAE*-C_{60} and *APHF-CO*. This absorption is a characteristic signature of the fullerene in the -1 charge state and corresponds to dipole allowed transitions between the new t_{1u}-derived occupied HOMO and the unoccupied t_{1g}-derived LUMO of *APHF* or *NIPHF*. We note that although the spectral shape (including the vibrational sidebands) and intensity of this feature is similar to C_{60}^- (in $TDAE^+C_{60}^-$), the t_{1u}–t_{1g} absorption peak of the $APHF^-$ is blue-shifted from 1080 nm in C_{60}^- to 1020 nm in $APHF^-$, suggesting an increased HOMO-LUMO spacing in the latter. On the other hand the absorption spectra of *APHF* and pure C_{60} in the visible and UV spectral range (190–500 nm) are virtually indistinguishable (see insert to Fig. 2), indicating that the electronic structure of *neutral APHF* is *very* similar to C_{60}.

Evidence for a *full* CT comes from vibrational spectroscopy. The mid-infrared transmittance spectra of *APHF-CO* as well as neutral Cp_2Co and *APHF* in the region of the four main C_{60}-derived IR-active vibrational modes of *APHF* are shown in Fig. 3. In addition to the four C_{60}-derived modes at 526, 573, 1184 and 1427 cm^{-1}, *APHF* shows a large number of additional weaker vibrational modes, which are assumed to be partly a result of the broken icosahedral symmetry of the fullerene by the adduct and partly due to the adduct itself. As expected, in

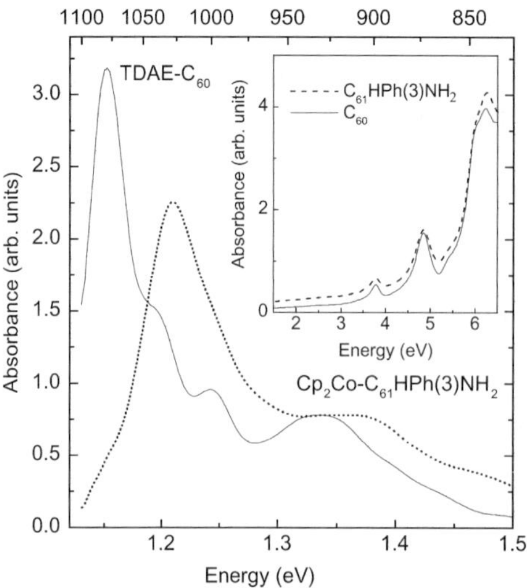

Fig. 2. The optical spectra of *APHF-CO* in comparison with *TDAE*-C_{60}. The peak near 1.2 eV corresponds to the HOMO-LUMO transition. The insert shows that the spectra of the monoadducts themselves are virtually identical

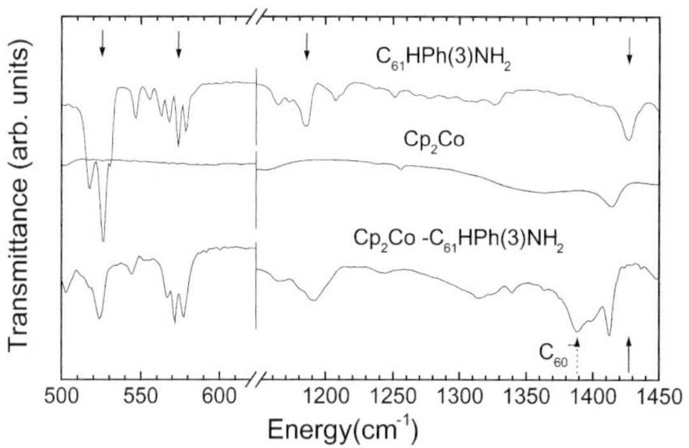

Fig. 3. The infrared vibrational spectra of *APHF*, cobaltocene and *APHF-CO*. The characteristic bands of the fullerene are shown by arrows

APHF-CO the four main fullerene-based modes are distinctly frequency shifted to 524, 572, 1191, and 1388 cm^{-1} respectively (Fig. 3), in excellent agreement with shifts in C_{60}^{-} anions in other doped fullerenes [11]. Careful examination of the *APHF-CO* spectra reveals no presence of the 1427 cm^{-1} peak corresponding to *neutral APHF.* This implies that there are no remaining unpaired electrons on the donor molecules and the ferromagnetic properties can be attributed entirely to the fullerene spins in these compounds.

4. Magnetic Phenomena

When the magnetic properties of *TDAE*-C_{60} were first measured at low temperatures [1], the researchers found a rather unusual cusp in the susceptibility with an onset at $T_c = 16$ K, indicating that below this temperature some kind of spin ordering was taking place. The behaviour was not typical for a ferromagnet, however, and only later, when single crystal data could be compared, it was shown that the unusual behaviour could be attributed to the nano-crystalline nature of the powder samples. The problem with powder samples is that the surface-to-volume ratio of the nano-crystalline particles is very large: up to 30% of all the buckyballs could be in the first few layers of each nano-particle (these were found by electron-microscopy to be typically of the order of 50 nm in diameter). This could lead to some unusual magnetic properties, resembling superparamagnetism. Since the dopants are sensitive to air, the surface could also be chemically degraded. As a result it is difficult to determine whether the observed behaviour is intrinsic or not.

Subsequent measurements have confirmed the existence of a magnetically ordered state at low temperatures: it was established by low-field radio-frequency ESR [12] that in the absence of external field there is a small, but finite internal field which is present only below T_c. The presence of such an internal field was confirmed later by muon spin resonance [13] and so the fact that there is a net ferromagnetic component to the spin order was established.

The magnetic behaviour of powder samples of *APHF-CO* and *NIPHF-CO* is qualitatively similar as with *TDAE*-C_{60} nanopowder. There are no single-crystal data yet available for the metallocene-doped fullerenes, but they may be expected to show similar differences between powder samples and single crystals as *TDAE*-C_{60}.

The magnetization of all the mentioned metallocene-doped powder samples measured so far show a non-linear curve as a function of magnetic field such as is found in soft ferromagnetic materials. (In these, there is little or no remnant magnetization in the absence of external field.) These properties together with an observed increase of T_c with magnetic field – another property typical of ferromagnets – lead to the belief that these materials are proper ferromagnets.

The low-temperature magnetic susceptibility $\chi(T)$ of *APHF-CO* measured by an AC susceptometer is shown in Fig. 4a). $\chi(T)$ shows an abrupt increase near 19 K and saturates rapidly, reaching a near-constant value $\chi \sim 2.7$ emu/mol at 17 K. A peculiar feature of the material is that the critical temperature of the transition to a ferromagnetic state appears to decrease with time and eventually disappears after a year or so (Fig. 4b). Although this could be attributed to chemical instability, the fact that T_c decreases is not trivial: such degradation would be expected to lead to a lower magnetisation, but not to such a marked change in the critical temperature.

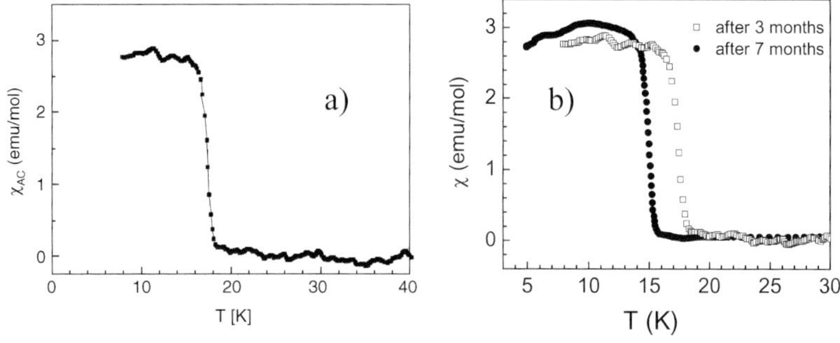

Fig. 4. a) The AC susceptibility χ_{AC} of *APHF-CO* as a function of temperature. b) χ_{AC} as a function of time after synthesis. T_c is clearly reduced with time

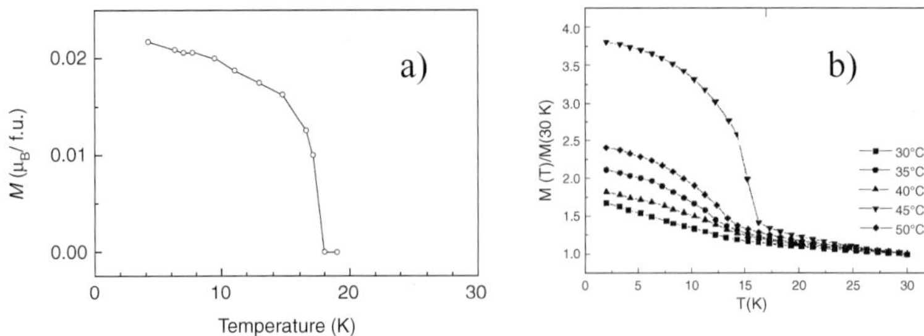

Fig. 5. a) The magnetisation of *APHF-CO* as a function of temperature measured by SQUID magnetometer. b) The magnetisation *M(T)* for *APHF-CO* synthesised at different temperatures. The highest M and T_c are achieved with synthesis at 45°C

It is an indication that the exchange coupling is strongly dependent on the structure and chemistry of the material.

The DC magnetization $\mathbf{M}(T)$ measured with a SQUID at $H = 30$ Oersted (Fig. 5a) also shows a rather characteristic rise below 18.5 K. The typical saturation moment of $\sim 0.021\,\mu_B/\text{f.u.}$ measured by SQUID magnetometer (Figs. 5a and 6a) and $\sim 0.045\,\mu_B/\text{f.u.}$ in AC magnetization measurements (Fig. 6b), shows that typically only a few percent of the sample is ferromagnetically ordered. The effect of synthesis temperature (at the doping stage) is shown in Fig. 5b. A temperature dependence of the magnetisation was measured in a static magnetic field of 50 Oe for five *APHF-CO* samples synthesised at five different temperatures between 30°C and 50°C. The magnetisation of the samples differs markedly both in magnitude and in critical temperature T_c. The magnetisation is highest when the synthesis is performed in the vicinity of 45°C and falls off rapidly on either side of that temperature. The critical temperatures range from 13 K and 17 K. The low-temperature magnetisation of the samples in a weak external field (the spontaneous magnetisation) can vary approximately by a factor of 3 among the different samples. The results shown are normalised by the value of magnetisation at 30 K (the temperature at which all samples were in the paramagnetic phase and so their magnetic moment at low temperature is directly proportional to the quantity of magnetic material present).

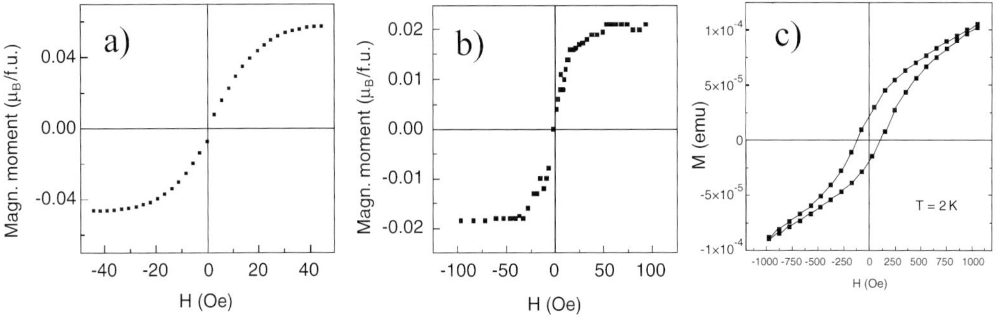

Fig. 6. The magnetisation curves of *APHF-CO* measured by a) AC susceptometer, b) SQUID magnetometer, c) SQUID measurement shows a hysteresis loop for samples synthesised under certain conditions (see text)

The magnetization curves measured directly using a SQUID magnetometer (Fig. 6b) and indirectly with an AC susceptometer and calculated from $M(H) = \int \chi_{AC}\, dH$ (Fig. 6a) both show complete saturation at $\mathbf{M} \sim 0.045\, \mu_B/\text{f.u.}$ at a relatively low field of $H = 40$ Oersted. These data appear to exclude the possibility of spin-canted antiferromagnetism, since application of higher fields above 40 Oe does not increase **M**, as expected for a proper ferromagnet.

In Fig. 6c we show the magnetisation data for *APHF-CO* synthesised by a different route [8] – for the sample with the highest low-field/low-temperature magnetisation, the magnetisation curve at 2 K in fields $-1\,\text{kOe} < H < 1\,\text{kOe}$ is shown. The magnetisation curve shows clear hysteretic behaviour with a coercitive field of $H_c \sim 100$ Oe, and a remanent magnetisation M_r which is about 0.1 per cent of the expected saturation magnetisation M_s. The magnetisation did not show saturation in fields up to 1 kOe. Repeating the measurement at 2 K up to fields of 50 kOe also did not produce saturation. A sample of approximate weight 10 mg and a molar mass of 1000 g/mol when magnetically saturated should have a magnetic moment of ~0.05 emu. The measured value in the highest experimentally accessible field is ~50 times smaller. Thus, in common with most other organic ferromagnets *APHF-CO* shows a very small, or no hysteresis, which is clearly very sample-dependent and is shown to depend on the synthesis conditions. The critical parameters determining this behaviour are not known at present.

In the X-band ESR spectra of *APHF-CO* above 26 K, a *single* strong symmetric line is observed with $g = 2.00$, whose intensity is consistent with one spin per formula unit, confirming the absence of spins on the donor. No other line is visible at high temperature, which indicates that there is no neutral $\text{Co}Cp_2$ with $s = 1/2$ present. The width of the ESR line decreases monotonically from 16 Oe at 300 K, to 2 Oe at 26 K, and shows no visible anomalies at intermediate temperatures. Below $T \sim 26$ K, an additional line appears, which grows in intensity upon reducing the temperature (Fig. 7). The susceptibility χ calculated from the intensity of this line is plotted as a function of T in Fig. 8. χ_0 shows a slight increase already at $T = 25 \sim 30$ K. In addition to the increase in static susceptibility, the T-dependent ESR line also shifts and broadens at low temperature. The shift in resonance field

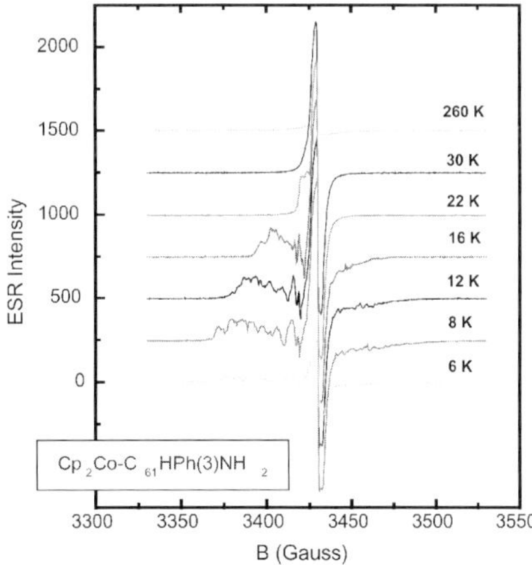

Fig. 7. The ESR signal of *APHF-CO* at different temperatures

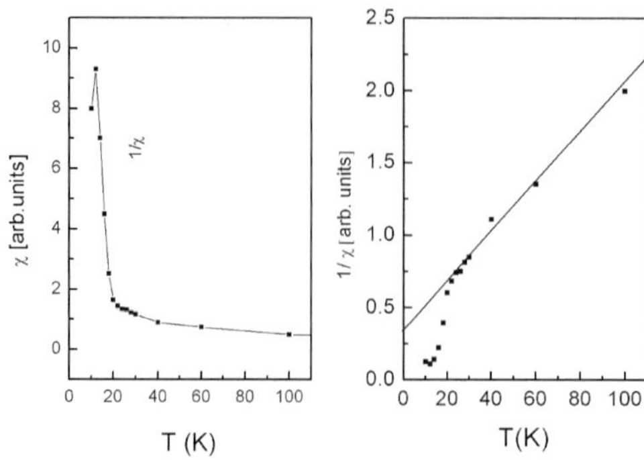

Fig. 8. a) The static susceptibility χ_0 as a function of T from the ESR intensity. b) The inverse susceptibility $1/\chi_0$ as a function of T shows a negative intercept implying AF spin correlations

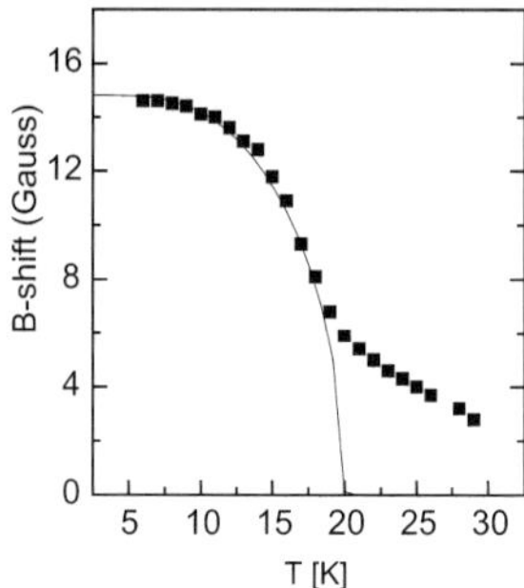

Fig. 9. The shift of the ESR resonance field with temperature is attributed to the existence of an internal field

(Fig. 9) is interpreted to be due to the appearance of an internal field below T_c. Although the broadening is difficult to measure accurately above T_c, a lineshift is clearly evident well above T_c, at least up to 30 K, which together with the increase in χ_0 could be interpreted as an onset of FM well above T_c. However, since no such increase in M or χ_{AC} is observed above T_c which are done at zero applied field, we attribute this effect to the external 3.4 kOe field of the ESR apparatus. (Such a field-induced enhancement in T_c and broadening of the transition is indeed expected for a ferromagnet.) The inverse ESR susceptibility $\chi_0{}^{-1}(T)$ of the FM line, shown as a function of temperature shown in Fig. 8b, suggests *Curie-Weiss* (*CW*) behavior above T_c with a negative temperature intercept of $\chi_0{}^{-1}(T)$ at $T \sim 30$ K. Below 26 K, χ_0 starts to depart significantly from the *CW* law, suggesting a changeover from AFM to FM spin correlations.

The low-temperature magnetic properties of 1-(3-nitrophenyl)-1*H*-methanofullerene C_{60} doped with cobaltocene (*NIPHF-CO*) are very similar to those of the amino compound. A measurement of the temperature dependence of the magnetization (Fig. 10) reveals an onset of the magnetic ordering at temperatures around 15 K. The magnetisation curve at 2 K (insert of Fig. 5) has non-linear behaviour with hysteresis which is narrower than in the amino compound but still clearly observable. Similarly as for the amino compound, M_s does not saturate in fields up to 50 kOe.

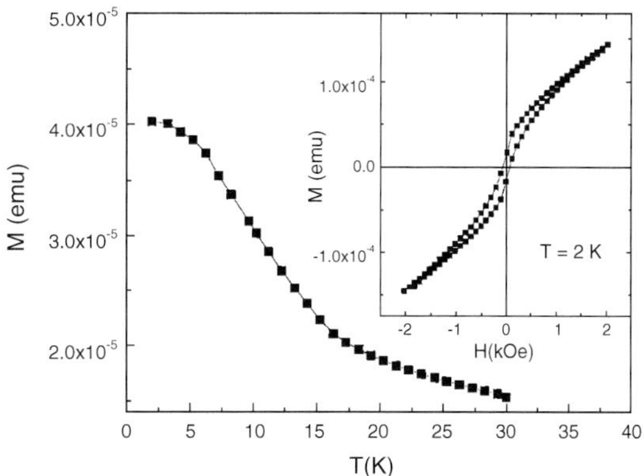

Fig. 10. The magnetisation $\mathbf{M}(T,H)$ as a function of temperature and magnetic field (insert) for *NIPHF-CO*

5. Discussion

There are some distinct and important differences in the magnetic behavior of *APHF-CO* compared to *TDAE*-C_{60}. First, *APHF-CO* shows a clear field-saturation of **M** below T_c in some samples, whereas *TDAE*-C_{60} powder samples typically do not show full saturation [14]. Secondly, the susceptibility χ_{AC} below T_c is temperature-*independent* in *APHF-CO* for $H = 0$, while *TDAE*-C_{60} shows a distinct spin-glass-like cusp at low temperatures [15] and finally in *APHF-CO* there is a conspicuous absence of anomalies in the ESR spectra for $100 < T < 200$ K which are a characteristic feature in *TDAE*-C_{60} [16]. Some of these effects in *TDAE*-C_{60} have been attributed to the molecular rotation of the C_{60} molecules [17, 18], which are believed to lead to a competition between spin-glass and FM interactions at low temperatures. *APHF-CO* and *NIPHF-CO* have inhibited rotational degrees of freedom because of the amino-phenyl or nitro-phenyl adduct, which may explain the absence of the glass-like behaviour and absence of anomalies related to the orientational ordering at intermediate temperatures.

To emphasize the importance of the structure for achieving a FM state, we mention that CoCp_2 salts of *ortho*- *para*-isomers of *APHF* also exhibit a full CT and, judging by their vibrational, NIR and UV-Vis spectra compared to the *meta*-isomer, show no detectable difference in electronic structure. However, repeated attempts at obtaining a FM phase with these isomers have so far proved unsuccessful, the *ortho*- and *para*-isomers showing antiferromagnetic ordering down to 4 K.

The magnetic properties of *APHF-CO* and *NIPHF-CO*, such as the transition temperature, values of the spontaneous magnetisation and the coercive field strongly depend on the details of the preparation procedure, different for each material. The effect of ageing shown in Fig. 4b is particularly unusual and cannot be understood in a simple way. All samples studied so far show a characteristic low value of magnetisation and in some cases an absence of magnetisation. Further, the magnetisation curve is non-linear and hysteretic at low fields (few kOe) and linear at higher applied fields, but this linear part can be small and clearly depends on synthesis conditions. Full saturation has been obtained in *APHF-CO*, with a magnetisation which is saturated already in very low applied fields ($<50\,\mathrm{Oe}$) but with no hysteresis. This was not found to be easily reproducible however. It is too early to speculate whether these compounds are proper ferromagnets with a full saturation of one μ_B per doped spin in crystalline form. In the case of *TDAE*C_{60} this was proven only after high quality single crystals were synthesized.

A systematic study of the choice of synthesis procedure of the fullerene based ferromagnets in correlation of their magnetic properties clearly shows that crystallisation conditions – for example temperature – play a crucial role in determining the magnetic properties.

To conclude, the saturation of the magnetization in both AC susceptibility and SQUID magnetization measurements together with infrared absorbance measurements enables us to classify *APHF-CO* and *NIPHF-CO* as π-electron molecular ferromagnets whose magnetic lattice is composed of single molecular species. *APHF-CO* and has the highest *Curie* temperature $T_c = 19\,\mathrm{K}$ to date for this kind of compound. In contrast to *TDAE*-C_{60}, no spin-glass properties have so far been observed in these materials, but rather distinct characteristics of a proper soft ferromagnet. The antiferromagnetic nature of other cobaltocene doped isomers of *APHF* appears to suggest that the main challenge of synthesizing new organic FM compounds is related to achieving appropriate self-assembly of the D-A pairs into a molecular crystal structure favoring FM over AFM interactions at low temperatures.

6. Outlook

There are very good reasons for finding new organic ferromagnets with higher T_c. Organic materials in general do not require high-temperature synthesis and are soluble in organic solvents, making processing easier for magnetic storage media for example. Moreover they are very light, which makes them potentially very interesting for applications. At present it seems that molecular magnets utilising high spin states of transition-metal ions embedded in an organic matrix are more likely to meet that challenge than fullerene materials.

However, in spite of the fact that common knowledge supposes that the spin–spin interactions are weak in organic materials, there is no *a priori* reason to suppose that ferromagnetism in molecular organic materials based on fullerenes should exist only at low temperatures. The recent report of ferromagnetism at high temperatures in a fullerene compound [19] show that the aims of achieving high T_cs in fullerene-based ferromagnets may indeed be achievable.

The fundamental challenge remains of understanding a spin correlated state composed of entirely *p* electrons still remains. It is clear that a good understanding of the origin of the inter-molecular exchange interactions in disordered molecular materials is essential for designing new organic magnetic materials.

Acknowledgement

I wish to acknowledge *Ales Mrzel* for carefully reading the manuscript.

References

[1] Allemand PM et al. (1991) Science **253**: 301
[2] Li Y et al. (1993) Sol Stat Comm **86**: 475; Ata M et al. (1994) Jap J Appl Phys **33**: 1865; Wang HL, Zhu DB (1994) J Phys Chem Sol **55**: 437
[3] Klos H, Rystau I, Schutz W, Gotschy B, Skiebe A, Hirsch A (1994) Chem Phys Lett **224**: 333; Gotschy B (1996) Fullerene Science and Technology **4**: 677
[4] Kveder et al. (2001) JETP Letters **74**: 422; Konarev DV et al. (2001) Synthetic metals **121**: 1127; Konared DV et al. (2000) J Mol Str **526**: 25; ibid (2000) J Mater Chem **10**: 803
[5] Tanaka K et al. (1992) Int Journal of Mod Phys B **6**: 3953; ibid (1993) Sol Stat Comm **85**: 69
[6] Mrzel A et al. (1998) Carbon **36**: 603
[7] Mrzel A et al. (1998) Chem Phys Lett **298**: 329
[8] Umek P, Omerzu A, Mihailovic D, Tokumoto M (2000) Chem Phys **253**: 361
[9] Blinc R et al. (2002) **88**: 086402
[10] Pokhodnia KI et al. (1999) Journal of Chemical Phys **110**: 3606
[11] Winter J, Kuzmany H (1994) Phys Rev B **49**: 15879
[12] Blinc R et al. (1998) Phys Rev Lett **80**: 1529
[13] Lappas A et al. (1995) Science **267**: 1799
[14] Suzuki A et al. (1994) Chem Phys Lett **223**: 517
[15] Venturini P et al. (1993) Int J Mod Phys B **6**: 3947; Cevc et al. (1994) Sol Stat Comm **90**: 543; Sato T (1997) Phys Rev B **55**: 11055
[16] Mrzel A, Cevc P, Omerzu A, Mihailovic D (1996) Phys Rev B **53**: R2922
[17] Mihailovic D, Arcon D, Venturini P, Blinc R, Omerzu A, Cevc P (1995) Science **268**: 400; Narymbetov B et al. (2000) **407**: 883
[18] Tanaka K et al. (1996) Chem Phys Lett **259**: 574
[19] Makarova TL et al. (2001) Nature **413**: 716

Invited Review

High Spin and Anisotropic Molecules Based on Polycyanometalate Chemistry

Valérie Marvaud*, **Juan M. Herrera**, **Thomas Barilero**, **Fabien Tuyeras**, **Raquel Garde**, **Ariane Sculler**, **Caroline Decroix**, **Martine Cantuel**, and **Cédric Desplanches**

Laboratoire de Chimie Inorganique et Matériaux Moléculaires, CNRS UMR 7071, Case 42, Université Pierre et Marie Curie, 75252 Paris Cedex 05, France

Received July 19, 2002; accepted July 23, 2002
Published online December 19, 2002

Summary. This paper points out some recent achievements in the chemistry and physics of high spin and anisotropic molecules based on polycyanometalate complexes. Following a step by step synthetic strategy and using a localized electron orbital model, isotropic high spin molecules were obtained with ground spin states ranging from $S = 9/2$ to $27/2$. In the same way, anisotropic molecules with various nuclearities (bi, tri, tetra, hexa, and hepta-nuclear complexes) have been synthesized. Mixing these two approaches, it has been possible to obtained anisotropic high spin molecules that behave as single molecule magnets. The paper reviews some of the steps that lead to these findings and some of the prospects opened in the field of single molecule magnets.

Keywords. High spin molecules; Magnetic dendrimers; Molecular magnetism; Polycyanometalate chemistry; Photomagnetism; Single molecule magnet.

1. Introduction

In the field of molecular magnetism [1, 2], the search for new polynuclear molecules displaying high spin ground state raises the interest of synthetic chemists [3–8] and physicists [9–11] involved in nanomagnetism, because it provides new magnetic objects with specific and controlled characteristics. The properties of these complexes may be considered from both classical and quantum approach [12]. Possessing *a priori* well defined and uniform volume, shape, magnetic moment and anisotropy, these large spin molecules exhibit also original magnetism such as single molecule magnet behavior [13–18] (long relaxation time for the magnetization

* Corresponding author. E-mail: marvaud@ccr.jussieu.fr

below a so-called blocking temperature, T_B) or magnetic quantum tunnelling effect [19, 20].

Our research approach has been always based on the synthesis of active molecules able to fulfill a particular function, such as molecular switches through the description of mixed valence species [21, 22], artificial enzymes based on porphyrins dimers and trimers [23, 24] or inorganic dendrimers acting as electrons reservoirs [25, 26]. In the same way, our present work is devoted to the synthesis of anisotropic high spin molecules in order to get single molecule magnets. It is briefly reviewed in the present paper.

2. Single Molecule Magnet

The investigation field is related to the so-called bottom-up approach of nanomagnetism: starting with mononuclear building blocks and reacting them together allows to get well characterized polynuclear complexes with high spins ground state and anisotropy. These species are of great interest to study the reversal of the magnetization in a magnetic field. The topic is both very fundamental (study of single-molecule magnets, evidence and study of the macroscopic quantum tunneling effect,...) and turned to applications (due to a possible enhancement of the information storage at the molecular scale).

The scheme in Fig. 1 allows to present simply the main physical parameters. First, in a polynuclear complex, the coupling between the spins S_M and S_{Mi} might be described by a phenomenological Hamiltonian,

$$\mathbf{H} = -J_{MM'}\left[\sum_{i,j} S_M(i) S_{M'}(j)\right] \quad (1)$$

leading to a spin state diagram. At low temperature, only the ground state is populated and the energy of the first excited state is proportional to the exchange

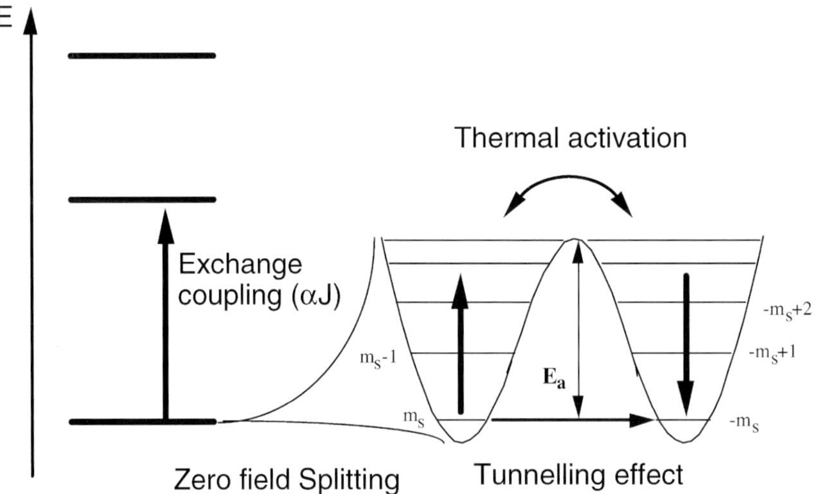

Fig. 1. Schematic plot of single molecule magnet experiencing axial zero field splitting

coupling, J. In presence of an uniaxial anisotropy, a zero-field splitting of the ground spin state is observed and the degeneracy of the ground state might be viewed as a two wells energy potential curve. The activation energy of the anisotropic barrier, E_a, between the spin states describing the two orientations of the magnetization, is a direct function of D (uniaxial anisotropy) and S (ground state spin value) following the equation $E_a = DS^2$ as indicated on Fig. 1.

To obtain a single molecule magnet, *i.e.*, to avoid the relaxation process, the spin as well as the absolute value of the anisotropy have to be as high as possible and D has to be negative so that the state of highest M_S lies below in energy. On the other hand, the exchange coupling, J, between two spin carriers has to be important to well separate the ground spin state from the excited states; the relative high value of the blocking temperature is dependent on it. Playing also an essential role, the intermolecular interaction, J'_{inter}, has to be negligible, necessary condition to avoid three-dimensional ordering and to observe the properties of an isolated nano-scale object. Very few examples of clusters responding to all these criteria have been described in the literature, among them Mn_{12} [15, 16, 27], Mn_4 [11, 14], Fe_8 [10, 28] and more recently a series of molecules by Winnpenny [29] and Murrie [30].

Thus, in order to get a single molecule magnet, four parameters have to be taken into consideration in the design, the conception and the synthesis of the target molecules: the ground state spin value, S, the anisotropy, D, the exchange coupling, J, and the intermolecular magnetic interaction, J'_{inter}. Concerning the spin value, this is a matter of fact that, up to now, in most cases, serendipity has purchased nice and attractive high spin structures of great interest but without an effective control by the chemistry [8]. However, since our understanding increases, there exists undoubtedly a demand for more accurate design of clearly identified target molecules. As part of our research activities devoted to molecular magnetism, we are interested in synthesizing polynuclear compounds showing both large spin ground state value and anisotropy. In the present paper, the spin value, the anisotropy, the exchange coupling and the intermolecular interaction are discussed through the description of polynuclear complexes based on polycyanometalate precursors.

3. Rational Strategy to Obtain High Spin and Anisotropic Molecules

The synthetic strategy that has been initiated by *Mallah* in the laboratory [4, 31] and developed by ourselves [32–35] for the last few years, is based on a step by step modular approach. The key idea is the use of polycyanometalate precursors, as polyfunctional core and of specifically designed mononuclear complexes with only one accessible coordination site, the other sites being blocked by a polydentate chelate ligand. Thus, the reaction between theses two precursors, a *Lewis* base and acid, respectively, leads to polynuclear complexes with a radial symmetry (Fig. 2).

The main interest of such strategy is that the spin value is always high whatever the nature of the magnetic exchange interaction is (ferromagnetic or antiferromagnetic). Of further interest is the high solubility of the target species and their large charge, useful parameters to separate well the molecules from each other and then

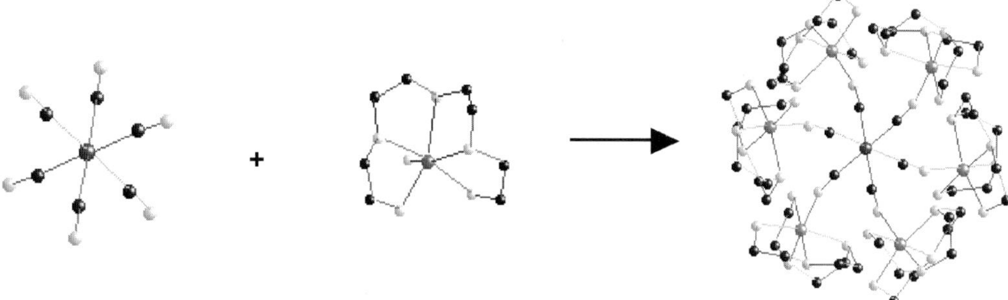

Fig. 2. Synthetic strategy

to minimize the intermolecular interactions. The other important point is the number of synthetic parameters to play with: the nature of the polycyanometalate cores, the polydentate ligand, the metallic ions, the stoichiometry, the counter anions, the solvent, etc... Acting with these different criterions, it has been possible to synthesize a large amount of polynuclear complexes with various nuclearity, structural anisotropy and/or high spin value.

3.1. Control of the spin value [33]

The nature of the metallic ions is the first parameter to modulate the spin value of the target polynuclear complex. Starting with an hexacyanochromate(III) complex and adding mononuclear species $[M(L)(H_2O)]^{2+}$ (with L polydentate ligand, and $M = Cu^{II}$, Ni^{II}, Mn^{II}), it has been possible to synthesize a series of heptanuclear complexes, $[Cr(CN\text{–}M\text{–}L)_6]^{9+}$, ($L =$ *tren*, *tetren*, $M = Cu^{II}$, Ni^{II}, Mn^{II}), noted $CrCu_6$, $CrNi_6$ and $CrMn_6$, characterized in solution by mass spectrometry and in solid state by X-ray crystallography on single crystal (Fig. 3).

The magnetic measurements have been performed on these three compounds, indicating the ferromagnetic nature of the interaction for $CrCu_6$ ($J = +45\,cm^{-1}$) and $CrNi_6$ ($J = +17.3\,cm^{-1}$) and the anti-ferromagnetic interaction for $CrMn_6$

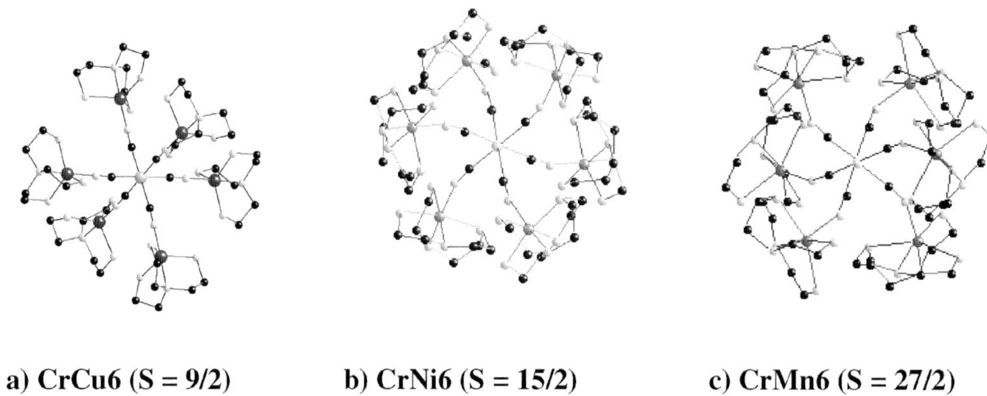

Fig. 3. Perspective view of three heptanuclear complexes: a) $CrCu_6$; b) $CrNi_6$; c) $CrMn_6$

($J = -9\,cm^{-1}$). The first magnetization curves allow us to determine the ground state spin value for the three complexes and the experimental results are in good agreement with the expected value: $CrCu_6$ ($S = 9/2$), $CrNi_6$ ($S = 15/2$) and $CrMn_6$ ($S = 27/2$). A specific work is in progress in order to get heptanuclear complexes based on lanthanide ions and a particular attention is given to a potential $CrGd_6$ complex that might display a spin ground state equal to 45/2.

At low temperature ($T < 2$ K), despite the high spin values, the three compounds $CrCu_6$, $CrNi_6$ and $CrMn_6$, do not behave as single molecule magnet. This is well understood by the isotropy of these heptanuclear complexes and the absence of the anisotropic barrier DS^2.

3.2. Control of the structural anisotropy [34, 35]

The results obtained on the isotropic heptanuclear complexes indicate the necessity of introducing a structural anisotropy, as well as a high spin value. Changing the polydentate ligand and/or the counter anions, we succeeded in getting complexes of various nuclearity based on hexacyanometalate chemistry. Starting from hexacyanocobaltate(III), a thermodynamically stable diamagnetic center, and different Ni(II) mononuclear species, dinuclear (CoNi), *trans*-trinuclear ($CoNi_2$), *fac*-tetranuclear ($CoNi_3$), hexanuclear ($CoNi_5$) has been synthesized and characterized by X-ray crystallography (Fig. 4).

It appears that by using adapted ligands, it is possible to induce a specific nuclearity. Thus, with *dienpy2* and *dipropy2* ligands, trinuclear and tetranuclear complexes have been obtained selectively. The counter ions play also an important role in the formation of well defined polynuclear complexes. Thus, starting with *tetren* ligand in presence of nickel, chloride ions and hexacyanocobaltate(III), a trinuclear complex, $[Co(CN)_4(CN{-}Ni(\mathit{tetren}))_2]Cl$, $CoNi_2$(*tetren*), appears to be the more thermodynamically stable product, instead of the hexanuclear species, $[Co(CN)(CN{-}Ni(\mathit{tetren}))_5](ClO_4)_7$, $CoNi_5$, obtained in similar conditions but in presence of perchlorate ions. Similarly, the CoNi anion is formed when a large cation, *i.e.* tetraphenyl phosphate or tetraphenyl arseniate is introduced in the reaction medium. These results have been also observed with other transition metal ions, such as cobalt(II). All the polynuclear complexes obtained from hexacyanochromate(III) and hexacyanocobaltate(III) are summarized in Table 1.

Following a similar synthetic route $[Cr(CN)_5(CN{-}Ni(\mathit{dienpy2}))](P(\mathit{Phi})_4)$, CrNi ($S = 5/2$), $[Cr(CN)_4(CN{-}Ni(\mathit{tetren}))_2]Cl$, $CrNi_2$(*tetren*) ($S = 7/2$) and $[Cr(CN)_4(CN{-}Ni(\mathit{dienpy2}))_2](ClO_4)$, $CrNi_2$ (*dienpy2*) ($S = 7/2$) have been isolated as single crystals and characterized. Investigations on the magnetic properties of the CrNi and $CrNi_2$ complexes have been performed, indicating in all cases the ferromagnetic nature of the interaction between Cr(III) and Ni(II) through the cyanide ligand, as predicted by *Kahn*'s model based on the orthogonality of orbitals (*vide infra*). On the contrary, anti-ferromagnetic interactions are expected in case of orbital overlap.

The experimental data of the thermal dependence of $\chi_M T$ (χ_M being the magnetic susceptibilty) have been fitted by the *Van Vleck* equation, leading to the intramolecular exchange coupling constants. Concerning the dinuclear CrNi and the trinuclear compounds, $CrNi_2$(*tetren*) and $CrNi_2$(*dienpy2*), the resulting J values

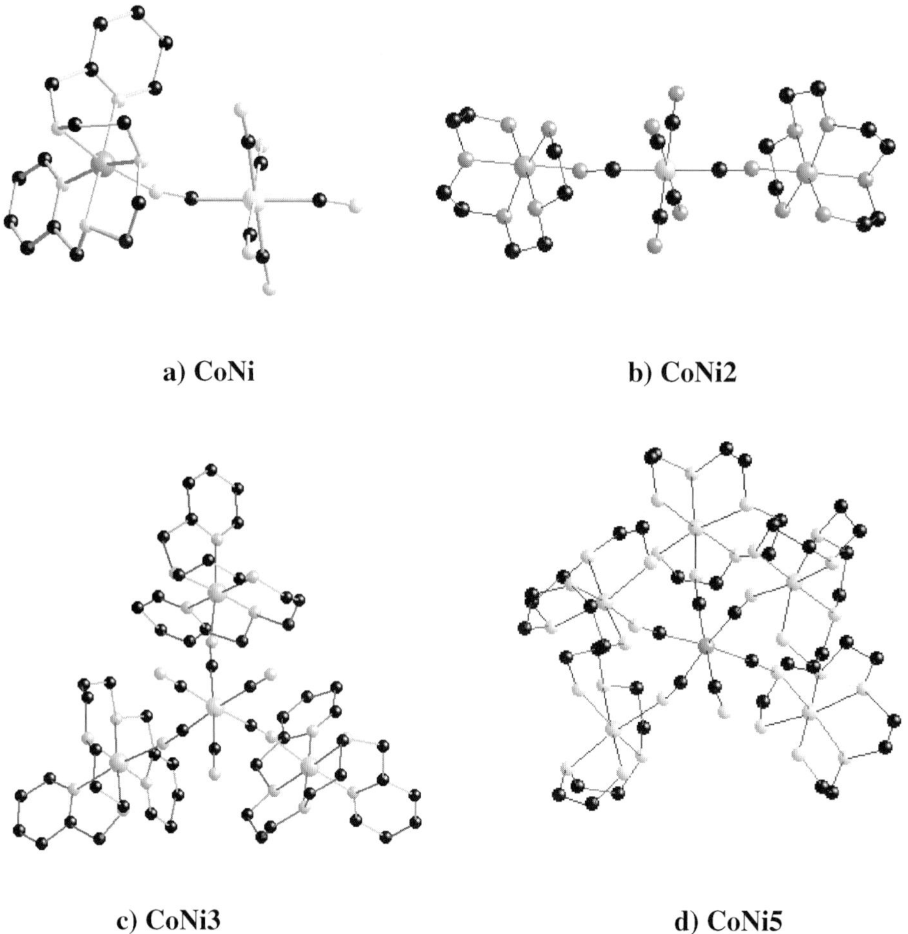

Fig. 4. Perspective view of four polynuclear complexes with different geometry : a) CoNi; b) $CoNi_2$; c) $CoNi_3$; d) $CoNi_5$

are $+18$, $+10$ and $+5\,cm^{-1}$ respectively with an agreement factor of the order of 10^{-4}. The result can be compared to the one obtained for $CrNi_6$ heptanuclear complex ($J = +17.3\,cm^{-1}$). The difference between the four coupling constants is due to the different values of the C–N–Ni angle (*vide infra*).

The magnetic measurements as well as EPR experiments allow us to determine the anisotropy constants, D, for all the polynuclear $CrNi_x$ complexes ($x = 1, 2, 6$). The first results indicate that, on one hand, heptanuclear complexes are isotropic high spin molecules ($D = 0\,cm^{-1}$), and on the other hand, dinuclear, CrNi and trinuclear $CrNi_2$ are anisotropic species ($|D| = 0.3\,cm^{-1}$) but with a modest spin value. Our intent is then to use the tools that have been developed in the previously described synthetic route to obtain simultaneously anisotropy and high spin value. Different strategies will be subsequently described.

3.3. *Magneto-structural correlation [35]*

Seven different "$CrNi_x$" ($x = 1, 2, 6$) compounds have been isolated as single crystal and characterized including X-ray diffraction. The magnetic studies that

Table 1. Geometry, symmetry, spin ground state and interaction in polynuclear complexes synthesized from [B(III)(CN)$_6$] precursors [B(III) = Cr(III), Co(III)]

Nuclearity	Complex	Geometry	Symmetry	Spin	Interaction[a]	J (cm^{-1})
Bi	CrNi		$D_{\infty h}$	5/2	F	+18
	CoNi		$D_{\infty h}$	1		
Tri	CrNi$_2$-**A**	trans	$D_{\infty h}$	7/2	F	+10
	CrNi$_2$-**B**	trans	$D_{\infty h}$	7/2	F	+10
	CrNi$_2$-**C**	trans	$D_{\infty h}$	7/2	F	+10
	CrNi$_2$-**D**	trans	$D_{\infty h}$	7/2	F	+5
	CrNi$_2$-**E**	trans	$D_{\infty h}$	7/2	F	+5
	CoCo$_2$	trans	$D_{\infty h}$	0	af	~0
	CoCu$_2$	cis	$C_2v/D_{\infty h}$	0	af	~0
	CoNi$_2$-**A**	trans	$D_{\infty h}$	0	af	~0
	CoNi$_2$-**B**	trans	$D_{\infty h}$	0	af	~0
Tetra	CrNi$_3$	facial	C_{3v}	9/2	F	+18
	CoCo$_3$	Facial	C_{3v}	frustration	af	~0
	CoNi$_3$	Facial	C_{3v}	frustration	af	~0
	CoCu$_3$	mer		1/2	af	~0
Hexa	CoNi$_5$		C4v	1	af	~0
Hepta	CrCu$_6$-**A**	octahedral	O_h	9/2	F	+45
	CrCu$_6$-**B**	octahedral	O_h	9/2	F	+45
	CrNi$_6$	octahedral	O_h	15/2	F	+17
	CrMn$_6$	octahedral	O_h	27/2	AF	−9
	CoCu$_6$-**A**	octahedral	O_h	0	af	~0
	CoCu$_6$-**B**	octahedral	O_h	0	af	~0
	CoMn$_6$	octahedral	O_h	0	af	~0
	CrNiMn$_5$	octahedral	O_h	10	F/AF	in progress
	CrNi2Mn$_4$	octahedral	O_h	13/2	F/AF	in progress
	CrNi2Ni$_4$	octahedral	O_h	15/2	F	in progress

[a] F = ferro, AF = antiferro, af = weak antiferromagnetic interaction

have been performed on all species indicate in all cases a ferromagnetic interaction between the spin carriers as predicted by orbital models. The exchange coupling value, J, varies from $+4.5\,\mathrm{cm}^{-1}$ to $+18\,\mathrm{cm}^{-1}$ according to the nature of the product and more precisely as a function of the distortion of the cyanide bridge (the C–N–Ni angle).

Two models exist in order to foresee the $|J|$ value based on orbitals consideration: the *Hoffmann* (orthogonalized magnetic orbitals) and *Kahn* (non orthogonalized magnetic orbitals) models. Both predict that orthogonal orbitals give rise to ferromagnetism and that overlaps give rise to antiferromagnetism. One expression summarizes *Kahn*'s models in the case of two electrons 1 and 2 on two sites, described by two identical orbitals a and b. The singlet-triplet energy gap, J ($J = E_S - E_T$) is given by Eq. (2):

$$J = 2k + 4\beta S \tag{2}$$

in which k is the bielectronic exchange integral (positive) between the two non orthogonalized magnetic orbitals a and b; β is the corresponding monoelectronic resonance (or transfer) integral (negative) and S the monoelectronic overlap integral (positive) between a and b. When the two a and b orbitals are different, no rigorous analytical treatment is still available; a semi-empirical relation was proposed by *Kahn* [1, 36]:

$$J = 2k + 2S\,(\Delta^2 - \delta^2)^{1/2} \tag{3}$$

where δ is the initial energy gap between a and b orbitals, Δ is the energy gap between the molecular orbitals built from them.

In such approach, the exchange interaction, J, results from the addition of the two terms ($J = J_F + J_{AF}$), with a positive term, J_F, favoring a parallel alignment of the spins and ferromagnetism, and a negative term, J_{AF}, favoring an antiparallel alignment of the spins and short range antiferromagnetism.

In case of Cr(III) in an O_h symmetry (d^3, $(t_{2g})^3$) and Ni(II) in an octahedral surrounding (d^8, $(e_g)^2$), the magnetic orbitals are orthogonal and ferromagnetic interaction is expected. Two different situations might be considered: (i) if the cyano bridge is close to linearity, the monoelectronic overlap integral is zero, $J = 2k$ and then the absolute value of the exchange coupling is maximum; (ii) on the other hand, the distortion of the cyano bridge induces a weak overlap of the orbitals ($S \neq 0$) leading to a J value that corresponds to Eq. (3). In such a case the $|J|$ value decreases when the overlap increases.

The linear plot of the exchange coupling constant as function of the N–C–Ni angle (Fig. 5) allows to determine that the maximum J value might be expected for strict orthogonality ($J_{max} = +35\,cm^{-1}$) and to predict that below 145 degrees an antiferromagnetic interaction will occur.

3.4. How to obtain anisotropic high spin molecules?

We have discussed previously the selective formation of isotropic high spin molecules through the description of $CrCu_6$ ($S = 9/2$), $CrNi_6$ ($S = 15/2$) and $CrMn_6$ ($S = 27/2$), as well as the synthesis of anisotropic complexes, such as CrNi ($S = 5/2$) and $CrNi_2$ ($S = 7/2$). Several synthetic strategies are presently developed in the laboratory in order to get anisotropic high spin molecules: (i) the synthesis of

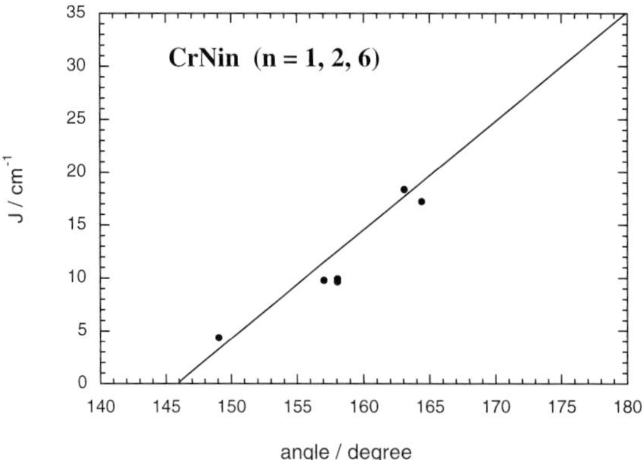

Fig. 5. Magneto structural correlation: $J = f(\theta)$

tetranuclear and hexanuclear complexes based on hexacyanochromate(III) chemistry, $CrNi_3$ and $CrNi_5$ for instance; (ii) the use of anisotropic metallic ions, such as Co(II) and Mn(III); (iii) the formation of heterotrimetallic complexes, *i.e.* $CrNi_2Mn_4$ and (iv) the use of anisotropic polycyanometalate cores.

3.4.1 Tetranuclear complex based on hexacyanochromate(III) chemistry [37]

The synthesis of the tetranuclear complex, $CrNi_3$, has been recently performed in the laboratory, starting from $[Ni(\textit{dipropy2})(H_2O)]^{2+}$ isolated as crystalline perchlorate salt, in presence of the hexacyanochromate(III) precursor. No X-ray structure on single crystal has been solved but the other characterizations (infrared spectroscopy, mass spectrometry, elemental analysis and powder diffraction) and the similarities with the well defined $CoNi_3$ complex converge and confirm the nature of the product. The magnetic measurements on the $CrNi_3$ compounds indicate a ferromagnetic interaction between the spin carriers and corroborate the ground state spin value: $S = 9/2$. First AC measurements and the presence of an hysteresis loop at low temperature (30 mK) attest the presence of anisotropy in the compound that at first sight behave as a single molecule magnet. This result comforts us in our strategy to get, in a rational way, high spin and anisotropic molecules.

3.4.2 Polynuclear complexes based on anisotropic metallic ions [38]

We have seen previously the importance of the anisotropy to obtain a single molecule magnet. The structural anisotropy is the first important and necessary factor. The local anisotropy of the metallic ions is the second significant parameter. It has to be as high as possible: Mn(III), Co(II) and Ni(II) are good candidates for such an approach. The polynuclear complexes based on nickel(II) have been already synthesized and the preliminary results on $CrNi_3$ are in conformity with the expected single molecule magnet behavior. We planned then to build similar complexes based on Co(II) precursors. In presence of cobalt(II), we observed in most of the cases, during the synthesis, the hydrolysis of the hexacyanochromate(III) precursor and the formation of the thermodynamically stable $[Co(III)(CN)_6]^{3-}$. Then, the hexacyanocobaltate(III) reacts with the excess of the mononuclear cobalt(II) species present in solution inducing the formation of $CoCo_2$, $CoCo_3$ or $CoCo_6$ complexes depending on the synthetic conditions. Disappointing for the "single molecule magnets" approach, such compounds might be interesting from other points of view and specially the $CoCo_3$ complex that present an interesting *Landau-Zener* effect. Our observations point out the importance of the inertness of the molecular precursors in the success of our strategy: the bricks must be both stable and inert to control kinetic and entropic hindrances. So that, to obtain high spin and anisotropic molecules based on cobalt(II) peripheric metallic complexes, we are exploring new synthetic conditions. We are also dealing with promising Mn(III) derivatives.

3.4.3 Hetero-trimetallic complexes [39]

In order to get high spin and anisotropic complexes, we have recently developed a new synthetic strategy based on the formation of hetero-trimetallic complexes. The key idea is to use anistropic complexes such as CrNi or $CrNi_2$ as new precursors for the synthesis of polynuclear complexes. Thus, starting with $CrNi_2$ and adding a mononuclear Mn(II) complex to increase the spin value, it has been possible to isolate as single crystal a $CrNi_2Mn_4$ species that has been characterized by X-ray crystallography. The ligand *tetren* used for the nickel and the manganese is the same and no difference between the two metallic ions can be done on the crystallographic structure. Nevertheless, several evidences (cell parameters, mass spectrometry and X-ray fluorescence analysis) show unambiguously the formation of the expected compound. The magnetic studies performed on the compound indicate a high spin value for the ground state, $S = 13/2$. This result might be explained by a ferromagnetic interaction between Cr(III) and Ni(II) and antiferromagnetic interaction between Cr(III) and Mn(II), in agreement with the orbital model and the results obtained previously on $CrNi_6$ and $CrMn_6$. Preliminary results performed at low temperature indicate that, unfortunately, no significant improvement of the anisotropy and no single molecule magnet behaviors have been observed. The high symmetry of the heptanuclear complex and more especially the contribution of local anisotropy of the nickel ions explain this. Indeed, as shown previously, the total anisotropy of the polynuclear complex might be described by the sum of the local (D_i) and the coupling (D_{ij}) anisotropic factors associated to a specific coefficient following the equation:

$$D = \sum_i c_i D_i + \sum_{ij} c_{ij} D_{ij}$$

The calculation performed in collaboration with *Dante Gatteschi* indicates that, despite the high D_{Ni} value, the c_{Ni} parameter is very small and much lower than the c_{Mn} and c_{Cr} coefficients. It explains the small anisotropy of the whole polynuclear $CrNi_2Mn_4$ complex. Furthermore, it allows to determine the more promising anisotropic hetero di- or trimetallic compounds. Thus was built a $CrNi_2Ni_4$ complex (where the nickel(II) ions are surrounded with two different polydentate ligands) that behave as single molecule magnet and comfort us in the multi steps synthetic strategy.

3.4.4 Polynuclear complexes based on anisotropic cores [40, 41]

Most of the results described previously have been obtained with hexacyanometalate cores, such as $[Cr(CN)_6]^{3-}$ or $[Co(CN)_6]^{3-}$. We have discussed the limitation of these precursors, due to their octahedral symmetry, leading to isotropic heptanuclear complexes, with high spin value but no anisotropy. In the aim of building anisotropic high spin molecules, it is possible to use other polycyanometalate precursors, playing the role of *Lewis* base for the synthesis of polynuclear complexes with well-defined nuclearity. Starting with anisotropic cores, $[Cr(\textit{acen})(CN)_2]^+$, $[Cr(Cp)(CN)_3]^+$, $[Cr(\textit{phen})(CN)_4]^-$ or $[Cr(NO)(CN)_5]^{3-}$ for instance, and applying the synthetic strategy used successfully for the heptanuclear complex, it is planed to get selectively tri-, tetra-, penta- and hexa-nuclear complexes. The feasibility of such an approach has been demonstrated through the formation of $NiCu_4$ and $Fe(\textit{phen})Cu_4$ complexes formed from $[Ni(CN)_4]^{2-}$ and $[Fe(\textit{phen})(CN)_4]^{2-}$ starting materials and characterized by X-ray crystallography (Fig. 6) [40].

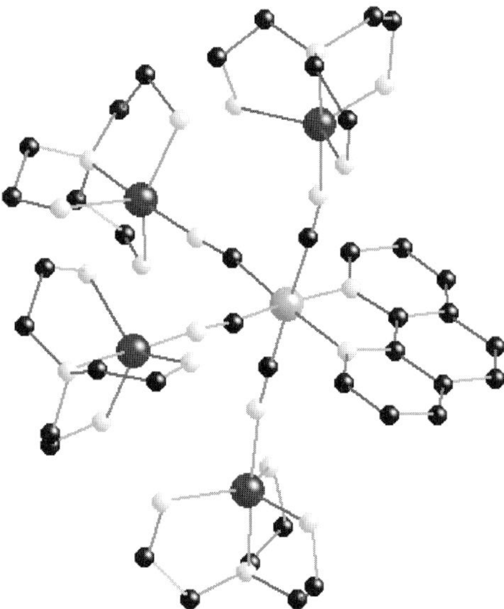

Fig. 6. Perspective view of a pentanuclear complex: Fe(phen)Cu_4

The octacyanometalate precursors might be viewed as an interesting family because of their anisotropic symmetry and the large number of accessible cyanide ligands. This might be of great interest for the spin value as well as the anisotropy. Furthermore, the more diffuse orbitals of these precursors might influence the exchange coupling value. The first attempts to obtain polynuclear complexes based on niobium(IV), molybdenium(V) and tungstate(V) complexes failed due to the hydrolysis of the niobium compounds and the reduction of the two other precursors. Nevertheless, heptanuclear complexes based on octacyanometalate chemistry, $[M(CN)_2(CN{-}M'(L))_6](ClO_4)_8$, ($M = W^{IV}$, Mo^{IV}, $M' = Cu^{II}$, Ni^{II}, and Mn^{II}), WCu_6, WNi_6, WMn_6, $MoCu_6$, $MoNi_6$ and $MoMn_6$, have been synthesized and fully characterized, including by X-ray crystallography. Assuredly, the compounds are made with a diamagnetic core and behave as six isolated paramagnetic species, but these compounds are of great interest for their photo-magnetic properties as discussed further [41].

4. New Achievements and Prospects

4.1. Supramolecular chemistry

The polynuclear complexes based on polycyanometalate chemistry that have been prepared and discussed previously might be used as new starting materials in the synthesis of more elaborate architectures. A large number of species with accessible cyanide ligands are now available, such as CrNi, $CrNi_2$, $CrNi_3$, $CoNi_5$, $MoCu_6$, etc... They are presently used in presence of an assembling mononuclear building block in order to get complexes of high nuclearity and higher spin value. The tools of supramolecular chemistry are widely used, such as the template effect or high dilution experimental conditions. Till now, these synthetic tools have been already used successfully including by ourselves [23] but they are not commonly employed in the molecular magnetic field. Following the routes opened by *Long* [42] very promising works are in progress.

4.2. Magnetic dendrimers

The chemistry of dendrimers, a novel field of research, is of real interest to build identical molecules (strictly monodisperse entities in the language of nanoscience) with a high number of branches. The flexibility of molecular chemistry allows not only to chose the chemical strategy (convergent or divergent) but also to design the branches so that the robust organic skeleton can complex metallic ions and catch at each generation a large and different number of metallic cations. However, very few examples of magnetic dendrimers are published in the literature and most of them are based on radicals [43]. It appears that the design of such architecture achieving at the same time (i) magnetic exchange interaction between the spin carriers, (ii) a high spin value of the ground state, and (iii) convincing characterization is a real challenge that we would like to take off. Our experience in the "dendrimer chemistry" [25, 26] and the molecular magnetism allows us to appreciate the difficulties of the approach, difficulties linked to the synthesis itself that has to involve quantitative yield reactions related to the construction of the skeleton and specific problems imposed by the magnetic requirements. Several synthetic strategies are nevertheless developed by our team in order to obtain magnetic dendrimers. Among them, we are dealing with i) design and synthesis of specific polydentate ligands, including radicals, ii) the design and the synthesis of organic dendrimers including specific coordination sites allowing exchange coupling between the spin bearers, and iii) design and synthesis of polynuclear complexes, their functionalization and a final assembling.

The polynuclear complexes $CrCu_6$ ($S=9/2$), $CrNi_6$ ($S=15/2$) and $CrMn_6$ ($S=27/2$) described previously might be viewed as a first generation of magnetic dendrimers. Using the step by step synthetic strategy that has been developed in our laboratory and specific polydentate blocking ligand made with radicals, it might be possible to get multi-shell polynuclear complexes. In such architectures, the chromium plays the role of a polyfunctional core surrounded by the six metallic ions forming the first generation and the radicals are viewed as the peripheric shell, forming the second generation of spin carriers. The synthesis of compounds of that type has been done by *Rey* and *Verdaguer* allowing the formation of $[Fe_2Ni_3(Rad)_6]$ viewed as a first example of high spin poly-shell complexes [44].

4.3. Photomagnetic high spin molecules

The molecular switch effect, extendedly used in molecular electronics, has always been part of our research activities, with the synthesis of designed species able to control the electron transfer. In molecular magnetism, the synthesis of photo-active materials are also present such as Co–Fe Prussian blue analogues [45–47], spin transition complexes or molecules presenting LIESST [48] or LDLIST [49] effect. But till now no one has succeeded to get a photo-active single molecule magnet. In our group, a very recent field of novelty is the design and the synthesis of photomagnetic high spin molecules. Our aim is then to modify the magnetic properties

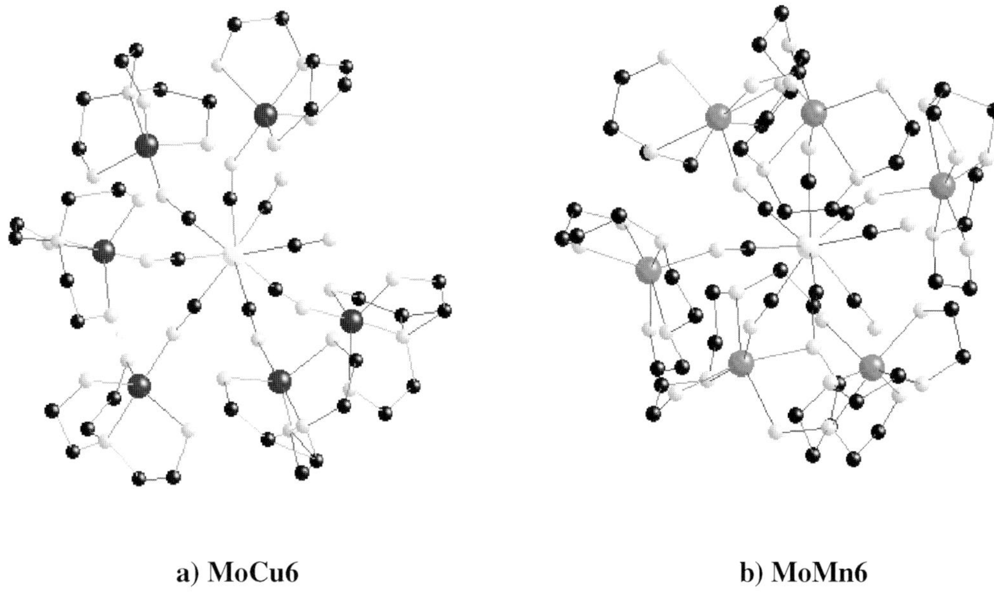

Fig. 7. Perspective view of two heptanuclear complexes: a) $MoCu_6$; b) $MoMn_6$

of the molecular entity using an external perturbation, such as a photonic irradiation. In continuation of the synthetic strategy herein described, we have selectively obtained heptanuclear complexes based on octacyanometalate diamagnetic precursors, Mo(IV) and W(IV). Then $MoCu_6$, WCu_6, $MoNi_6$, WNi_6, $MoMn_6$ and WMn_6 have been obtained and characterized by X-ray crystallography on single crystal and mass spectrometry (Fig. 7).

In collaboration with *C. Mathonière* (Bordeaux) who has previously suggested a possible electron transfer in MoCu Complexes [50], we have studied the photomagnetic properties of $MoCu_6$ complexes and demonstrated that before irradiation, the $Mo(IV)Cu_6$ complexes behave as six isolated paramagnetic centers, as expected by the presence of the diamagnetic center. After irradiation and a supposed electron transfer, the molecule behaves as a high spin molecule ($S = 3$) with ferromagnetic interaction between the spin carriers and the formation of paramagnetic centered species: $Mo(V)Cu(I)Cu(II)_5$. To our knowledge, the $MoCu_6$ appears as the first photomagnetic high spin molecule [41]. One has to consider that the phenomenon is reversible. Work is in progress (i) to evaluate the photomagnetic properties of the other complexes of the octacyanometalate family and (ii) in order to get not only photomagnetic high spin molecules, but also photomagnetic single molecule magnets.

5. Conclusion

We have developed a synthetic strategy and obtained selectively polynuclear complexes of high nuclearity and high spin value. A good understanding of the parameters involved in the single molecule magnet behavior permits us to design specific complexes and the step by step synthetic route allows us to get the tailor-made species with predicted and foreseen properties. The number of compounds obtained are of great interest for chemists to compare and to evaluate the best compounds to prepare, but also for physicists that have a complete family of products with interesting and diversified magnetic properties. The new fields of magnetic dendrimers or photomagnetic high spin molecules appear as promising areas of research for the future.

Acknowledgements

We would like to thank M. Verdaguer for his help and constant support, in this long term work. We thank Corine Mathonière (Bordeaux) for the photomagnetic studies, Wolfgang Wernsdorfer for micro-Squid experiments and Carley Paulsen for low temperatures magnetic studies. We thank the CNRS, the University Pierre et Marie Curie (Paris), the European Community (HPRNCT19990012 and ERBFMRXCT980181) and European Science Foundation for financial support ("Molecular Magnets" Programme).

References

[1] Kahn O (1993) Molecular Magnetism, Wiley, New York
[2] Miller JS and Drillon M (eds) (2001) Magnetism: Molecules to Materials, Wiley, Weinheim
[3] Powell AK, Heath SL, Gatteschi D, Pardi L, Sessoli R, Spina G, del Giallo F, Piralli F (1995) J Am Chem Soc **117**: 2491
[4] Mallah T, Auberger C, Verdaguer M, Veillet P (1995) J Chem Soc Chem Commun 61–62
[5] Zhong ZJ, Seino H, Mizobe Y, Hidai M, Fujishima A, Ohkoshi S, Hashimoto K (2000) J Am Chem Soc **122**: 2952–2953
[6] Pilkington M, Decurtins S (2000) Chimia **54**(10): 593–601
[7] Parker RJ, Spiccia L, Berry KJ, Fullon GD, Moubaraki B, Murray KS (2001) Chem Commun 333
[8] Winpenny REP (2002) J Chem Soc Dalton Trans 1–10
[9] Werndorfer W, Sessoli R (1999) Science **284**: 133–135
[10] Barra AL, Bencini F, Caneschi A, Gatteschi D, Paulsen C, Sangregorio C, Sessoli R, Sorace L (2001) Chem Phys Chem **2**: 523–531
[11] Werndorfer W, Aliaga-Alcade N, Hendrickson DN, Christou G (2002) Nature **416**: 406–409
[12] Leuenberger MN, Loss D (2001) Nature **410**: 789–793
[13] Christou G, Gatteschi D, Hendrickson DN, Sessoli R (2000) MRS Bull **25**(11): 66–71 and reference therein
[14] Hendrickson DN et al. (1992) J Am Chem Soc **114**: 2455–2471
[15] Sessoli R, Gatteschi D, Caneschi A, Novak MA (1993) Nature **365**: 141–143
[16] Sessoli R, Tsai H-L, Schake AR, Wang S, Vincent JB, Folting K, Gatteschi D, Christou G, Hendrickson DN, J Am Chem Soc **115**: 1804–1816
[17] Gatteschi D, Caneschi A, Pardi L, Sessoli R (1994) Science **265**: 1054
[18] Aubin SMJ, Wemple MW, Adams DM, Tsai H-L, Christou G, Hendrickson DN (1996) J Am Chem Soc **118**: 7746–7754
[19] Thomas L, Lionti F, Ballou R, Gatteschi D, Sessoli R, Barbara B (1996) Nature **383**: 145–147
[20] Friedman JR, Sarachik MP, Tejada J, Maciejewski J, Ziolo R (1996) Phys Rev Lett **76**: 3830–3833
[21] Marvaud V, Launay JP (1993) Inorg Chem **32**: 1376
[22] Lainé P, Marvaud V, Gourdon A, Launay J-P, Argazzi R, Bignozzi C-A (1996) Inorg Chem **35**: 711
[23] Marvaud V, Vidal-Ferran A, Webb SJ, Sanders JKM (1997) J Chem Soc Dalton Trans 985
[24] Bampos N, Marvaud V, Sanders JKM (1998) Chem Eur J **4**(2): 335–343

[25] Marvaud V, Astruc D (1997) Chem Commun 773
[26] Sartor V, Djakovitch L, Fillaut J-L, Moulines F, Neveu F, Marvaud V, Guittard J, Blais J-C, Astruc D (1999) J Am Chem Soc **121**: 2929
[27] Lis T (1980) Acta Cryst B **36**: 2042–2046
[28] Gatteschi D, Sessoli R, Cornia A (2000) Chem Commun 725–732
[29] Cadiou C, Murrie M, Paulsen C, Villar V, Wernsdorfer W, Winpenny REP (2001) Chem Commun 2666–2667
[30] Murrie M, Stoeckli-Evans H, Güdel HU, Angew Chem Int Ed **40**: 1957–1960
[31] Scuiller A, Mallah T, Verdaguer M, Nivorozkhin A, Tholence J-L, Veillet P (1996) New J Chem **20**: 1–3
[32] Verdaguer M, Bleuzen A, Marvaud V, Vaissermann J, Seuleiman M, Desplanches C, Scuiller A, Train C, Gelly G, Lomenech C, Rosenman I, Veillet P, Cartier dit Moulin C, Villain F (1999) Coord Chem Rev **190**: 1023–1047
[33] Marvaud V, Scuiller A, Decroix C, Vaissermann J, Guyard-Duhayon C, Gonnet F, Verdaguer M (2002) Chem Eur J (in press)
[34] Marvaud V, Decroix C, Scuiller A, Tuyèras F, Vaissermann J, Guyard-Duhayon C, Gonnet F, Verdaguer M (2002) Chem Eur J (in press)
[35] Scuiller A, Marvaud V, Vaissermann J, Guyard-Duhayon C, Fabrizi de Biani F, Verdaguer M (2002) Chem Eur J (in press)
[36] Girerd JJ, Journeaux Y, Kahn O (1981) Chem Phys Lett **82**: 534
[37] Tuyèras F, Marvaud V et al. (manuscript in preparation)
[38] Marvaud V et al. (manuscript in preparation)
[39] Barilero T, Marvaud V et al. (manuscript in preparation)
[40] Garde R, Marvaud V et al. (manuscript in preparation)
[41] Herrera J-M, Marvaud V et al. (manuscript in preparation)
[42] Berseth PA, Sokol JJ, Shores MP, Heinrich JL, Long JR (2000) J Am Chem Soc **122**(40): 9655
[43] Rajca A, Wongsriratanakul J, Rajca S, Cerny R, Angew Chem Int Ed Engl **37**(8): 201
[44] Vostrikova KE, Luneau D, Wernsdorfer W, Rey P, Verdaguer M (2000) J Am Chem Soc **122**(4): 718–719
[45] Sato O, Iyoda T, Fujishima A, Hashimoto K (1996) Science **272**: 704
[46] Verdaguer M (1996) Science **272**: 698
[47] Bleuzen A, Lomenech C, Escax V, Villain F, Varret F, Cartier dit Moulin C, Verdaguer M (2000) J Am Chem Soc **122**(28): 6648
[48] a) Decurtins S, Gütlich P, Köhler CP, Spiering H, Hauser A (1984) Chem Phys Lett **105**: 1. b) Gütlich P, Hauser A (1990) Coord Chem Rev **97**: 1
[49] Boillot M-L, Roux C, Audière J-P, Dausse A, Zarembowitch J (1996) Inorg Chem **35**: 3975
[50] Rombaut G, Verelst M, Golhen S, Ouahab L, Mathonière C, Kahn O (2001) Inorg Chem **40**(6): 1151

Invited Review

Spin Crossover Properties of the [Fe(*PM-BiA*)$_2$(NCS)$_2$] Complex – *Phases I* and *II*

Jean-François Létard[1,*], **Guillaume Chastanet**[1], **Olivier Nguyen**[1], **Silvia Marcén**[1], **Mathieu Marchivie**[1], **Philippe Guionneau**[1,*], **Daniel Chasseau**[1], and **Philipp Gütlich**[2]

[1] Groupe des Sciences Moléculaires, Institut de Chimie de la Matière Condensée de Bordeaux, UPR CNRS No. 9048, F-33608 Pessac, France

[2] Institut für Anorganische Chemie und Analytische Chemie, Universität Mainz, D-55099 Mainz, Germany

Received August 26, 2002; accepted August 30, 2002
Published online November 21, 2002

Summary. In the present review, we reexamine the photomagnetic properties of the [Fe(*PM-BiA*)$_2$(NCS)$_2$], *cis*-bis(thiocyanato)-bis[(*N*-2′-pyridylmethylene)-4-(aminobiphenyl)]iron(II), compound which exhibits, depending on the synthetic method, an exceptionally abrupt spin transition (*phase I*) with a very narrow hysteresis ($T_{1/2}\downarrow = 168$ K and $T_{1/2}\uparrow = 173$ K) or a gradual spin conversion (*phase II*) occurring at 190 K. In both cases, light irradiation in the tail of the ^{1}MLCT-LS absorption band, at 830 nm, results in the population of the high-spin state according to the light-induced excited spin-state trapping (LIESST) effect. The capacity of a compound to retain the light-induced HS information, estimated through the *T*(LIESST) experiment, is determined for both phases. Interestingly, the shape of the *T*(LIESST) curve is more gradual for the *phase II* than for the *phase I* and the *T*(LIESST) value is found considerably lower in the case of the *phase II*. The kinetics parameters involved in the photo-induced high-spin → low-spin relaxation process are estimated for both phases. From these data, the experimental *T*(LIESST) curves are simulated and the particular influence of the cooperativity as well as of the parameters involved in the thermally activated and tunneling regions are discussed. The Light-Induced Thermal Hysteresis (LITH), originally described for the strongly cooperative *phase I*, is also reinvestigated. The quasi-static LITH loop is determined by recording the photostationary points in the warming and cooling branches.

Keywords. Spin crossover; Iron(II); Photomagnetism; Kinetic relaxation.

* Corresponding authors. E-mail: letard@icmcb.u-bordeaux.fr

1. Introduction

Today, a promising strategy in the development of information technology is offered by the small-upward (bottom-up) approach where molecules or assemblies of molecules are used for information processing [1]. In this context, molecular bistability, that is the ability of a molecular system to occur in two different electronic states, is a particularly interesting property since logic operations can be executed. One of the most fascinating example of molecular bistability is certainly given by the spin crossover (SC) phenomenon discovered by *Cambi et al.* in 1931 on iron(III) dithiocarbamate complexes [2], and some 30 years later by *Baker* and *Bobonich* on the first iron(II) coordination compound [3]. Many examples of SC complexes, particularly of iron(II) and iron(III) became shown thereafter [4–6]. In fact, it is only during the 1980's that researchers realized that SC compounds could be used as active elements in memory devices [7] and the iron(II) transition metal ion was particularly studied [8]. In this latter metal case, at the molecular scale, the SC phenomenon corresponds to an intra-ionic transfer of two electrons occurring in the nanosecond scale between the e_g and t_{2g} orbitals. The phenomenon may be induced by a change of temperature, of pressure, or by light irradiation [8].

The first photomagnetic effect on an iron(II) SC compound was reported by *Decurtins et al.* in 1984 [9]. The authors demonstrated the possibility to convert a low-spin (LS) state into a metastable high-spin (HS) state at sufficiently low temperatures ($\leq$50 K) by using green light irradiation. This phenomenon was named the Light-Induced Excited Spin-State Trapping (LIESST effect). Later on, *Hauser* proved that red light switches the system back from the HS to the LS state (reverse-LIESST) [10]. Today, the LIESST effect has been observed in a large number of iron(II) systems [8, 11]. From all these examples, we have learned that due to the large increase in metal-ligand bond lengths during the light-induced population of the HS state at low temperatures, the role of the elastic interactions are particularly important for the interpretation of the photomagnetic data [12]. The resulting elastic interactions may be pictured as an internal pressure which is proportional to the concentration of the low-spin species. In diluted systems, the HS $\rightarrow$ LS relaxation is single exponential whereas in concentrated systems the curves follow a sigmoidal behavior due to a self-accelerating process resulting from the cooperative interactions.

Another important feature of the LIESST effect is that trapping the system in a metastable HS state requires very low temperatures. Usually, above 50 K, the metastable HS state decays within minutes and the "information" is erased. The relaxation process follows the non-adiabatic multi-phonon theory with a temperature independent HS $\rightarrow$ LS relaxation below $\sim$50 K [13] and a thermally activated process at elevated temperature which can be regarded as a tunneling from thermally populated vibrational levels of the HS state [14]. Nevertheless, in some unexplained cases, long-lived metastable HS states have been reported even at elevated temperatures [15, 16], suggesting that some unknown secondary reactions in the nearby lattice surroundings of a light-generated HS complex molecule are able to stabilize the metastable HS state. In this context, we have introduced the notion of critical LIESST temperature, *T*(LIESST), defined as the temperature for which the light-induced HS information is erased [17]. Our idea was to constitute a data base which collects the photomagnetic properties of a large number of iron(II) SC compounds. Today, more than thirty SC compounds have been characterized with the same *T*(LIESST) procedure. It seems already that a direct relation exists between the T(LIESST) value and the thermal spin transition temperature, ($T_{1/2}$); *i.e.* $T(\text{LIESST}) = T_0 - 0.3T_{1/2}$ with T_0 estimated at $T \rightarrow 0$ [18, 19]. The influence of the coordination sphere (nature of the organic core, of the anion, of the hydration degree) and of the cooperativity on the magnitude of the T(LIESST) is currently investigated [19].

In the present paper, we review the particular case of the [Fe(*PM-BiA*)$_2$(NCS)$_2$], *cis*-bis(thiocyanato)-bis[(*N*-2′-pyridylmethylene)-4-(aminobiphenyl)] iron(II), compound. This system has been selected by reason of the presence of two phases exhibiting different magnetic and photomagnetic properties [17, 20–27]. The thermal SC regimes is either gradual (*phase II*) or exceptionally abrupt (*phase I*) with a narrow hysteresis of ca. 5 K ($T_{1/2}\downarrow = 168$ K and $T_{1/2}\uparrow = 173$ K) and, consequently, the particular role of the cooperativity on the T(LIESST) value will be investigated.

In Section 2, some highlights concerning the thermal spin crossover properties of the [Fe(*PM-BiA*)$_2$(NCS)$_2$] compound will be recalled. In Section 3, on the basis of new results the photomagnetic properties of both phases will be reexamined. The Light-Induced Thermal Hysteresis (LITH) [17], originally described for the strongly cooperative *phase I*, will also be reinvestigated. The various kinetics parameters involved in the HS $\rightarrow$ LS relaxation will be determined and the T(LIESST) curve will be simulated. In Section 4, the kinetics parameters of the strongly cooperative *phase I* will be compared to the weak cooperative *phase II*. The influence of these parameters on the T(LIESST) value will be analyzed.

2. Highlights on the [Fe(*PM-BiA*)$_2$(NCS)$_2$] Complex

2.1. Synthesis

The first description of the synthesis of the [Fe(*PM-BiA*)$_2$(NCS)$_2$] complex was reported in 1997 and the analysis of the SC properties recorded for the powder sample form gave a gradual spin transition [20]. Later on in 1998, we have reported that a change of the synthetic conditions allowed to obtain a new phase with an exceptionally abrupt spin transition [17]. The abrupt form was arbitrarily named *phase I* and the gradual spin conversion *phase II*. Today, the synthesis of the two phases has been performed several times. From all these experiments, we have learned that the main difference in the synthesis procedure concerns the speed to obtain the [Fe(*PM-BiA*)$_2$(NCS)$_2$] complex. In some intermediate case, the final powder compound corresponds to a mixture of both phases and the magnetic data revealed the summation of both behavior.

The procedure to obtain the *phase I* was as following. Under nitrogen atmosphere, 2.7 g of iron(II) sulfate heptahydrate, $Fe(SO_4)_2 \cdot 7H_2O$, (9.7 mmol) and 1.8 g of potassium thiocyanate, KNCS, (19 mmol) were dissolved in 20 cm^3 of freshly distillated methanol [17]. The presence of ascorbic acid was used to prevent the iron(II) oxidation. The colorless solution of $Fe(NCS)_2$ was separated from the white precipitate of potassium sulfate by filtration, and added dropwise to a stoichiometric amount of *PM-BiA*, *N*-2′-pyridyl-methylene-4-aminobiphenyl (5 g, 19 mmol) in 20 cm^3 of methanol. The green [Fe(*PM-BiA*)$_2$(NCS)$_2$] precipitate was then slowly formed. The *PM-BiA* ligand was synthesized from the 2-pyridinecarbaldehyde and the aminobiphenyl.

The *phase II* was obtained by a fast precipitation of the [Fe(*PM-BiA*)$_2$(NCS)$_2$] complex [20, 21]. For this, we have used an excess of *PM-BiA* ligand (1 g, 3.88 mmol) in regard to $Fe(SO_4)_2 \cdot 7H_2O$, (0.27 g, 0.97 mmol) and KNCS (0.189 g, 1.94 mmol) constituents and moreover the methanol quantity was limited to 5 cm^3.

2.2. Thermal Spin Transition Behavior

Figure 1 compares the spin transition properties recorded for the two phases of the [Fe(*PM-BiA*)$_2$(NCS)$_2$] compound [17, 20–22]. At room temperature, the $\chi_M T$ product (χ_M stands for the molar magnetic susceptibility and T the temperature) is close to 3.5 cm^3 K mol^{-1}, corresponding to what is expected for a quintet HS ground state. When the temperature is lowered, the magnetic signal of the *phase II* gradually decreases while the $\chi_M T$ product of the *phase I* drops suddenly within 2 K around 168 K. Below 150 K, the magnetic responses of both phases are close to zero, indicating the presence of solely a singlet LS ground state. In the warming mode, the $\chi_M T$ product of the *phase II* smoothly increases and the magnetic data perfectly correspond to those obtained in the cooling mode, showing that a complete gradual spin transition occurs at $T_{1/2} = 198$ K without thermal hysteresis (Fig. 1a). For *the phase I*, the magnetic data recorded during the warming mode abruptly increase, within 1 K, at 173 K. The compound exhibits a complete abrupt spin transition with a thermal hysteresis loop of 5 K ($T_{1/2}\downarrow = 168$ K and $T_{1/2}\uparrow = 173$ K, see Fig. 1b).

The thermal spin crossover properties of the *phase I* has also been investigated by *Mössbauer* spectroscopy. Figure 2 shows a representative series of *Mössbauer* spectra recorded at various temperatures. From the spectra recorded at 200 K (and above) and at 140 K (and below), it has been checked that the thermal spin transition of the *phase I* is complete. For instance, the spectrum at 200 K presents a unique doublet characteristic of iron(II) in HS state, the isomer shift δ being 0.91(1) mm/s (relative to α-iron) and a quadrupole splitting E_Q being 2.82(1) mm/s, while the spectrum at 140 K shows only the doublet characteristic of the iron(II) LS state with

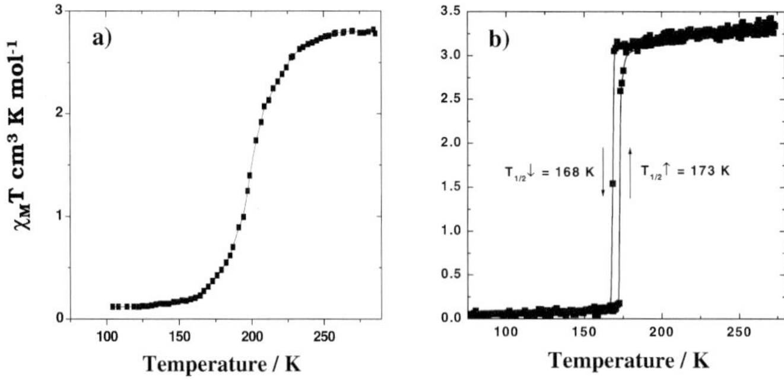

Fig. 1. $\chi_M T$ product *vs.* T (χ_M stands for the molar magnetic susceptibility and T the temperature) in the 80–300 K region of [Fe(*PM-BiA*)$_2$(NCS)$_2$] in both warming and cooling modes. (a) *phase II*; (b) *phase I*. The thermal spin transition has been followed with a Manics DSM-8 fully automatized *Faraday*-type magnetometer equipped with an DN-170 Oxford Instruments continuous-flow cryostat and a BE 15f Bruker electromagnet operating at *ca.* 0.8 Tesla and in the 80–300 K temperature range

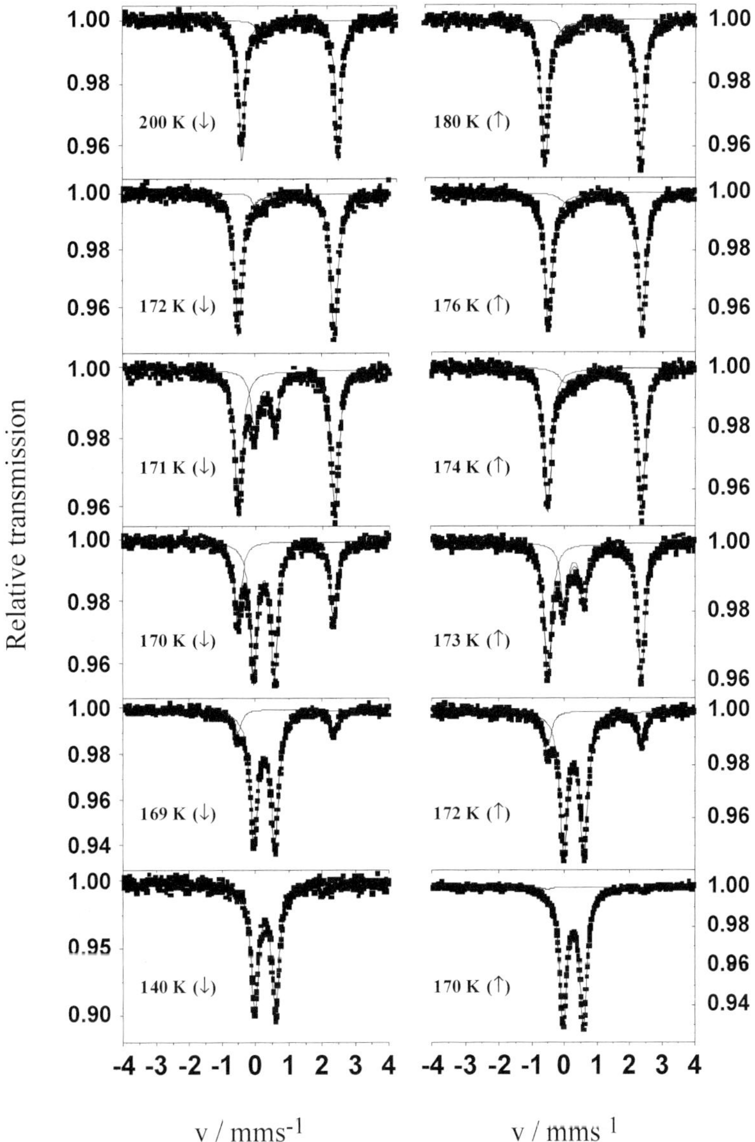

Fig. 2. *Mössbauer* spectra of *phase I* at different temperatures

$\delta = 0.28(1)$ mm/s and $E_Q = 0.62(1)$ mm/s. The spectra recorded in the vicinity of the 168–172 K region show the coexistence of the two spin isomers with strongly temperature-dependent relative intensities. Figure 3 reports the calculated HS area fractions as function of the temperature during the cooling and warming modes. This result has confirmed the exceptional abrupt character of the thermal spin transition and the presence of a thermal hysteresis of 5 K obtained by magnetic susceptibility measurements.

In the particular case of the strongly cooperative *phase I*, pressure experiments have also been performed [22]. The aim was to study the effect of an external pressure on the hysteresis loop. Figure 4 recalls the magnetic properties obtained

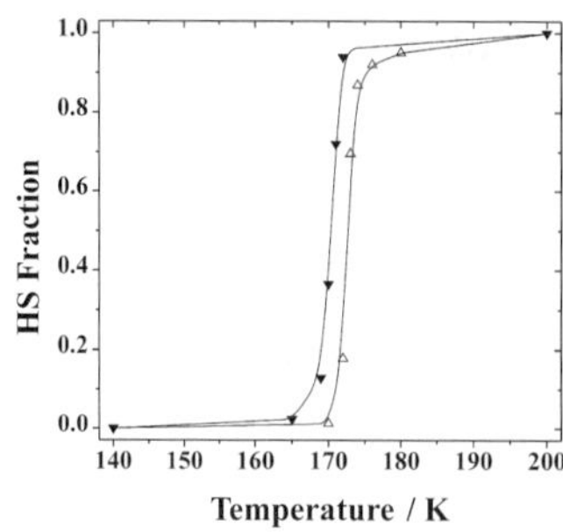

Fig. 3. Plot of the HS fraction *vs.* T of *phase I* determined by *Mössbauer* spectroscopy in the warming and the cooling modes

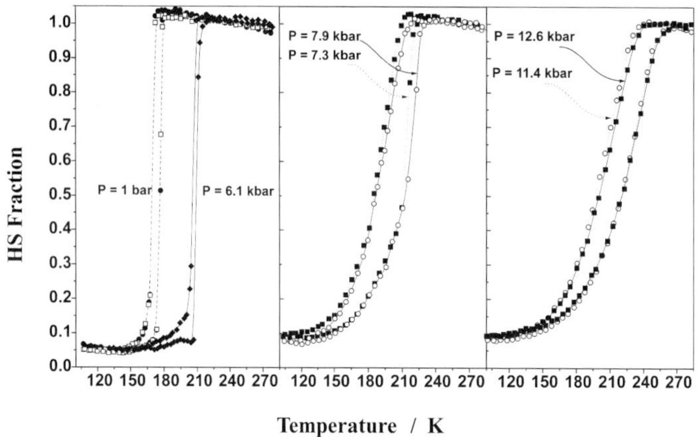

Fig. 4. HS fraction under hydrostatic pressures for the *phase I*

under hydrostatic pressure. When the pressure increases from 1 bar to 6.1 kbar, the transition temperature $T_{1/2}$ was observed to shift upwards and the width of the hysteresis was reduced. At higher pressures, 7.3 kbar and 7.9 kbar, the hysteresis width became large (25 K), and beyond 10 kbar, the spin crossover was found to become more gradual and the hysteresis width slightly diminished. The increase of the hysteresis loop in the vicinity of 7–8 kbar has been interpreted as the presence of a new phase transition [22]. Nevertheless, whatever the nature reached in the high pressure domain, these data show that the width of the thermal hysteresis remains smaller than 25 K. It was not possible to drastically affect the cooperativity of the *phase I*, linked to the supramolecular organization of the aromatic ligand, by an external pressure perturbation.

2.3. X-Ray Investigation

The [Fe(*PM-BiA*)$_2$(NCS)$_2$] compound is a member of the [Fe(*PM-L*)$_2$(NCS)$_2$] family with *PM-L* the *N*-2′-pyridylmethylene-4-aminoaryl ligand. Interestingly, this family of complexes offers a large diversity of the spin conversion features together with, in a first approach, similarities in the structural properties [23–27]. A careful comparative examination of the crystal structures determined both in the

HS and the LS states sheds light on some differences within the crystal packing. These differences have turned out to be directly connected to the diversity of the spin conversion features. For instance, the paramount role of hydrogen bonds has been pointed out.

The crystal structures of the *phase I* have been determined at 293 K, 140 K [17] and 30 K [25] and those of *phase II* at 293 K and 120 K [27]. The temperature dependence of the cell parameters has also been determined for both phases and the thermal expansion tensors discussed [23–27]. It is important to note that the single crystals of *phase I* and *phase II* are hardly differentiable by microscope observation as they show no significant differences in morphology. In order to identify the nature of a batch of samples (in powder or single crystal form), a quick X-ray powder experimental diffractogram have been compared to the ones calculated from the single crystal X-ray crystal structures of the two phases. Indeed, the *phase I* and *phase II* diffractograms appear to be notably different (Fig. 5).

At room temperature, *phase I* crystallises in the orthorhombic space group Pccn ($a = 12.949(7)$ Å, $b = 15.183(2)$ Å, $c = 17.609(5)$ Å, $V = 3462(2)$ Å^3) and *phase II* in the monoclinic space group $P2_1/c$ ($a = 17.358(5)$ Å, $b = 12.602(2)$ Å, $c = 17.570(5)$ Å, $\beta = 115.68(1)°$, $V = 3464(2)$ Å^3). Interestingly, the unit cell volumes of both phases are identical and the unit cell contains four entities. In the orthorhombic phase, iron atoms lie on a two-fold axis and consequently the molecule is symmetrical, which is not the case in the monoclinic phase. The main intramolecular differences in the *phase II* concerns one of the two external phenyl groups of the complex which displays very large atomic displacement parameters. Otherwise, the geometry of the FeN_6 octahedron is very similar in both phases at room temperature. For instance, the Fe–N bond lengths as well as the associated distortion parameter Σ of the two phases are very similar. Likewise, despite the difference in space group, the crystal packing is also similar (Fig. 6). One of the differences comes from the perfect alignment of the iron atoms, for symmetrical reasons, in *phase I* while these atoms form a zigzag arrangement in *phase II*. The packing is driven by π–π interactions between neighbouring complexes as well as intermolecular hydrogen like contacts involving the NCS branches. These intermolecular interactions show some

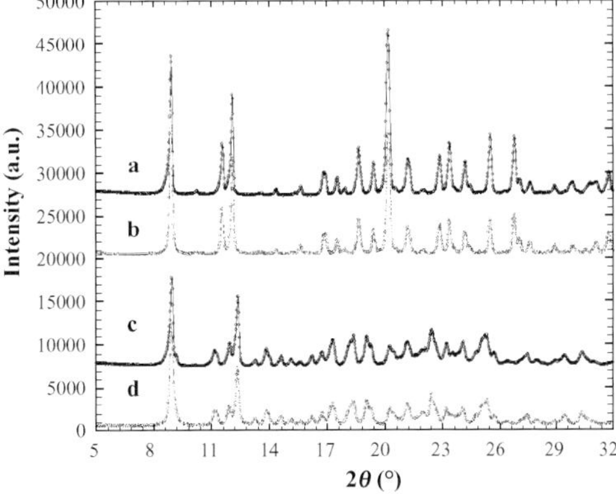

Fig. 5. Simulated and experimental powder diffractograms of [Fe(*PM-BiA*)$_2$(NCS)$_2$] in *phase I* (a, b respectively) and in *phase II* (c, d respectively)

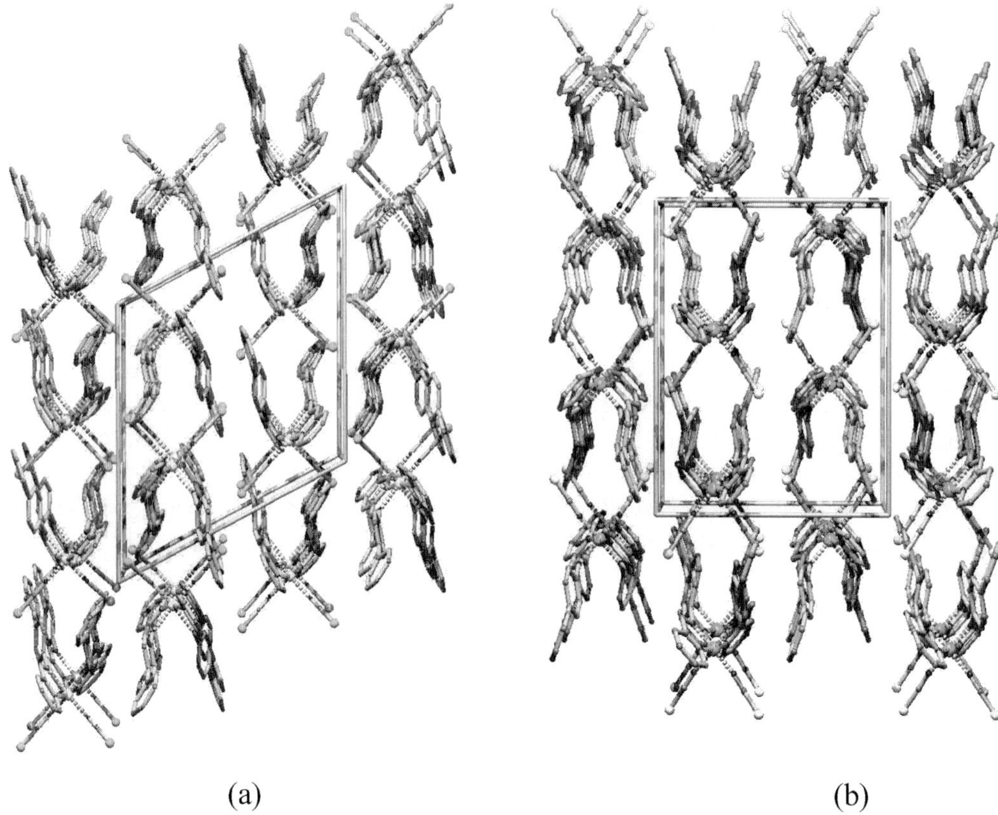

Fig. 6. Crystal packing and unit cell of [Fe(*PM-BiA*)$_2$(NCS)$_2$] in the monoclinic phase (a) and in the orthorhombic phase (b)

small differences in both structures but the main difference involves the sulphur atoms. Indeed, the intermolecular hydrogen like contacts S···HC is significantly shorter in *phase I* (S···C: 3.41(1) Å) than in *phase II* (S···C: 3.54(1) Å).

At low temperature, below the spin crossover, no essential modification occurs in the crystal structures. From 293 K to 140 K, the amplitude of contraction of the unit cell is slightly larger in *phase II* (149 Å^3) than in *phase I* (125 Å^3). On going from HS to LS, the amplitude of the geometrical modifications of the FeN_6 octahedron are comparable: the average variation of bond lengths is $\Delta r = 0.218$ Å in *I* and $\Delta r = 0.200$ Å in *II* and the Σ parameter decreases by 39° in *I* and 37° in *II*. The FeN_6 octahedron volume is reduced by 3 Å^3 at the spin crossover in both phases like for all the Fe(II) complexes investigated up to now [26]. Interestingly, the length of the intermolecular hydrogen like contacts S···HC, discussed above, becomes identical for both phases in the low spin state: S···C: 3.44(1) Å for *phase I* and 3.45(1) Å for *phase II*.

3. Light-Induced Excited Spin-State Trapping (LIESST)

3.1. UV-Visible Absorption

The absorption spectra of [Fe(*PM-BiA*)$_2$(NCS)$_2$] have been performed for both phases at room temperature under various experimental conditions, in KBr pellet, in cellulose acetate film, as well as in solution of acetonitrile and of butyronitrile. All the spectra are similar. In acetonitrile, the electronic absorption spectrum contains mainly three absorption bands at 256 nm ($\varepsilon = 38000\,l\,mol^{-1}\,cm^{-1}$), at 337 nm ($\varepsilon = 22600\,l\,mol^{-1}\,cm^{-1}$) and at 585 nm ($\varepsilon = 1230\,l\,mol^{-1}\,cm^{-1}$). This spectrum is typical for an iron(II) spin-crossover compound where the metal ion is conjugated with an aromatic ring [28, 29]. The electronic transitions below 350 nm can be assigned to the $\pi \rightarrow \pi^*$ transitions of the *PM-BiA* ligand and those around 590 nm to the metal-to-ligand charge-transfer bands of the LS state (^{1}MLCT-LS).

The evolution of the UV-visible spectra as function of temperature has been followed in butyronitrile solution and in cellulose acetate film. In both cases, the decrease of the temperature results in an enhancement of the ^{1}MLCT-LS band. At low temperature, the general aspect of the UV-Visible spectra recorded at room temperature is conserved and the color of the compound remains unchanged. This particular finding contrasts with what happens for tetrazole and triazole SC derivatives, where the spin transition is accompanied by a drastic color change between violet (or pink) in the LS state and white in the HS state [7]. In fact, in the [Fe(*PM-BiA*)$_2$(NCS)$_2$] compound, the intense and broad ^{1}MLCT-LS band situated in the visible range totally overlaps with the forbidden d–d transitions of the LS and HS states and any temperature dependence of the d–d transitions remains insignificant.

3.2. LIESST Experiments

In 1998, it was reported that the metastable HS state can be generated at 10 K by irradiating the *phase I* with a Kr^+ Laser ($\lambda = 647.1$–676.4 nm), through the well-known LIESST effect. One year later, a similar behavior was described for the *phase II*. In both cases, the photoinduced HS fraction at 10 K was estimated to be around 20%. Figure 7 shows the photomagnetic properties recorded on both phases by irradiating with a diode laser emitting at 830 nm. Obviously, the magnetic signal at 10 K under irradiation rapidly increases. The $\chi_M T$ product reached after saturation

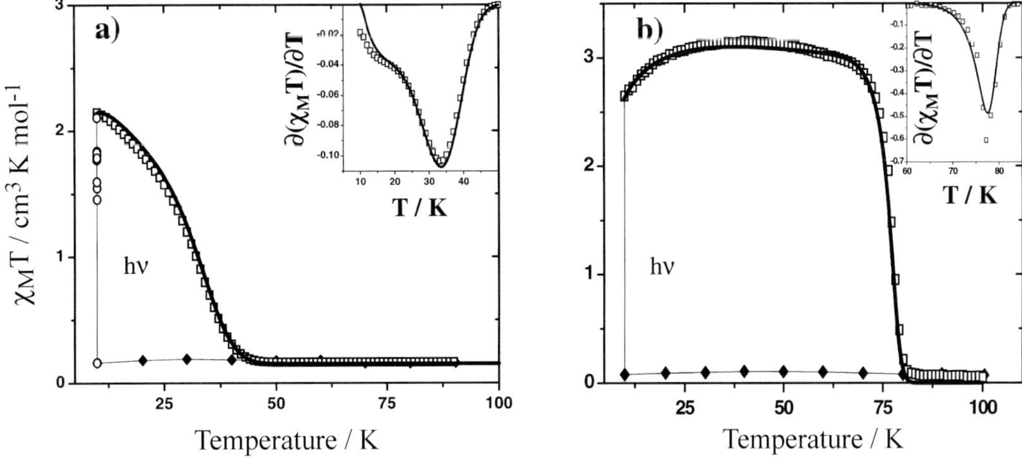

Fig. 7. Evolution of $\chi_M T$ product *vs.* T before (◆), during (o) and after (□) irradiation at 830 nm. (a) *phase II*; (b) *phase I*. The minimum of the derivative of $\chi_M T$ *vs.* T (inserts) gives the T(LIESST) value. The solid lines represent the simulation obtained according to Eq. (7)

was close to 2.5 $cm^3\,K\,mol^{-1}$ for the *phase I* and close to 2.1 $cm^3\,K\,mol^{-1}$ for the *phase II*. Figure 7 also reports on the *T*(LIESST) experiments. At the saturation of the magnetic signal reached under light irradiation (830 nm) at 10 K, the light was turned off and the temperature was slowly increased at a rate of 0.3 $K\,min^{-1}$. Interestingly, these *T*(LIESST) curves are strongly different. The magnetic response of the *phase II* continuously decreases with increasing temperature and the *T*(LIESST) value is 34 K. On the contrary, the $\chi_M T$ product of the *phase I* is almost independent of temperature up to *ca.* 70 K and the *T*(LIESST) value is estimated to be 78 K. The magnetic response of the *phase I* only decreases in the vicinity of 75–79 K.

From these photomagnetic data it follows: i) The increase of the magnetic signal recorded at 10 K under irradiation at 830 nm suggests that direct LIESST process occurs for both phases. This differs from the reverse-LIESST effect previously observed in these experimental conditions for tetrazole derivatives. In fact, this behavior can be understood by the presence of an intense ^{1}MLCT-LS electronic transition which totally masks the d–d transition of the HS state expected around 800–850 nm. Consequently, irradiation at 830 nm, in the ^{1}MLCT-LS absorption band, induced the population of the metastable HS state through the direct LIESST process. ii) The second remark refers to the comparison of the $\chi_M T$ product determined at the maximum of the *T*(LIESST) curves and the value recorded at room temperature. For the *phase I*, the light irradiation at 830 nm induces a nearly quantitative LS $\rightarrow$ HS photoconversion while light irradiation at 647.1–676.4 nm or at 532 nm only generates 20% of photoinduced HS fraction. This points out the particular role of the ^{1}MLCT-LS absorption band. At 830 nm the irradiation occurs on the borderline of the ^{1}MLCT-LS absorption band while at 647.1–676.4 nm or at 532 nm it is much around the maximum. Consequently, the bulk attenuation of the light intensity is lower at 830 nm than in the visible range where the opacity of the sample linked to the strong ^{1}MLCT-LS absorption prevents the light penetration. iii) The last comments concern the *T*(LIESST) experiment. From Fig. 7, it is clear that the shape of the *T*(LIESST) curves is more gradual for the *phase II* than for the *phase I*. The *T*(LIESST) experiment seems to be the mirror image of the thermal spin transition (Fig. 1). A similar comment has been recently given in the case of the $[Fe(bpp)_2]X_2{\cdot}nH_2O$ series (*bpp* = 2,6-bis(pyrazol-3-yl) pyridine) [19] exhibiting gradual and abrupt thermal spin transitions. In parallel, it is also interesting to note that the *T*(LIESST) value of the *phase II* is lower than that of the *phase I*. Basically it is expected that the *T*(LIESST) value decreases with increasing $T_{1/2}$, in agreement with the inverse-energy-gap law [14] and with the recently observed relation $T(\text{LIESST}) = T_0 - 0.3T_{1/2}$ [18, 19]. However, the $T_{1/2}$ difference between both phases is only 28 K and certainly can not alone explain such drastic changes. Figure 8 shows a comparison of the relaxation kinetics recorded for both phases in the same time window.

3.3. HS $\rightarrow$ LS Relaxation

An analysis of the relaxation curves recorded for the *phase II* (Fig. 8a) shows that the kinetics can be satisfactorily fitted by using a single exponential. This behavior is, in fact, typical of a system exhibiting a weak cooperativity, as reflected by the gradual spin transition of the *phase II*. Figure 9a shows the plot of the

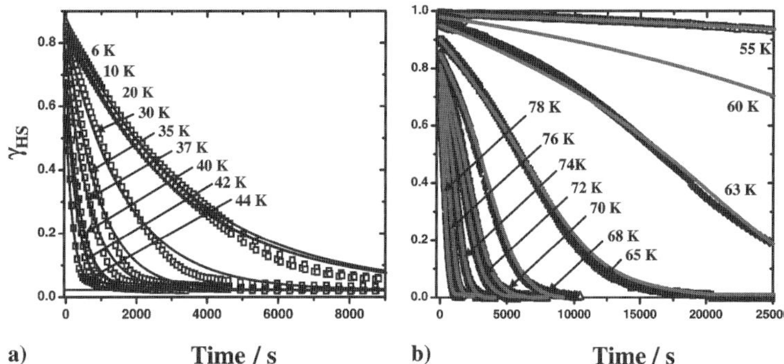

Fig. 8. Relaxation kinetics of the HS fraction *vs.* time at different temperatures. (a) *phase II*; (b) *phase I*. The solid lines represent the simulation obtained according to Eqs. (1) and (3)

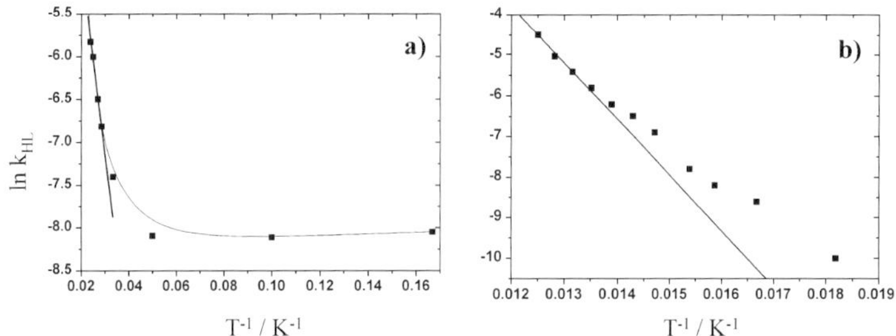

Fig. 9. *Arrhenius* plot of $\ln k_{HL}(T)$ *vs.* $1/T$ for *phase II* (a) and for the *phase I* (b)

logarithm of the rate constant as function of the inverse temperature. Below 20 K, the process is found nearly temperature independent in agreement with a quantum mechanical tunneling [13] and above 20 K, the relaxation curves follow an *Arrhenius* behavior which can be regarded as a tunneling from thermally populated vibrational levels of the HS state [14]. The deduced rate constant associated to the temperature independent domain, $k_{HL}(T \rightarrow 0)$, is estimated at $3 \cdot 10^{-4}\,s^{-1}$. The preexponential factor in the thermally activated region, $k_\infty = 0.6\,s^{-1}$, and the activation energy, $E_a = 150\,cm^{-1}$, are deduced from Eq. (1),

$$k_{HL}(T) = k_\infty \exp(E_a/k_B T) \tag{1}$$

Figure 8b shows the kinetics of the *phase I*. Below 55 K, the relaxation rates are so slow that a time dependence can not be correctly estimated. For instance, at 50 K the change in HS fraction is less than 6% within 10 hours. The HS → LS relaxation kinetics becomes measurable in the time window of our SQUID setup in the 63–78 K temperature range. The most striking feature of these relaxation curves is the strong deviation from single exponential. This result perfectly agrees with the strong cooperativity of the *phase I*. In fact, it has been pointed out that elastic interactions in cooperative spin-crossover compounds act as an "internal" pressure in reason of the large difference in metal-ligand bond lengths between HS and LS states [8, 12, 30]. The height of the energy barrier is an increase (or a decrease) function of γ_{HS} (or γ_{LS}). In this self-acceleration process, the relaxation rate $k^*_{HL}(T, \gamma_{HS})$ depends exponentially on both γ_{HS} and T (Eqs. (2) and (3)), and $\alpha(T)$ is the acceleration factor at a given temperature,

$$\frac{d\gamma_{HS}}{dt} = -k^*_{HL}\,\gamma_{HS} \quad (2)$$

$$k^*_{HL}(T,\gamma_{HS}) = k_{HL}(T)\exp[\alpha(T)(1-\gamma_{HS})] \quad (3)$$

A least-squares fit of the experimental data with $k_{HL}(T)$ and $\alpha(T)$ as free parameters is quite satisfactory. Figure 9b reports the plot of the logarithm of the rate constant, $\ln[k_{HL}(T)]$, as function of the inverse temperature. Clearly, it is difficult to deduce the kinetic parameters (activation energy and preexponential factor) involved in the thermally activated region. The true values would be only obtained at higher temperatures, in the limit regime of the thermal activation [12, 13]. The straight line of $\ln[k_{HL}(T)]$ *vs.* $1/T$ plot found in the 70–78 K region only gives access to the deduced apparent activation energy, $E_a = 1100\,cm^{-1}$, and the apparent preexponential factor, $k_\infty = 2\cdot 10^6\,s^{-1}$. Obviously, these parameters, which are certainly underestimated, are already considerably higher than those obtained for the *phase II*.

Concerning the determination of the acceleration factor $\alpha(T)$, *Hauser* has pointed out that this parameter depends on $E^*_a/k_B T$ in the thermally activated region and reaches a limiting value, $\alpha(T\to 0)$ in the tunneling region [12]. E^*_a represents the additional activation energy linked to the cooperativity of the system. From our data, it is difficult to exactly deduce E^*_a and $\alpha(T\to 0)$. The various kinetics recorded on the 70–78 K region suggest that the magnitude of E^*_a should be in the range of $120\,cm^{-1}$. Therefore, by considering that the tunneling region acts until 40 K, the deduced limiting value $\alpha(T\to 0)$ is around 4.3. Note that the cooperativity parameter deduced from the photomagnetic data is lower than the magnitude obtained using the mean-field model proposed by *Slichter* and *Drickamer* [31] during the thermal spin transition ($\Gamma = 280\,cm^{-1}$ [32]).

An attempt to estimate the $k_{HL}(T\to 0)$ parameter for the *phase I* may lead to erroneous results in view of the long lifetimes recorded at low temperature in the tunneling region. Nevertheless, if we consider in first approximation that the k_{HL} value extrapolated from the data recorded at 63 K, where a full relaxation occurs within 11 hours, represents the upper limit of the $k_{HL}(T\to 0)$ rate constant, the deduced value is $2\cdot 10^{-5}\,s^{-1}$. Interestingly, this rate constant, which may be an overestimation of the $k_{HL}(T\to 0)$ value, is already lower than the rate constant recorded for the *phase II* ($3\cdot 10^{-4}\,s^{-1}$). This value is also lower than the previously reported values for $[Zn_{1-x}Fe_x(pic)_3]Cl_2\cdot MeOH$, $k_{HL}(T\to 0) = 9\cdot 10^{-3}\,s^{-1}$ with $T_{1/2} = 140\,K$, and $[Zn_{1-x}Fe_x(mepy)_3(tren)](PF_6)_2$, $k_{HL}(T\to 0) = 1.4\cdot 10^{-1}\,s^{-1}$ with $T_{1/2} = 210\,K$ [33].

3.4. T(LIESST) Simulation

One way to test the validity of the various kinetic parameters, $k_{HL}(T\to 0)$, k_∞, E_a and $\alpha(T)$, is to reproduce the *T*(LIESST) curve. In fact, this measurement is, at least within the investigated temperature domain, a global analysis which intrinsically reflects the effects of time, temperature and cooperativity. In a measurement of a *T*(LIESST) curve, the sample was first irradiated at 10 K, then without further irradiation the temperature was slowly warmed at the rate of approximately $0.3\,\mathrm{K\,min}^{-1}$. Rigorously, the temperature during a *T*(LIESST) measurement, in a SQUID cavity, is changed in 1 K steps. At each temperature T_i, the time for the signal acquisition is 60 s and the time to reach the next temperature is 120 s. In reality, it is difficult to reproduce the relaxation connected to the stabilization of the temperature. In a first approximation, we have decided to neglect this subtle effect and the relaxation is calculated at each temperature T_i during the global time of 180 s. The γ_HS fraction obtained after 180 s of relaxation at T_i is used as started value for the next temperature, $T_{\mathrm{i}+1}$; *i.e.* $(\gamma_\mathrm{HS})_{t=180}^{T=T_\mathrm{i}} = (\gamma_\mathrm{HS})_{t=0}^{T=T_{\mathrm{i}+1}}$.

Another important point in the simulation of the *T*(LIESST) curve is that such an experiment combines the relaxation of both the tunneling and the thermally activated regions. Rigorously, this imposes to use the theory of the non-adiabatic multi-phonon process in the strong vibronic coupling limit [13]. In a first approximation, we have assumed that the evolution of the HS fraction is just a summation of the two effects. In absence of any significant cooperative effects, as for the *phase II*, the time dependence of the HS fraction at the temperature T_i is given by Eq. (4). For a cooperative compound exhibiting sigmoidal relaxation curves, as encountered for the *phase I*, Eq. (5) can be used.

$$\left(\frac{\mathrm{d}\gamma_\mathrm{HS}}{\mathrm{d}t}\right)_{T_\mathrm{i}} = -\gamma_\mathrm{HS}\{k_\mathrm{HL}(T\to 0) + k_\infty \exp(-E_\mathrm{a}/k_\mathrm{B}T_\mathrm{i})\} \tag{4}$$

$$\left(\frac{\mathrm{d}\gamma_\mathrm{HS}}{\mathrm{d}t}\right)_{T_\mathrm{i}} = -\gamma_\mathrm{HS}\{k_\mathrm{HL}(T\to 0) + k_\infty \exp(-E_\mathrm{a}/k_\mathrm{B}T_\mathrm{i})\}\exp[\alpha(T_\mathrm{i})(1-\gamma_\mathrm{HS})] \tag{5}$$

The *T*(LIESST) curve is finally calculated by taking into account the anisotropy of the HS iron(II) ion in an octahedral surrounding. This phenomenon called zero-field splitting is associated to the spin-orbit coupling between the ground state and the excited state in a zero applied magnetic field. For an iron(II) ion in HS configuration, the S = 2 ground state is split into three levels. The magnetic contribution of each state is determined by their energy separation, *D*, and their thermal population [34]. For a powder sample, the $(\chi_\mathrm{M}T)_\mathrm{ZFS}$ product associated to the zero-field splitting and the $\chi_\mathrm{M}T$ of the *T*(LIESST) curve are, respectively, given by Eqs. (6) and (7),

$$(\chi_\mathrm{M}T_\mathrm{i})_\mathrm{ZFS} = \frac{\chi_\mathrm{M}T_{\mathrm{i}_{//}} + 2\chi_\mathrm{M}T_{\mathrm{i}_\perp}}{3} \tag{6}$$

$$\text{with } \chi_\mathrm{M}T_{\mathrm{i}_{//}} = \frac{3}{4}g^2\frac{\mathrm{e}^{-D/k_\mathrm{B}T_\mathrm{i}} + 4\mathrm{e}^{-4D/k_\mathrm{B}T_\mathrm{i}}}{1 + 2\mathrm{e}^{-D/k_\mathrm{B}T_\mathrm{i}} + 2\mathrm{e}^{-4D/k_\mathrm{B}T_\mathrm{i}}}$$

$$\text{and } \chi_\mathrm{M}T_{\mathrm{i}_\perp} = \frac{1}{4}g^2\frac{k_\mathrm{B}T_\mathrm{i}}{D}\frac{9 - 7\mathrm{e}^{-D/k_\mathrm{B}T_\mathrm{i}} - 2\mathrm{e}^{-4D/k_\mathrm{B}T_\mathrm{i}}}{1 + 2\mathrm{e}^{-D/k_\mathrm{B}T_\mathrm{i}} + 2\mathrm{e}^{-4D/k_\mathrm{B}T_\mathrm{i}}}$$

$$\chi_\mathrm{M}T_\mathrm{i} = (\chi_\mathrm{M}T_\mathrm{i})_\mathrm{ZFS}\gamma_\mathrm{HS}(t, T_\mathrm{i}) \tag{7}$$

Figure 7 shows the calculated *T*(LIESST) curves of the two phases. The rate constant of the tunneling region, $k_\mathrm{HL}(T\to 0)$, the preexponential factor, k_∞, the activation energy of the thermally activated region, E_a, and the acceleration factor

$\alpha(T_i)$ used are those obtained from the experimental kinetics (see Section 3.3). The agreement with the experimental data is very good. The shape and the capacity of the compound to retain the light-induced HS information, estimated through the determination of the T(LIESST), are well described. Indirectly, this demonstrates the validity of our fitting procedure as well as our estimation of the various kinetics parameters, at least, in this temperature range. The relative importance of all these kinetics parameters on the T(LIESST) will be analyzed in Section 4. The last part of this Section 3 deals with the presentation of the light-induced thermal hysteresis.

3.5. Light-Induced Thermal Hysteresis

In 1998, it was reported that under permanent irradiation the *phase I* of the [Fe(*PM-BiA*)$_2$(NCS)$_2$] complex displays a new kind of thermal hysteresis in the vicinity of the T(LIESST) temperature [17]. This phenomenon was named the Light-Induced Thermal Hysteresis (LITH) [17]. Later on *Varret et al.* report a similar effect on the [Fe_xCo_{1-x}(*btr*)$_2$(NCS)$_2$] · H_2O system and a non-linear macroscopic master equation was proposed based on the competition between the constant photo-excitation and the self-accelerated thermal relaxation process [35].

The LITH phenomenon on the *phase I* of the [Fe(*PM-BiA*)$_2$(NCS)$_2$] was originally observed with a Kr^+ Laser ($\lambda = 647.1$–676.4 nm) and only 20% of the light-induced HS state was populated [17]. Figure 10 displays the LITH loop obtained by irradiating the sample with a photodiode emitting at 830 nm. The sample was irradiated at 10 K until saturation of the magnetic signal, then under permanent irradiation the temperature was slowly warmed up to 100 K at the rate of 0.3 K min^{-1} and cooled back to 10 K at the rate of 0.2 K min^{-1}. In regard to the magnetic signal recorded at elevated temperatures, this experiment represents for

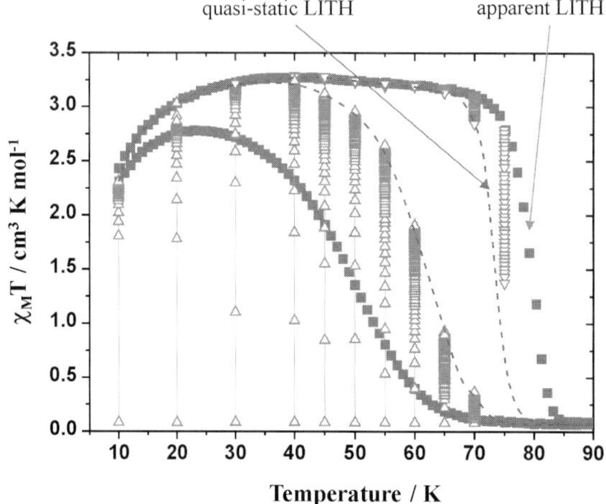

Fig. 10. Evolution of the $\chi_M T$ product *vs.* T of *phase I* under permanent irradiation at 830 nm (■) for a warming speed of 0.3 K min^{-1} and a cooling speed of 0.2 K min^{-1}. This gives the apparent LITH. (Δ) irradiation at 820 nm of the LS state for different temperatures. (∇) Irradiation at 820 nm of the HS state for different temperatures. Each irradiation leads to a photostationary point which describes the quasi-static LITH (---)

the *phase I* the first LITH loop with quantitative photoconversion. Obviously, the steepness of the warming branch is higher than of the cooling branch. We call this thermal hysteresis the *apparent* LITH loop. The *real* LITH loop will be obtained by recording the hysteresis loop with an infinite time, that is, when the steady state between the population and the relaxation is achieved. Otherwise, we are dealing with a dynamic hysteresis.

In order to approach the final form of the *real* LITH loop, we have determined some photostationary points on both the warming and the cooling branches. The procedure used was the following:

- On the cooling branch, the sample in the dark was slowly cooled down to the given temperature, T_{h_ν}. The sample was then irradiated and the $\chi_M T$ product was recorded. The experiment was stopped when the photostationary limit, equilibrium between population and depopulation, was reached. The temperature was finally warmed to 100 K and maintained at this temperature for 5 minutes to assure that any photoinduced HS fraction was erased. The T_{h_ν} was alternatively fixed at 70, 65, 60, 55, 50, 45, 40, 30, and 20 K.
- On the warming mode, the system was irradiated at 10 K until saturation, then warmed to T_{h_ν} where a new photostationary point was recorded. The T_{h_ν} was alternatively fixed at 10, 20, 30, 40, 45, 50, 55, 60, 65, and 70 K. Between each T_{h_ν}, the temperature was maintained 5 minutes at 100 K in order to erase any photoinduced HS fraction, then the sample was irradiated at 10 K until reaching the photostationary limit.

Figure 10 collects the different photostationary points and the proposed profile of the *real* LITH curve. As expected, this loop is much more narrow than the *apparent* one, *i.e.* ~10 K broad as compared to the previous 35 K. It is interesting to note that the asymmetry character of the LITH loop remains. The cooling branch is much more gradual than the warming branch. Further work is currently in progress to understand this particular behavior. The influence of various parameters, such as light intensity, as absorption of the LS/HS states, have been recently analyzed [36, 37]. It appears to be clear that the cooperativity is at the origin of the LITH phenomenon. This point is particularly well illustrated by the absence of the LITH effect in the *phase II* of the [Fe(*PM-BiA*)$_2$(NCS)$_2$] compound. The photomagnetic signals recorded during the warming and the cooling modes are very similar.

4. Comparative Analysis of the Kinetic Parameters Involved in the *Phases I* and *II*

4.1. Cooperative Effects

Figure 11a reports the calculated *T*(LIESST) curves by using the kinetic parameters of the *phase I* as function of the additional activation energy associated to the cooperativity, E_a^*. Clearly, the shape of the *T*(LIESST) curve becomes more gradual as the cooperativity decreases. This behavior agrees with the experimental data recorded for the *phases I* and *II*. The *T*(LIESST) curve was found to be gradual for the weakly cooperative *phase II* and abrupt in the case of *phase I*.

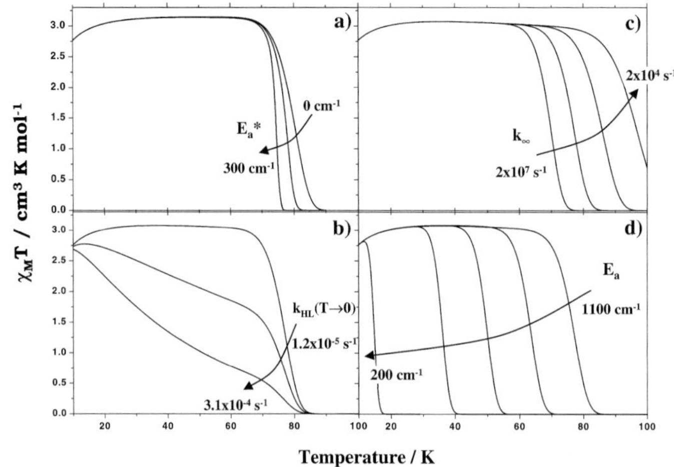

Fig. 11. Influence of the kinetics parameters on the *T*(LIESST) curve. (a) Variation of E_a^*; 0, 120, and 300 cm^{-1}. (b) Variation of $k_{HL}(T \to 0)$; $1.2 \cdot 10^{-5}$, $1.2 \cdot 10^{-4}$, and $3.1 \cdot 10^{-4}$ s^{-1}. (c) Variation of k_∞; $2 \cdot 10^7$, $2 \cdot 10^6$, $2 \cdot 10^5$, and $2 \cdot 10^4$ s^{-1}. (d) Variation of E_a; 1100, 900, 700, 500, and 200 cm^{-1}

From Figure 11a, it can also be noticed that the calculated *T*(LIESST) value is predicted to slightly increase when the cooperativity is reduced. This effect is linked to the increase of the height of the activation energy for a given HS fraction in Eq. (5) when E_a^* is reduced. Obviously, experimental data are inverse. The *T*(LIESST) value was clearly higher for the cooperative *phase I* than for the weakly cooperative *phase II*, suggesting that the main difference between the *T*(LIESST) values of *both phases* is not linked to a cooperativity effect. This finding is in line with the experimental results obtained from the comparison of a series of $[Fe(bpp)_2]X_2 \cdot nH_2O$ compounds, where independently to the cooperativity, the photomagnetic properties are found to follow the $T(\text{LIESST}) = T_0 - 0.3T_{1/2}$ law with $T_0 = 150$ [19].

4.2. *Tunneling and Thermally Activated Regions*

It is now well established that the HS → LS relaxation in general is determined by a nearly temperature-independent tunneling rate at low temperatures and an activated behavior at elevated temperatures. On the basis of a comparison between the photomagnetic properties of diluted SC compounds, *Hauser et al.* have shown that the magnitude of the low temperature tunneling rates as well as the activated region can be understood in terms of nonadiabatic multiphonon relaxation [33]. Figure 11b shows the influence of $k_{HL}(T \to 0)$ on the *T*(LIESST) curve by using the kinetics parameters of the *phase I* with a cooperative factor fixed at zero. Interestingly, from the values of the *phase I* ($k_{HL}(T \to 0) = 1.2\,10^{-5}\,s^{-1}$) and of the *phase II* ($3.1\,10^{-4}\,s^{-1}$), the shape of the *T*(LIESST) curve becomes more gradual as experimentally observed. Nevertheless, the *T*(LIESST) value, defined by the minimum in the derivative of the $\chi_M T$ *vs.* T plot, remains around 78 K. In the temperature-independent region, the decrease of the *T*(LIESST) curve is, in fact, relative to the $k_{HL}(T \to 0)$ constant and the derivative of the $\chi_M T$ *vs.* T plot gives a straight line. The minimum in the derivative of the $\chi_M T$ *vs.* T plot occurs only in the vicinity of the thermally activated region.

Figures 11c and 11d report the influence of the parameters of the thermally activated region on the *T*(LIESST) curve. A decrease of the preexponential factor, k_∞, increases the stabilization of the photoinduced HS fraction and rises the *T*(LIESST) value. This tendency is opposite to the experimental data and suggests that the change of the preexponential factor between the *phase I* ($2\,10^6\,s^{-1}$) and the *phase II* ($0.6\,s^{-1}$) can not explain the decrease of the *T*(LIESST) value. The role of the activation energy seems to be, in contrary, more important. A change between $1100\,cm^{-1}$ to $200\,cm^{-1}$ dramatically reduces the *T*(LIESST) value. At $200\,cm^{-1}$, the *T*(LIESST) value is in the expected region experimentally recorded for the *phase II*.

These results suggest that the main differences between the photomagnetic properties of the two phases are linked to the thermally activation energy. From a structural point of view, it is difficult to attribute the increase of the activation energy to a given modification of the FeN_6 core. Between both phases, the distortion of the octahedron is very similar and the crystal packing is also very close. The main difference seems to be the average of the bond lengths between the HS and the LS states which are higher for the *phase I* ($\Delta r = 0.218$ Å) than for the *phase II* (0.200 Å). Such a change is, in fact, consistent with the decrease of the activation energy on going from *phase I* ($1100\,cm^{-1}$) to *phase II* ($150\,cm^{-1}$) as well as to the observed increase of the tunneling rate constant, *i.e.* $k_{HL}(T \rightarrow 0) = 2 \cdot 10^{-5}\,s^{-1}$ for *I* instead of $3 \cdot 10^{-4}\,s^{-1}$ for *II*.

5. Conclusions

The study of the [Fe(*PM-BiA*)$_2$(NCS)$_2$] compound has shown the importance of the intense ^{1}MLCT-LS electronic transition in the photomagnetic experiments. The strong opacity of the [Fe(*PM-BiA*)$_2$(NCS)$_2$] compound in the visible range imposes to perform the photomagnetic experiments at the tail of the ^{1}MLCT-LS absorption band. Otherwise incomplete photoconversion may occur due to a bulk attenuation of light intensity. At 10 K, it was shown that light irradiation with a diode laser emitting at 830 nm induced an almost quantitative conversion to the HS state. A complete LITH loop has been described for the *phase I* of the [Fe(*PM-BiA*)$_2$(NCS)$_2$] compound. The various kinetic parameters involved in the HS $\rightarrow$ LS relaxation have been determined and used to simulate the *T*(LIESST) experiments. The influence of the cooperativity, of the tunneling and of the thermally activated kinetics parameters on the capacity of a compound to retain the metastable HS information, *i.e.* *T*(LIESST) value, have been discussed.

Acknowledgements

We are grateful for financial assistance from the European Commission for granting the TMR-Network "Thermal and Optical Switching of Spin States (TOSS)", Contract No. ERB-FMRX-CT98-0199 and from the ESF "Network Molecular Magnets" program.

References

[1] Balzani V, Credi A, Raymo FM, Stoddart JF (2000) Angew Chem Int Ed **39**: 3348
[2] Cambi L, Szegö L, Cassano A (1931) Accd Naz Lincei **13**: 809
[3] Baker WA, Bobonich HM (1964) Inorg Chem **3**: 1184

[4] König E (1968) Coord Chem Rev **3**: 471
[5] Goodwin HA (1976) Coord Chem Rev **18**: 293
[6] Gütlich P (1981) Struct Bonding (Berlin) **44**: 83
[7] Kahn O, Kröber J, Jay C (1992) Adv Mater **4**: 718
[8] Gütlich P, Hauser A, Spiering H (1994) Angew Chem Int Ed Engl **33**: 2024
[9] Decurtins S, Gütlich P, Köhler CP, Spiering H, Hauser A (1984) Chem Phys Lett **105**: 1
[10] Hauser A (1986) Chem Phys Lett **124**: 543
[11] Gütlich P, Garcia Y, Woike T (2001) Coord Chem Rev **219**: 839
[12] Hauser A, Jeftic J, Romstedt H, Hinek R, Spiering H (1999) Coord Chem Rev **190–192**: 471
[13] Bukhs E, Navon G, Bixon M, Jortner J (1980) J Am Chem Soc **102**: 2918
[14] Hauser A (1991) Coord Chem Rev **111**: 275
[15] Buchen Th, Gütlich P, Goodwin HA (1994) Inorg Chem **33**: 4573
[16] Hayami S, Gu ZZ, Einaga Y, Kobayasi Y, Ishikawa Y, Yamada Y, Fujishima A, Sato O (2001) Inorg Chem **40**: 3240
[17] Létard J-F, Guionneau P, Rabardel L, Howard JAK, Goeta AE, Chasseau D, Kahn O (1998) Inorg Chem **37**: 4432
[18] Létard J-F, Capes O, Chastanet G, Moliner N, Létard S, Real JA, Kahn O (1999) Chem Phys Lett **313**: 115
[19] Marcen S, Lecren L, Capes L, Goodwin HA, Létard J-F (2002) Chem Phys Lett **358**: 87
[20] Létard J-F, Montant S, Guionneau P, Martin P, Le Calvez A, Freysz E, Chasseau D, Lapouyade R, Kahn O (1997) Chem Commun 745
[21] Létard J-F, Daubric H, Cantin C, Kliava J, Bouhedja YA, Nguyen O, Kahn O (1999) Mol Cryst and Liq Cryst **335**: 495
[22] Ksenofontov V, Levchenko G, Spiering H, Gütlich P, Létard J-F, Bouhedja Y, Kahn O (1998) Chem Phys Lett **294**: 545
[23] Daubric H, Kliava J, Guionneau P, Chasseau D, Létard J-F, Kahn O (2000) J Phys Condens Mater **12**: 5481
[24] Guionneau P, Létard J-F, Yufit DS, Chasseau D, Bravic G, Goeta AE, Howard JAK, Kahn O (1999) J Mater Chem **9**: 985
[25] Guionneau P, Brigouleix C, Barrans Y, Goeta AE, Létard J-F, Howard JAK, Gaultier J, Chasseau D (2001) CR Acad Sci Paris, Chemistry **4**: 161
[26] Guionneau P, Marchivie M, Bravic G, Létard JF, Chasseau D (2002) J Mater Chem **12**: 2546
[27] Marchivie M, Guionneau P, Létard JF, Chasseau D, submitted to Acta Cryst B
[28] Ferguson J, Herren F (1982) Chem Phys Lett **89**: 371
[29] Hauser A, Adler J, Gütlich P (1988) Chem Phys Lett **152**: 468
[30] Gütlich P, Spiering H, Hauser A (1999) Inorganic Electronic Structure and Spectroscopy, In: Solomon EI, Lever ABP (eds) Vol II, Wiley, New York, p 575
[31] Slichter CP, Drickamer HG (1972) J Chem Phys **56**: 2142
[32] Capes L, Létard J-F, Kahn O (2000) Chem Eur **6**: 2246
[33] Hauser A, Vef A, Adler P (1991) J Chem Phys **95**: 8710
[34] Kahn O (1993) Molecular Magnetism, VCH: New York
[35] Desaix A, Roubeau O, Jeftic J, Haasnoot JG, Boukheddaden K, Codjovi E, Linarès J, Noguès N, Varret F (1998) Eur Phys J B **6**: 183
[36] Enachescu C, Constant-Machado H, Codjovi E, Linares J, Boukheddaden K, Varret F (2001) J Phys Chem Solids **62**: 1409
[37] Jeftic J, Matsarski M, Hauser A, Goujon A, Codjovi E, Linarès J, Varret F (2001) Polyhedron **20**: 1599

Invited Review

Spin Transition of 1D, 2D and 3D Iron(II) Complex Polymers The Tug-of-War between Elastic Interaction and a Shock-Absorber Effect

Petra J. van Koningsbruggen[1,*], **Matthias Grunert**[2], and **Peter Weinberger**[2]

[1] Institute for Inorganic and Analytical Chemistry, Johannes Gutenberg University, D-55099 Mainz, Germany

[2] Institute of Applied Synthetic Chemistry, Vienna University of Technology, A-1060 Vienna, Austria

Received April 8, 2002; accepted April 18, 2002
Published online September 19, 2002

Summary. The structures of linear chain Fe(II) spin-crossover compounds of α,β- and α,ω bis (tetrazol-1-yl)alkane type ligands are described in relation to their magnetic properties. The first threefold interlocked 3-D catenane Fe(II) spin-transition system, [μ-tris(1,4-bis(tetrazol-1-yl)butane-N1,N1′) iron(II)] bis(perchlorate), will be discussed. An analysis is made among the structures and the cooperativity of the spin-crossover behaviour of polynuclear Fe(II) spin-transition materials.

Keywords. Spin-crossover; High-spin; Low-spin; Fe(II); Tetrazole; Polynuclear compounds; Chain; Catenane.

Introduction

Spin-crossover materials are increasingly investigated due to their perceived technological importance, which is based on their possible application as molecular-based memory devices and displays [1–3]. Especially, Fe(II) spin-crossover compounds exhibit favourable response functions towards a change in temperature or pressure, and also upon light irradiation [1–15]: the thereby occurring interconversion from low-spin (LS; S = 0) and high-spin (HS; S = 2) represents the magnetic response, and moreover, it is frequently associated with a pronounced thermochromic effect. This is, for instance, the case for the extensively studied

* Corresponding author. E-mail: koning@iacgu7.chemie.uni-mainz.de

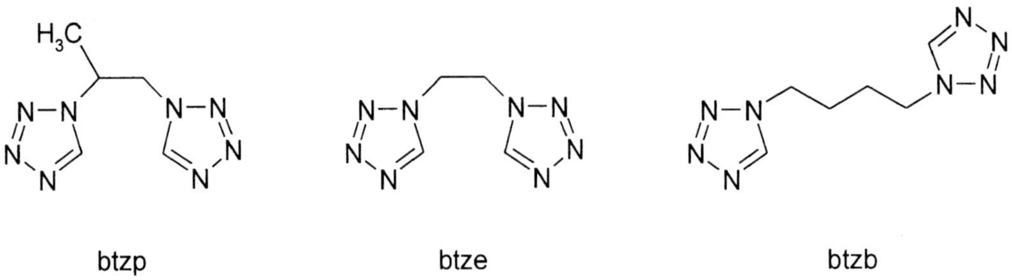

btzp btze btzb

Fig. 1. 1,2-bis(tetrazol-1-yl)propane (*btzp*), 1,2-bis(tetrazol-1-yl)ethane (*btze*) and 1,4-bis(tetrazol-1-yl)butane (*btzb*)

[Fe(1-propyl-tetrazole)$_6$](BF$_4$)$_2$ [12–17], which shows very abrupt spin transitions, a feature which may very well be described by the model of elastic interactions [18], and even thermal hysteresis, which is due to a first order crystallographic phase transition [19]. Generally, the occurrence of thermal hysteresis in mononuclear Fe(II) spin-crossover compounds may also be brought about by strong intermolecular interactions resulting from the presence of an important hydrogen bonding network [20, 21] or extended π–π interactions [22, 23]. Unfortunately, these features invoked to be responsible for thermal hysteresis are extremely difficult to control, hence alternative strategies involving polynuclear Fe(II) compounds have been applied during the last decade. This quest for polynuclear Fe(II) spin-crossover compounds has been motivated by the fact that an efficient propagation of the molecular distortions originating from the Fe(II) spin transition through the crystal lattice is enhanced by the direct covalent intramolecular bonds.

Our approach is based on the use of α,β- and α,ω-bis(tetrazol-1-yl)alkane type ligands. This paper deals with the comparison of the structural features in relation to the Fe(II) spin-crossover properties of various linear chain Fe(II) spin-crossover compounds obtained with the ligands 1,2-bis(tetrazol-1-yl)propane (abbreviated as *btzp*) and 1,2-bis(tetrazol-1-yl)ethane (abbreviated as *btze*) (Fig. 1). These structures will also be compared with the crystallographic data of a related Cu(II) linear chain of *btze*.

It will also be shown that increasing the length of the alkyl spacer in such a way as to yield 1,4-bis(tetrazol-1-yl)butane (abbreviated as *btzb*) (Fig. 1), proves to be a valuable tool in determining the dimensionality of the Fe(II) spin-crossover material. The Fe(II) spin-crossover properties of this polynuclear compound will be discussed and compared to these reported for other polynuclear Fe(II) spin-transition materials. Special emphasis is given to the factors leading to the cooperativity of the Fe(II) spin-crossover behaviour in these various systems.

Results

Spin-Crossover Behaviour and Structure of [Fe(btzp)$_3$](ClO$_4$)$_2$

[Fe(*btzp*)$_3$](ClO$_4$)$_2$ represents the first structurally characterized Fe(II) linear-chain compound exhibiting thermal spin-crossover [24]. It shows gradual spin-crossover behaviour with a transition temperature $T_{1/2}$, *i.e.* where equivalent amounts of spin

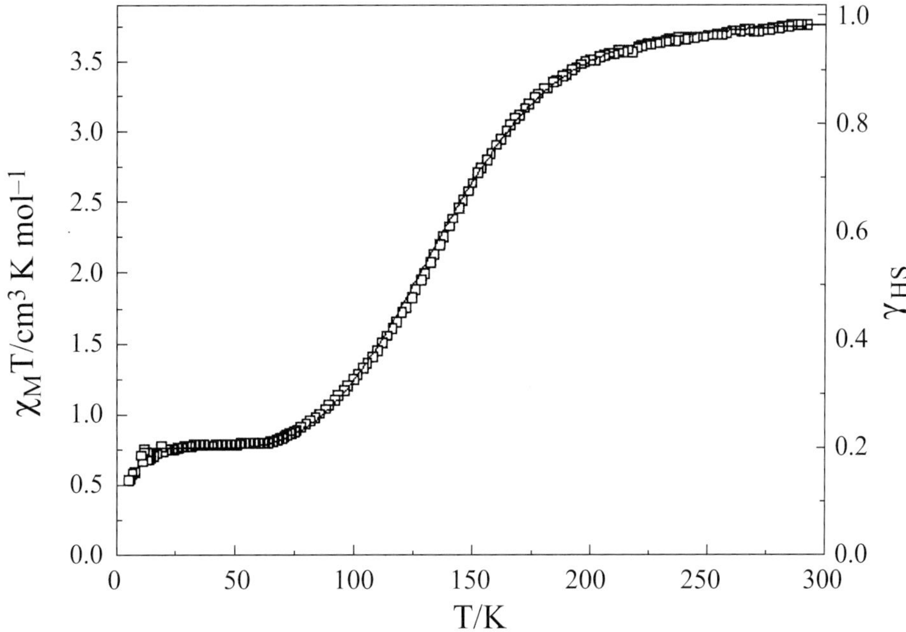

Fig. 2. $\chi_M T$ and γ_{HS} *versus* T plots both in the cooling and warming mode for $[Fe(btzp)_3](ClO_4)_2$ in the 4.2–300 K temperature range [24]

switching Fe(II) ions in low-spin and high-spin forms are present, of 148 K, as evidenced by variable temperature magnetic susceptibility measurements and ^{57}Fe *Mössbauer* spectroscopy. The magnetic behaviour of $[Fe(btzp)_3](ClO_4)_2$ is shown in Fig. 2 in the form of the $\chi_M T$ *versus* T plot, χ_M being the molar magnetic susceptibility per iron(II) ion and T the temperature. At higher temperatures, the spin-crossover is fairly complete yielding 100% of high-spin Fe(II) ions, whereas at 60 K a mixture of low-spin and high-spin Fe(II) ions, with the molar fractions 0.80 and 0.20, respectively, could be detected. The presence of Fe(III) could be ruled out based on the ^{57}Fe *Mössbauer* spectroscopy data [24]. The solid line shown in Fig. 2 shows the fraction of high-spin Fe(II) species γ_{HS} as function of the temperature, as derived from the regular solution model. The transition does not show any hysteresis, since the $\chi_M T$ *versus* T curves recorded at decreasing and increasing temperatures are identical.

Most interestingly, $[Fe(btzp)_3](ClO_4)_2$ undergoes light-induced excited spin-state trapping (LIESST effect). To the best of our knowledge, this is the first and only one-dimensional Fe(II) spin-crossover compound behaving this way.

The structure has been solved at 200 K and 100 K by single-crystal X-ray crystallography. A view of the cationic iron(II) linear chain is depicted in Fig. 3. The space group at 100 K and 200 K is $P\bar{3}c1$. The asymmetric unit consists of an iron(II) ion and one half of the *btzp* ligand. The C3 of the 1,2-propane linkage is crystallographically disordered and yielded two partially occupied positions. This originates from the use of the racemic mixture of the ligand in the synthesis of the Fe(II) compound. A disordered perchlorate anion completes this asymmetric unit. The Fe(II) ion lies on the threefold axis and has an inversion center. It is in an octahedral environment formed by six crystallographically related *N*4 coordinating

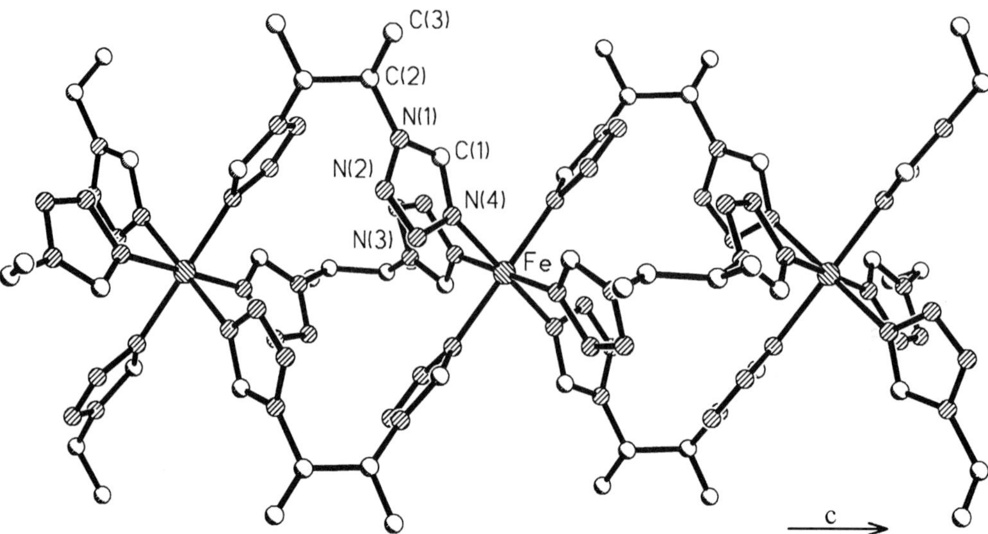

Fig. 3. View of the structure of $[Fe(btzp)_3](ClO_4)_2$ perpendicular to the c-axis at 100 K (adapted from [24])

1-tetrazole moieties. The almost perfect O_h symmetry for the FeN_6 core is therefore present in the high-spin and low-spin state. The latter feature could also be confirmed by the ^{57}Fe *Mössbauer* spectra, showing a singlet, the characteristic signature for one of the rare cases of cubic local symmetry for an Fe(II) ion in the low-spin state. The Fe–N4 distance of 2.164(4) Å at 200 K corresponds to the value expected for an Fe(II) ion in the high-spin state [5]. At 100 K, the Fe–N4 distance is equal to 2.038(4) Å, which is a typical value for an Fe(II) ion in the LS state [5]. The Fe–N4 distance decreases by 6% upon the spin conversion, which corresponds to the values found for other spin-crossover compounds [5]. The Fe(II) octahedron is very slightly distorted in the high-spin state with two sets of bond angles N–Fe–N of 91.1(2)° and 88.9(2)°. However, as expected, it is quasi regular in the low-spin state with bond angles N–Fe–N of 90.8(2)° and 89.2(2)°. The Fe(II) ions are linked by a bridge composed of three bis(tetrazole) ligands, leading to a regular linear chain running along the c-axis with Fe–Fe separations of 7.422(1) Å at 200 K and 7.273(1) Å at 100 K. The *btzp* ligand has a bent *syn* conformation, which is illustrated by the torsion angle N1–C1–C2^a–N1^a ($a = y, x, 0.5 - z$) of $-34(1)°$ at 200 K and of $-35(1)°$ at 100 K. A projection of the structure at 200 K perpendicular to the b-axis is shown in Fig. 4. The space-filling of the Fe(II) chains shows a hexagonal motif. In turn, the linear chains are packed in such a way as to form hexagonal cavities in the ab plane. The non-coordinated perchlorate anions reside in the voids of this molecular architecture. There are no intermolecular contacts between the linear chains. The apparent contacts between C3 atoms originating from different chains may be attributed to the statistical disorder of these C3 atoms.

The largest change in cell dimensions due to the spin transition amounts to 2.1% over the temperature range 100–200 K, and was found for the c-axis, *i.e.*, the chain axis. The changes in the a- and b-axes are considerably smaller (0.6%). The actual cell volume decreases by 3.3% in the temperature range from 200 to 100 K.

Fig. 4. View of the crystal structure of $[Fe(btzp)_3](ClO_4)_2$ down the *c*-axis at 100 K (adapted from [24])

The volume change per Fe(II) ion is 25.4 $Å^3$, which falls in the range normally observed [5].

Spin-Crossover Behaviour and Structure of $[Fe(btze)_3](BF_4)_2$

Continuing this research, we investigated in which way the variation of a 1,2-propane linkage towards a 1,2-ethane linkage between the tetrazole moieties influences the structural features. Recently, we reported [Fe(1,2-bis(tetrazol-1-yl) ethane)$_3$]$(BF_4)_2$, a comparable Fe(II) spin transition linear chain with a somewhat more abrupt spin transition centered at 140 K [25]. The variable temperature magnetic susceptibility measurements (Fig. 5), as well as the ^{57}Fe *Mössbauer* spectroscopy study gave evidence for an Fe(II) spin-crossover behaviour that is complete at higher temperatures, however, with a residual Fe(II) high-spin fraction of about 9.3% at lower temperatures. Superimposed on this spin-crossover behaviour of the Fe(II) ions, the ^{57}Fe *Mössbauer* spectra recorded over the whole temperature range also revealed the presence of a small fraction (7%) of high-spin Fe(III) ions.

The structure of [Fe(1,2-bis(tetrazol-1-yl)ethane)$_3$]$(BF_4)_2$ has been determined in the trigonal space group $P\bar{3}c1$ at 296, 200, 150 and 100 K, and corresponds to these described for [tris(1,2-bis(tetrazol-1-yl)propane)iron(II)] bisperchlorate [24]. Also in this case, paramount disorder has been encountered in the crystal structure, however, of a different nature than for the former Fe(II) linear chain compound.

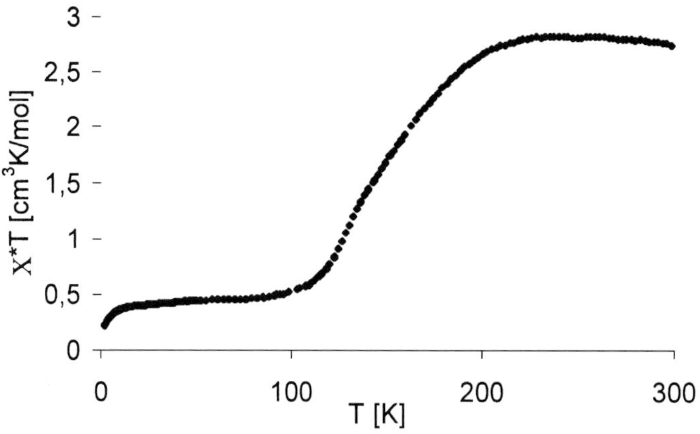

Fig. 5. Temperature dependent magnetic susceptibility of $[Fe(\textit{btze})_3](BF_4)_2$ [25]

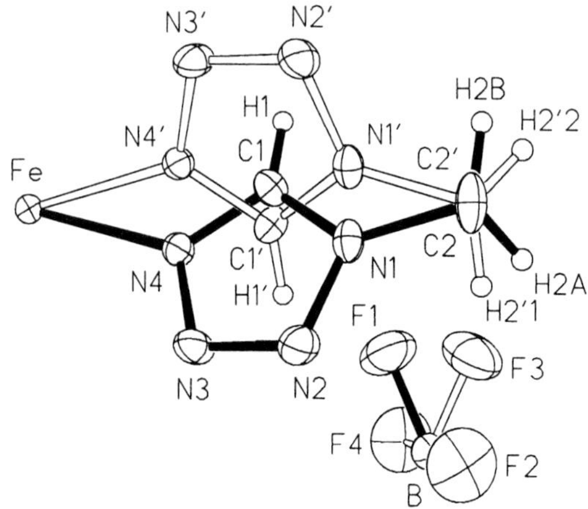

Fig. 6. Asymmetric unit of the structure of $[Fe(\textit{btze})_3](BF_4)_2$ ($T = 296$ K) showing two alternatively occupied sets of atoms linked by full/open bonds, as used in the final structure refinement [25]

This may best be viewed from Fig. 6 showing the asymmetric unit of the structure of $[Fe(\textit{btze})_3](BF_4)_2$. The asymmetric unit consists of an iron(II) ion, one half of the *btze* ligand and a tetrafluoroborate anion. The *btze* ligand adopts two orientations, characterized by two sets of atoms that are related by pseudo-symmetry and distinguished by unprimed and primed atoms. The refined site occupation factors for split positions were almost 0.5, *i.e.* 0.498(3). Also in this case, the geometry formed by the six N-donating *btze* ligands about the Fe(II) ion is almost perfectly octahedral. The Fe–N bond lengths are markedly temperature dependent: At room temperature the Fe–N bond length of 2.182(1) Å corresponds to a typical Fe(II) ion in high-spin state [5]. After a small contraction to 2.160(1) Å at 200 K, there is a significant decrease to 2.095(2) Å at 150 K and to 2.004(1) Å at 100 K. This is accompanied by a concomitant change in the colour of the crystal from colorless (296 and 200 K) to an intense pink (150 K and 100 K). The short Fe–N distance at 100 K is consistent with the transition to the low-spin state [5]. Three *btze* ligands link the Fe(II) centers to form cationic chains running parallel to the

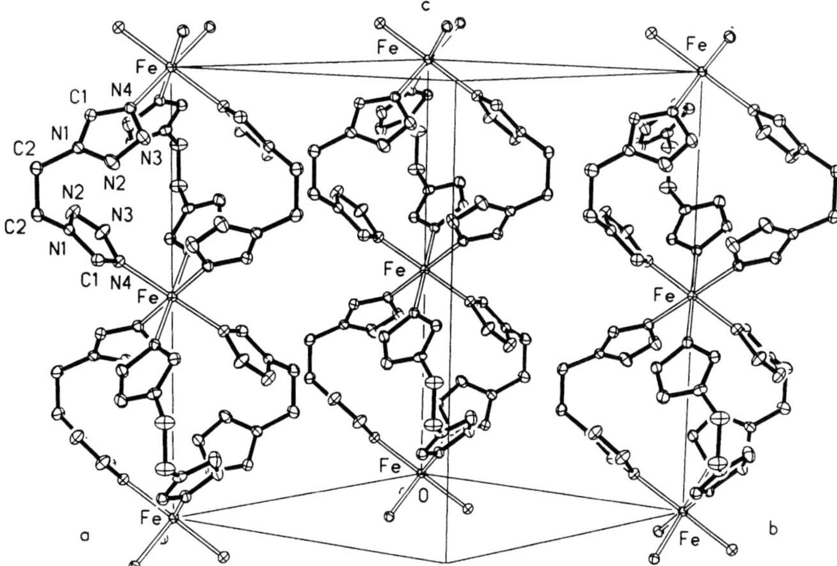

Fig. 7. Projection showing the unit cell of [Fe(*btze*)$_3$](BF$_4$)$_2$. Alternative orientation of the chains defined by primed atoms (see Fig. 6) is omitted for clarity [25]

crystallographic *c*-axis, as depicted in Fig. 7. Naturally, the Fe(II) spin-crossover behaviour is also reflected in the Fe...Fe separations, which are 7.477, 7.461, 7.376 and 7.293 Å at 296, 200, 150 and 100 K, severally. In the *ab* plane the linear chains are arranged in a hexagonal close-packed fashion, creating channel-like spaces between them and occupied by the tetrahedral BF_4^- anions (Fig. 8).

Structure of [Cu(btze)$_3$](ClO$_4$)$_2$

The crystal structure determination of [Cu(1,2-bis(tetrazol-1-yl)ethane)$_3$](ClO$_4$)$_2$ carried out at 298 K also revealed a linear chain structure [26], which, however, shows important differences with respect to the one for the Fe(II) tetrafluoroborate derivative [25]. Interestingly, the Cu(II) compound crystallises in the orthorhombic space group Pbcn, whereas both Fe(II) linear chain compounds crystallise in the trigonal space group P$\bar{3}$c1 [24, 25]. Therefore, the Cu(II) chain lacks the threefold symmetry about the chain axis, which implies that the perfect octahedral symmetry about the metal(II) center is not conserved anymore. Indeed, the Cu(II) ions are in a *Jahn-Teller* distorted octahedral environment (Cu(1)–N(11) = 2.034(2) Å, Cu(1)–N(21) = 2.041(2) Å and Cu(1)–N(31) = 2.391(2) Å). The N–Cu–N angles are close to 90°, varying from 88.07(7) to 91.93(7)°. Furthermore, in contrast to [Fe (*btze*)$_3$](BF$_4$)$_2$ there is only one crystallographic orientation for each *btze* ligand. The Cu(II) ions are linked by three *N*4, *N*4′ coordinating bis(tetrazole) ligands leading to a regular linear chain running along the *c*-axis, as displayed in Fig. 9. The *btze* ligands have a bent *syn* conformation which is shown by the torsion angles N(14)–C(16)–C(16)^2–N(14)^2 of $-51.1(2)°$ and N(24)–C(26)–C(36)^2–N(34)^2 of $-37.1(3)°$ (symmetry operation 2: $1-x, y, 3/2-z$). It can noticed that the long Cu–N distance is compensated by a smaller torsion angle, in order to

Fig. 8. Projection of the structure of $[Fe(\mathit{btze})_3](BF_4)_2$ down the *c*-axis [25]

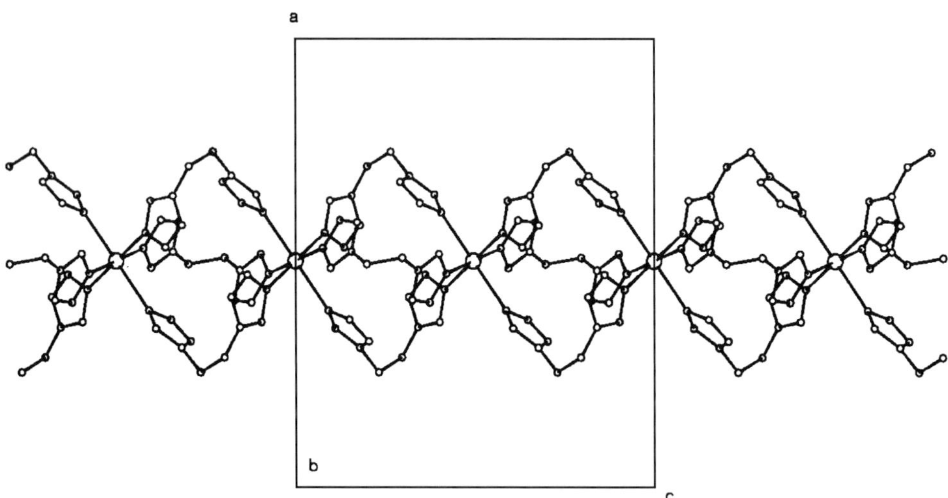

Fig. 9. Projection of the structure of $[Cu(\mathit{btze})_3](ClO_4)_2$ along the *c*-axis [26]

allow the three ligands to fit in the linkage between the Cu ions. The N(24)–C(26)–C(36)^2–N(34)^2 torsion angle of $-37.1(3)^\circ$ (symmetry operation 2: $1-x, y, 3/2-z$) is comparable to those observed for $[Fe(1,2\text{-bis(tetrazol-1-yl)propane})_3](ClO_4)_2$ of $-34(1)^\circ$ at 200 K and $-35(1)^\circ$ at 100 K [24]. These features lead to a Cu...Cu separation of 7.420(3) Å, *i.e.* only slightly shorter than the Fe...Fe separations in the Fe(II) linear chain spin-crossover compounds in the high-spin form [24, 25]. The crystal packing of all three linear chain compounds discussed above is essentially identical, notwithstanding the fact that there are neither significant intermolecular contacts between the linear chains, nor hydrogen bonding interactions present in the structures.

Spin-Crossover Behaviour and Structures of [M(btzb)$_3$](ClO$_4$)$_2$ (M(II) = Fe, Cu)

Continuing our strategy of applying the linkage of tetrazole moieties by alkyl groups in order to obtain polynuclear iron(II) spin-crossover materials, the length of the alkyl spacer was varied. This yielded a class of threefold interlocked 3-D catenanes of formula [M(1,4-bis(tetrazol-1-yl)butane)$_3$](ClO$_4$)$_2$ (M(II) = Fe, Ni, Cu) [27]. The highly thermochromic polynuclear compound [Fe(1,4-bis(tetrazol-1-yl)butane)$_3$](ClO$_4$)$_2$ has been obtained in the form of colourless crystals. This feature arises from the fact that the spin-allowed *d–d* transition of lowest energy of the compound in the high-spin state, $^5T_{2g} \rightarrow {}^5E_g$, occurs in the near infrared region. Upon cooling, the colour changes to an intense pink. This is due to the $^1A_{1g} \rightarrow {}^1T_{1g}$ *d–d* transition of the compound in the low-spin state. The results of the variable temperature optical measurements are displayed in Fig. 10. Upon cooling, a very abrupt high-spin → low-spin transition taking place at 155 K is observed. Subsequent heating shows the low-spin → high-spin transition at 180 K, yielding a thermal hysteresis of 25 K. Further heating-and-cooling cycles within the temperature range 77–298 K indicate that this hysteresis is retained. It is worth noting that these measurements provide an accurate determination of the transition temperatures, but do not give any information on the population of the active spin-crossover sites, *i.e.* the percentage of iron(II) ions involved in the spin transition. The magnetic susceptibility measurements revealed that only *ca.* 16% of the Fe(II) ions are involved in the spin transition, characterized by $T_{1/2}{\downarrow} = 150$ K and $T_{1/2}{\uparrow} = 170$ K. This hysteresis of 20 K has been reproduced along several thermal cycles. The

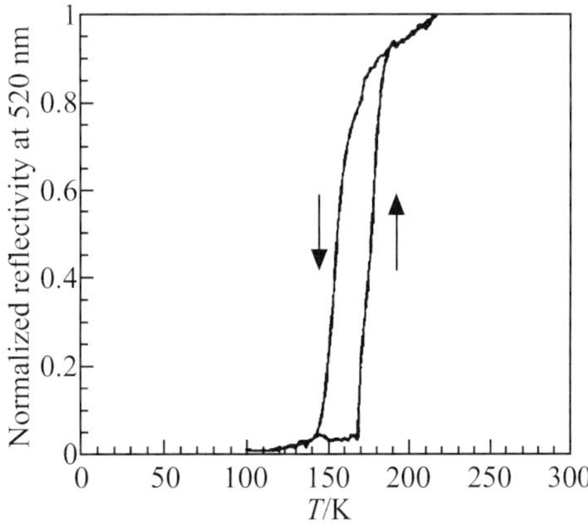

Fig. 10. Optical detection of the spin transition for [Fe(*btzb*)$_3$](ClO$_4$)$_2$ [27]

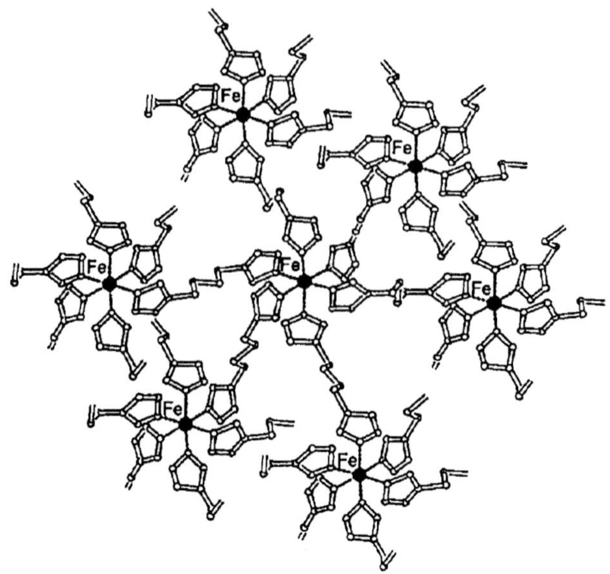

Fig. 11. Tentative 3-D model (150 K) for [Fe(*btzb*)$_3$](ClO$_4$)$_2$ [27]

slight discrepancy between the transition temperatures determined by optical and magnetic methods is most likely related to different sample – thermal response/temperature detector geometries in both applied techniques. The optical measurements focus on the colour, *i.e.* the surface of the sample, whereas the magnetic data reflect the physical behaviour of the bulk material [28, 29].

Furthermore, irradiation with green light at 30 K lead to the population of the metastable high-spin state for the thermally active iron(II) ions. Therefore, [Fe(btzb)$_3$](ClO$_4$)$_2$ represents the first 3-D Fe(II) spin-crossover material displaying the LIESST effect.

Unfortunately, only a tentative model of the 3-D structure of [Fe(*btzb*)$_3$](ClO$_4$)$_2$ at 150 K could be determined (Fig. 11). Each of the ligands is located on an inversion centre. This causes all central C–C linkages to be in the *anti* conformation. Of the six independent N–C–C–C torsions in the ligands, four are also in the *anti* conformation, but two fit the electron density best when brought into a *gauche* conformation. A detailed reanalysis of the crystallographic data has been carried out recently [30]. This revealed a model showing three symmetry related, interpenetrating, 3-D Fe–*btzb* networks. Iron atoms within one network are separated by the unit cell translations. The iron atoms of two non-connected networks approach each other as close as 8.3 and 9.1 Å according to this model. The crystal structure of the Cu(II) analogue confirmed this threefold interpenetrating 3-D catenane structure [31]. The driving force for the formation of these unprecedented supramolecular 3-D catenane materials lies in the conformation adopted by the alkyl spacer used to link the tetrazole moieties. Upon increasing of the spacer length the *anti* conformation, as has been found for the free *btzb* and for the Fe(II) catenane of *btzb* [27], is favoured over the bent *syn* conformation as found in the linear chains of ligands with smaller spacers [24–26].

Discussion

[Fe(*btzb*)$_3$](ClO$_4$)$_2$ is the first Fe(II) spin-crossover material with a supramolecular catenane structure consisting of three interlocked 3-D networks. Although, especially during the past few years several new polynuclear Fe(II) spin-crossover compounds have been reported (*vide infra*), the structure of the present compound is highly novel [27, 30]. Its structure is built up from three interlocked crystallographically dependent 3-D [Fe(*btzb*)$_3$]$^{2+}$ networks [30]. From the disorder met in this structure, it is evident that the crystal lattice does not involve the most efficient packing. In fact, it appears that in [Fe(*btzb*)$_3$](ClO$_4$)$_2$ four of the six crystallographically independent *btzb* ligands are in an *anti* conformation, whereas the remaining two are in a *syn* conformation. Interestingly, the crystal structure determination did not reveal any well-defined type of intra- or intermolecular interaction, which could be responsible for the stabilization of this unusual supramolecular structure.

A comparison may be made with the first supramolecular 2-D catenane exhibiting thermal spin-crossover behaviour [32]. The structure of [Fe(*tvp*)$_2$(NCS)$_2$] · CH$_3$OH (*tvp* = 1,2-di-(4-pyridyl)-ethylene) has been elucidated at room temperature and consists of two interpenetrating two-dimensional [Fe(*tvp*)$_2$(NCS)$_2$] sheets. The octahedrally coordinated Fe(II) ion contains two thiocyanate anions in *trans* positions, whereas the four surrounding bidentate N-coordinating *tvp* ligands link the Fe(II) ions into 2-D layers. The Fe...Fe separation through the *tvp* ligand is 13.66 Å. The Fe...Fe separations involving metal ions originating from interpenetrating layers are 22.59 and 15.36 Å. The difference in ligand dimensions between *tvp* and *btzb*, leads to a somewhat larger Fe...Fe separation over the direct *btzb* linkage, *i.e. ca.* 15 Å. However, the interweaving of three 3-D lattices gives rise to extremely short Fe...Fe separations of 8.3 and 9.1 Å between symmetry related non-connected lattices.

The same group pursued this research and reported on the threefold interlocked 2-D catenane system [Fe(*bpb*)$_2$(NCS)$_2$] · 0.5CH$_3$OH (*bpb* = 1,4-bis(4-pyridyl)-butadiyne) [33]. The structure consists of 2-D [FeL$_2$(NCS)$_2$] sheets comparable to the ones found in the former material, however, because of the increased rigidity and larger size of the present *bpb* ligand, now three mutually perpendicular nets are intercatenated. The crystal structure involves two symmetry independent networks leading to Fe...Fe separations over the direct *bpb* linkage of 16.628 Å and 16.393 Å, respectively.

The only other Fe(II) spin-crossover compound having a 3-D lattice with which comparisons may be made is [Fe(*btr*)$_3$](ClO$_4$)$_2$ (*btr* = 4,4′-bis-1,2,4-triazole) [34]. This compound may be regarded as a 3-D analogue of the 2-D spin-crossover materials [Fe(*btr*)$_2$(NC*X*)$_2$] (*btr* = 4,4′-bis-1,2,4-triazole; *X* = S [35], or Se [36]). The latter systems were first reported 15 years ago, and solely by incorporating the non-coordinating anion perchlorate for the N-coordinating thiocyanate anion, the 3-D Fe(II) spin-transition compound has been obtained. The structure of [Fe(*btr*)$_3$](ClO$_4$)$_2$ [34] was solved at 150, 190 and 260 K and comprises a 3-D network in which the crystallographically independent Fe1 and Fe2 ions are connected by single μ-*N*1,*N*1′-4,4′-bis-1,2,4-triazole bridges. Since the rather rigid *btr* ligand is smaller than the *btzb* ligand, the metal-metal separations over the bridging *btr* ligands are considerably shorter than for [Fe(*btzb*)$_3$](ClO$_4$)$_2$, *i.e.* Fe1...Fe2 (both in the HS form) = 8.67 Å at 260 K, Fe1 (HS)...Fe2 (LS) = 8.55 Å at 190 K, and Fe1...Fe2 (both in the LS form) = 8.42 Å at 150 K. Interestingly, these values are of the same order as the Fe...Fe spacings of 8.3 and 9.1 Å between different nets, *i.e.* through space, in the *btzb* material.

It is believed that the direct connectivity of the Fe(II) sites in polynuclear Fe(II) spin-transition compounds may have a favourable effect on the strength of the elastic interactions between the active Fe(II) spin-crossover centers, thereby increasing the cooperativity of the spin transition, leading to very abrupt spin-crossover behaviour. Indeed, this appears to be the case for the series of linear chain compounds of formula [Fe(4-*R*-*trz*)$_3$](anion)$_2 \cdot x$H$_2$O (4-*R*-*trz* = 4-substituted-1,2,4-triazole) [1–3, 7, 9–11, 28, 37–40], where the direct linkage of the Fe(II) spin-crossover centers by triple *N*1,*N*2-1,2,4-triazole bridges is assumed to account for the cooperative nature of the spin transition. When the ligand spacer linking the Fe(II) ions becomes more flexible, as is the case for [Fe(1,2-bis(tetrazol-1-yl)propane)$_3$](ClO$_4$)$_2$ [24] and [Fe(1,2-bis(tetrazol-1-yl)ethane)$_3$](ClO$_4$)$_2$ [25], the spin-crossover behaviour becomes more gradual. This is the signature for the negligible magnitude of the elastic interactions, which is most probably due to the 1,2-propane or 1,2-ethylene unit acting as some kind of shock absorber of the elastic interactions. This may be further illustrated by a comparison of the structures of [Cu(*hyetrz*)$_3$](ClO$_4$)$_2 \cdot$ 3H$_2$O (*hyetrz* = 4-(2′-hydroxy-ethyl)-1,2,4-triazole) [41] and [Cu(*btze*)$_3$](ClO$_4$)$_2$ [26]. The structure of [Cu(*hyetrz*)$_3$] (ClO$_4$)$_2 \cdot$ 3H$_2$O shows Cu(II) ions linked by triple *N*1,*N*2 1,2,4-triazole bridges yielding an unsymmetrical chain with two different alternating copper–copper distances, *i.e.* Cu1–Cu2 = 3.853(2) Å and Cu2–Cu3 = 3.829(2) Å, respectively. It is important to notice that even though the Cu(II) ions are in *Jahn-Teller* distorted octahedra, the chain shows only a relatively small deviation from linearity [41]. For [Cu(*btze*)$_3$](ClO$_4$)$_2$, in spite of the *Jahn-Teller* distorted geometry about the Cu(II) ions, the chain does not show any deviation from linearity [26]. Obviously, the flexibility of the ethylene linkage allows the preservation of the perfect linear chain structure, since the *Jahn-Teller* deformation of the Cu(II) coordination sphere has been successfully compensated by important variations in the N–C–C–N torsion angles of the *btze* ligands.

The same type of reasoning concerning the shock-absorbing properties of the direct bridging ligand may be applied to [Fe(*tvp*)$_2$(NCS)$_2$] $\cdot$ CH$_3$OH [32] and [Fe(*bpb*)$_2$(NCS)$_2$] $\cdot$ 0.5CH$_3$OH [33], which show only a very gradually proceeding and incomplete spin transition. Although the 1,2-di-(4-pyridyl)-ethylene and 1,4-bis(4-pyridyl)-butadiyne ligands themselves may *a priori* not be considered as flexible, it seems that the negligible magnitude of the elastic interactions is brought about by the flexibility of the 2-D network itself. This is illustrated by the absence of significant intra- and intermolecular interactions, which leads to the formation of large channels between the interlocked lattices.

On the other hand, both 3-D Fe(II) spin-crossover compounds show abrupt spin-crossover behaviour. The crystal structure of [Fe(*btr*)$_3$](ClO$_4$)$_2$ revealed the presence of two slightly different Fe(II) spin-crossover sites, each displaying its own magnetic behaviour [34]. This leads to a two-step spin conversion with 50% of the Fe(II) ions (site Fe1) undergoing a very abrupt spin transition with a small hysteresis of 3 K centered at 184 K, whereas the other 50% of Fe(II) ions (site Fe2) display rather gradual spin-crossover behaviour with $T_{1/2} = 222$ K.

The structural model for [Fe(*btzb*)$_3$](ClO$_4$)$_2$ showed three interlocked 3-D [Fe(*btzb*)$_3$]$^{2+}$ lattices [27, 30]. At 150 K, the Fe(II) ions of two non-connected networks approach each other as close as 8.3 and 9.1 Å. These distances involving no direct bridging ligands are comparable to the Fe1...Fe2 separation within [Fe(*btr*)$_3$](ClO$_4$)$_2$, albeit over a direct *btr* linkage [34]. Interestingly, the spin-crossover behaviour of [Fe(*btzb*)$_3$](ClO$_4$)$_2$ is far more abrupt with transition temperatures $T_{1/2}\uparrow = 170$ K and $T_{1/2}\downarrow = 150$ K, *i.e.* involving a rather large thermal hysteresis of about 20 K for a small fraction of *ca.* 16% of the Fe(II) ions involved in this transition; the remaining Fe(II) ions stay in the HS state [27]. It is worth noting that this is the largest thermal hysteresis observed up to now for iron(II) tetrazole derivatives. Apparently, the rigidity originating from the interweaving within this threefold 3-D interlocked lattice, is responsible for the efficient propagation of the elastic interactions leading to this type of cooperative spin-crossover behaviour. However, the same factors may also be invoked for explaining the small fraction of Fe(II) ions undergoing the spin transition. Most probably, the structural changes accompanying the Fe(II) spin transition modify the structure in such a way that the further spin-crossover of the high-spin Fe(II) ions upon cooling is severely hampered.

Conclusion

The comparison of these various polynuclear Fe(II) spin-crossover compounds has revealed that the predominant factor related to the cooperativity of the spin-crossover behaviour is neither the Fe...Fe separation, nor the dimensionality of the Fe(II) spin transition material, but rather the stiffness and rigidity with which the Fe(II) centers are maintained within the crystal lattice. Up to now, it has been observed that this rigidity may either arise from the stiffness of the direct bridges between the active spin-crossover centers as found in the linear chain materials [Fe(4-*R*-*trz*)$_3$](anion)$_2 \cdot x$H$_2$O [1–3, 7, 9–11, 28, 37–40], or from the efficient crystal packing of the lattice itself, as demonstrated for the 3-D spin-crossover material [Fe(*btzb*)$_3$](ClO$_4$)$_2$ [27, 30].

For the Fe(II) linear chain spin-transition materials, it is generally agreed upon that the cooperative Fe(II) spin-crossover behaviour of the Fe(II) 4*R*-1,2,4-triazole linear chains is related to the tight linking of the active Fe(II) centers by triple *N*1,*N*2 1,2,4-triazole bridges. On the other hand, the crucial factor responsible for the cooperative Fe(II) spin-crossover behaviour for the 3-D supramolecular catenane [Fe(*btzb*)$_3$](ClO$_4$)$_2$ [27] is the closed 3-D packing of the whole coordination polymer, and is certainly not caused by the rather flexible bridging *btzb* ligand itself.

It appears that the difference in dimensionality of the Fe(II) spin transition materials may be relevant to important variations in the response of the crystal lattice towards the Fe(II) spin-crossover. It becomes evident from the structural data of [Fe(*btzp*)$_3$](ClO$_4$)$_2$ [24] and [Fe(*btze*)$_3$](BF$_4$)$_2$ [25] that the main structural changes occurring during the spin transition are directed along the chain axis. Concomitant with this, the change in unit-cell dimensions predominantly takes place along the *c*-axis; therefore, the volume change upon spin-crossover is extremely anisotropic. The magnitude of the thermal contraction upon the Fe(II) spin transition could not be determined for the Fe(II) spin-crossover linear chain compounds of 4-*R*-1,2,4-triazole, since no structural data are available. Fe(II) spin transition materials of higher dimensionality may have more favourable thermal expansion characteristics upon Fe(II) spin-crossover. In these compounds, the change in unit-cell volume will occur more or less uniformly in all directions. Therefore, the elastic interactions are also expected to be more isotropic. Preliminary results have already confirmed this for another threefold interlocked 3-D supramolecular Fe(II) catenane of *btzb* [42].

Clearly, the family of α,β- and α,ω-bis(tetrazol-1-yl)alkane type ligands have been very valuable building blocks in the Fe(II) spin-crossover research. This approach already lead to the first structurally characterized 1-D iron(II) spin-crossover compound. This same $[Fe(btzp)_3](ClO_4)_2$ was also the first linear chain Fe(II) spin-crossover material to show the LIESST effect [24]. Furthermore, $[Fe(btzb)_3](ClO_4)_2$ represents the first supramolecular catenane Fe(II) spin-crossover compound with three interlocked 3-D networks. This thermochromic material can be switched by both temperature and light [27].

Further research on this type of polytetrazole systems is currently being carried out. Variation in the synthetic route provides a solid basis for tuning the spin-crossover behaviour by modifying the nature of the spacer between the tetrazole entities. In this way, a systematic tuning of the Fe...Fe separation is achieved, and more importantly, the dimensionality of the Fe(II) spin-crossover system can be changed.

Acknowledgements

For the supportive fruitful discussions we want to express our thank to *Kurt Mereiter* of the Institute of Chemical Technologies and Analytics of the Vienna University of Technology. This work was partly funded by the European Union within the TMR Research Network ERB-FMRX-CT98-0199 entitled "Thermal and Optical Switching of Molecular Spin States (TOSS)", the European Science Foundation program "Molecular Magnets" and the Hochschuljubiläumsfonds der Stadt Wien project H-65/2000.

References

[1] Kahn O, Jay-Martinez C (1998) Science **279**: 44
[2] Kahn O, Kröber J, Jay C (1992) Adv Mater **4**: 718
[3] Jay C, Grolière F, Kahn O, Kröber J (1993) Mol Cryst Liq Cryst **234**: 255
[4] Gütlich P (1981) Struct Bonding (Berlin) **44**: 83
[5] Zarembowitch J, Kahn O (1991) New J Chem **15**: 181
[6] König E (1987) Prog Inorg Chem **35**: 527
[7] Haasnoot JG (1996) In: Kahn O (ed) Magnetism: A Supramolecular Function. Kluwer, Dordrecht, 299
[8] Gütlich P, Garcia Y, Goodwin HA (2000) Chem Soc Rev **29**: 419
[9] Kahn O, Codjovi E (1996) Phil Trans R Soc London A **354**: 359
[10] Kahn O, Codjovi E, Garcia Y, van Koningsbruggen PJ, Lapouyade R, Sommier L (1996) In: Turnbull MM, Sugimoto T, Thompson LK (eds) Molecule-Based Magnetic Materials, Symposium Series No. 644, American Chemical Society, Washington, DC, 298

[11] Haasnoot JG (2000) Coord Chem Rev **200–202**: 131
[12] Gütlich P, Hauser A (1990) Coord Chem Rev **97**: 1
[13] Gütlich P, Hauser A, Spiering H (1994) Angew Chem Int Ed Engl **33**: 2024
[14] Gütlich P, Jung J, Goodwin HA (1996) NATO ASI Series. Kluwer, Dordrecht, p 327
[15] Gütlich P (1997) Mol Cryst Liq Cryst **305**: 17
[16] a) Franke PL, Haasnoot JG, Zuur AP (1982) Inorg Chim Acta **59**: 5; b) Müller WE, Ensling J, Spiering H, Gütlich P (1983) Inorg Chem **22**: 2074
[17] Wiehl L (1993) Acta Crystallogr **B49**: 289
[18] a) Sanner I, Meißner E, Köppen H, Spiering H (1984) Chem Phys **86**: 227; b) Willenbacher N, Spiering H (1988) J Phys C: Solid State Phys **21**: 1423; c) Spiering H, Willenbacher N. (1989) J Phys Condens Matter **1**: 10089
[19] Jung J, Schmitt G, Wiehl L, Hauser A, Knorr K, Spiering H, Gütlich P (1996) Z Phys B **100**: 523
[20] Sorai M, Ensling J, Hasselbach KM, Gütlich P (1977) Chem Phys **20**: 197
[21] a) Buchen T, Gütlich P, Sugiyarto KH, Goodwin HA (1996) Chem Eur J **2**: 1134; b) Sugiyarto KH, Weitzner K, Craig DC, Goodwin HA (1997) Aust J Chem **50**: 869
[22] a) Letard JF, Guionneau P, Codjovi E, Lavastre O, Bravic G, Chasseau D, Kahn O (1997) J Am Chem Soc **119**: 10861; b) Letard JF, Guionneau P, Rabardel L, Howard JAK, Goeta AE, Chasseau D, Kahn O (1998) Inorg Chem **37**: 4432
[23] a) Zhong ZJ, Tao JQ, Yu Z, Dun CY, Liu YJ, You XZ (1998) J Chem Soc Dalton Trans 327; b) Boca R, Boca M, Dlhán L, Falk K, Fuess H, Haase W, Jarošciak R, Papánková B, Renz F, Vrbová M, Werner R (2001) Inorg Chem **40**: 3025
[24] van Koningsbruggen PJ, Garcia Y, Kahn O, Fournès L, Kooijman H, Haasnoot JG, Moscovici J, Provost K, Michalowicz A, Renz F, Gütlich P (2000) Inorg Chem **39**: 891
[25] Schweifer J, Weinberger P, Mereiter K, Boca M, Reichl C, Wiesinger G, Hilscher G, van Koningsbruggen PJ, Kooijman H, Grunert M, Linert W (2002) Inorg Chim Acta (accepted)
[26] van Koningsbruggen PJ, Garcia Y, Bravic G, Chasseau D, Kahn O (2001) Inorg Chim Acta **326**: 101
[27] van Koningsbruggen PJ, Garcia Y, Kooijman H, Spek AL, Haasnoot JG, Kahn O, Linares J, Codjovi E, Varret F (2001) J Chem Soc Dalton Trans 466
[28] van Koningsbruggen PJ, Garcia Y, Codjovi E, Lapouyade R, Kahn O, Fournès L, Rabardel L (1997) J Mater Chem **7**: 2069
[29] a) Varret F, Constant-Machado H, Dormann JL, Goujon A, Jeftic J, Noguès M, Bousseksou A, Klokishner S, Dolbecq A, Verdaguer M (1998) Hyperfine Interact **113**: 37; b) Morscheidt W, Codjovi E, Jeftic J, Linarès J, Bousseksou A, Constant-Machado H, Varret F (1998) Meas Sci Technol **9**: 1311; c) Codjovi E, Morscheidt W, Jeftic J, Linarès J, Nogues M, Goujon A, Roubeau O, Constant-Machado H, Desaix A, Bousseksou A, Verdaguer M, Varret F (1999) Mol Cryst Liq Cryst **335**: 1295
[30] Mereiter K, Kooijman H, van Koningsbruggen PJ, Grunert M, Weinberger P, Linert W (2002) (unpublished results)
[31] van Koningsbruggen PJ, Bravic G, Chasseau D, Mereiter K, Grunert M, Weinberger P, Linert W (in preparation)
[32] Real JA, Andrés E, Munoz MC, Julve M, Granier T, Bousseksou A, Varret F (1995) Science **268**: 265
[33] Moliner N, Muñoz C, Létard S, Soans X, Menéndez N, Goujon A, Varret F, Real JA (2000) Inorg Chem **39**: 5390
[34] Garcia Y, Kahn O, Rabardel L, Chansou B, Salmon L, Tuchagues JP (1999) Inorg Chem **38**: 4663
[35] a) Vreugdenhil W, Haasnoot JG, Kahn O, Thuéry P, Reedijk J (1987) J Am Chem Soc **109**: 5272; b) Vreugdenhil W, van Diemen JH, de Graaff RAG, Haasnoot JG, Reedijk J, van der Kraan AM, Kahn O, Zarembowitch J (1990) Polyhedron **9**: 2971
[36] Ozarowski A, Shunzhong Y, McGarvey BR, Mislankar A, Drake JE (1991) Inorg Chem **30**: 3167

[37] a) Lavrenova LG, Ikorskii VN, Varnek VA, Oglezneva IM, Larionov SV (1986) Koord Khim **12**: 207; b) Lavrenova LG, Ikorskii VN, Varnek VA, Oglezneva IM, Larionov SV (1990) Koord Khim **16**: 654; c) Lavrenova LG, Yudina NG, Ikorskii VN, Varnek VA, Oglezneva IM, Larionov SV (1995) Polyhedron **14**: 1333

[38] Codjovi E, Sommier L, Kahn O, Jay C (1996) New J Chem **20**: 503

[39] a) Garcia Y, van Koningsbruggen PJ, Codjovi E, Lapouyade R, Kahn O, Rabardel L (1997) J Mater Chem **7**: 857; b) Garcia Y, van Koningsbruggen PJ, Lapouyade R, Fournès L, Rabardel L, Kahn O, Ksenofontov V, Levchenko G, Gütlich P (1998) Chem Mater **10**: 2426

[40] Garcia Y, van Koningsbruggen PJ, Lapouyade R, Rabardel L, Kahn O, Wieczorek M, Bronisz R, Ciunik Z, Rudolf MF (1998) C R Acad Sci Paris **II c**: 523

[41] Garcia Y, van Koningsbruggen PJ, Bravic G, Guionneau P, Chasseau D, Cascarano GC, Moscovici J, Lambert K, Michalowicz A, Kahn O (1997) Inorg Chem **36**: 6357

[42] Grunert M, Schweifer J, Weinberger P, Mereiter K, Boca M, Hilscher G, van Koningsbruggen PJ, Linert W (in preparation)

Invited Review

Is There a Need for New Models of the Spin Crossover?

Roman Boča[1,*] and **Wolfgang Linert**[2]

[1] Department of Inorganic Chemistry, Slovak University of Technology, SK-812 37 Bratislava, Slovakia
[2] Institute of Applied Synthetic Chemistry, Vienna University of Technology, A-1060 Vienna, Austria

Received April 10, 2002; accepted (revised) April 17, 2002
Published online October 7, 2002

Summary. The existing models of the low-spin to high-spin transition (spin crossover) are briefly reviewed. Experimental data pointing to a need of new models are displayed. A statistical model with the distribution of the solid-state cooperativeness is outlined. A modeling is shown as well as its application to a spin crossover system $[Fe(bzimpy)_2](ClO_4)_2 \cdot 0.25H_2O$. This shows an abrupt spin crossover at temperature as high as 403 K with a hysteresis width of 12 K. The angled walls of the hysteresis loop can be followed by the outlined statistical model.

Keywords. Spin crossover; Domain model; Solution model; Statistical model; Cooperativeness.

Introduction

Several transition metal complexes, especially those with d^4 to d^7 metal ion configuration, can exist either in the low-spin (LS) or high-spin (HS) states. Exceptionally, they can exist also in an intermediate-spin (IS) state. When the high-spin state is the ground-one, this is not altered by the temperature variation. However, when the ground state is the low-spin, a spin transition to the high-spin state can occur (Fig. 1).

There are two conditions for the spin crossover:

1. the enthalpy change (that includes the electronic and the vibrational contribution) should be positive

$$\Delta H = (E_{HS}^{el} + \varepsilon_{HS}^{vib}) - (E_{LS}^{el} + \varepsilon_{LS}^{vib}) > 0 \quad (1)$$

* Corresponding author. E-mail: boca@cvtstu.cvt.stuba.sk

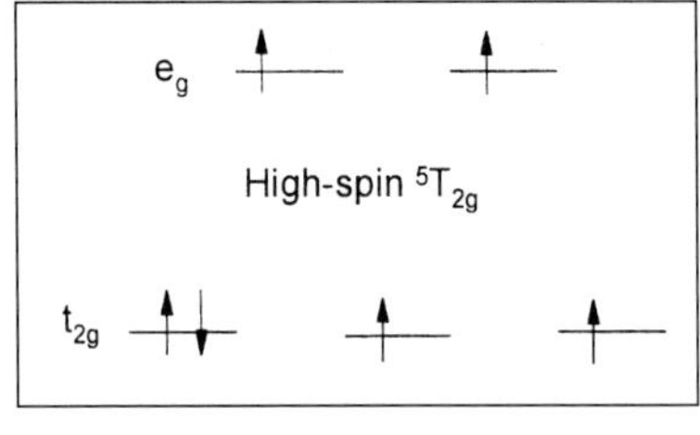

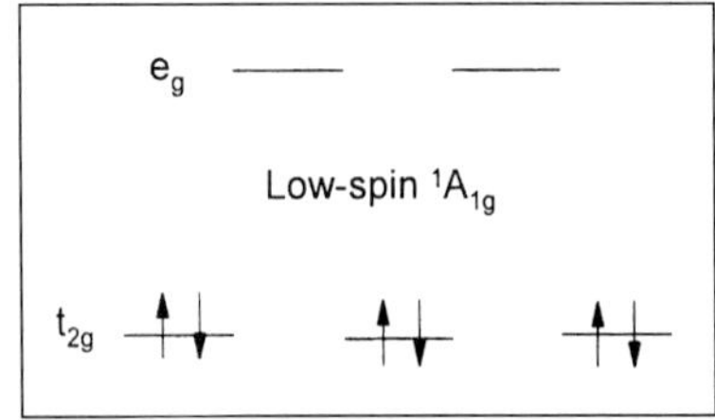

Electron configuration for LS and HS of an octahedral d^6 system

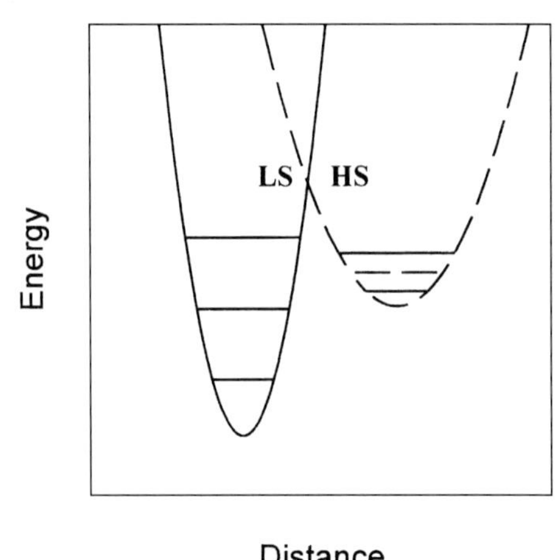

Total energy (adiabatic potential)

Fig. 1. Circumstances of the spin crossover

2. the entropy change (in the first, very crude approximation given by the electronic state degeneracy g^{el}) should be positive

$$\Delta S = k \ln g_{HS}^{el} - k \ln g_{LS}^{el} = k \ln\left(\frac{g_{HS}^{el}}{g_{LS}^{el}}\right) > 0 \tag{2}$$

Consequently the *Gibbs* energy $\Delta G = \Delta H - T \cdot \Delta S$ passes through the zero at the transition (critical) temperature (Fig. 2).

$$T_c = \Delta H / \Delta S \tag{3}$$

and the *vant' Hoff* plot, *i.e.* $\ln K$ vs. $(1/T)$, is a straight line intercepting zero at the transition temperature

$$\ln K = \ln \frac{x_{HS}}{x_{LS}} = \ln \frac{x_{HS}}{1 - x_{HS}} \tag{4}$$

$$\ln K = -\frac{\Delta G}{RT} = -\frac{\Delta H}{RT} + \frac{\Delta S}{R} \tag{5}$$

The slope of the *vant' Hoff* plot determines the enthalpy change whereas the intercept with the abscissa (when $1/T \to 0$) determines the entropy change. All these estimates are valid for a perfect fulfillment of the *Boltzmann* statistics when the deviations (the solid state cooperativeness) are negligible. When the entropy change would vanish, the system does not show the spin crossover: it stays low-spin.

The above requirements are well fulfilled for d^6 systems – iron(II) complexes. The ground low-spin state is $^1A_{1g}$ and this transforms to the high-spin excited state $^5T_{2g}$. The enthalpy change is positive owing to the promotion of electrons from

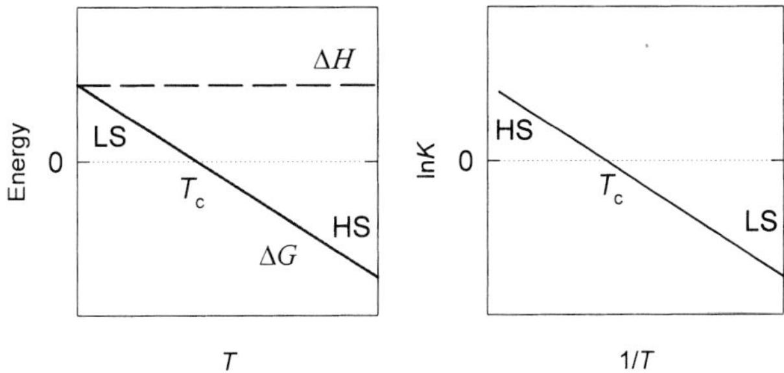

Fig. 2. Temperature variation of the *Gibbs* energy (left) and the *vant' Hoff* plot (right) for a spin crossover system

Table 1. Experimental monitoring of the spin crossover

Experimental method	Monitored property LS $\xrightarrow{T}$ HS
Magnetic measurements (MM)	magnetic susceptibility, effective magnetic moment, low $\xrightarrow{T}$ high
Mössbauer spectra (MS)	quadrupole splitting, low $\xrightarrow{T}$ high for Fe(II)
Vibrational spectra (IR)	M-L stretching wavenumber, high $\xrightarrow{T}$ low
Electron spectra (ES)	excitation energy, low $\xrightarrow{T}$ high
Calorimetric measurements (DSC)	heat capacity, low $\xrightarrow{T}$ high, a lambda-peak
X-ray diffraction	cell parameters, volume of the unit cell, low $\xrightarrow{T}$ high
Extended X-ray absorption fine structure (EXAFS)	metal-ligand distances, low $\xrightarrow{T}$ high
Nuclear magnetic resonance (NMR) in solutions	paramagnetic shift, effective magnetic moment, low $\xrightarrow{T}$ high
Volumetric measurements (VM)	partial molar volume, low $\xrightarrow{T}$ high
Electron spin resonance (ESR)	absorption, g-factor, no $(S = 0) \xrightarrow{T}$ yes

non-bonding t_{2g} orbitals to the antibonding e_g-ones (the energetically unfavorable process). The entropy change is positive as $g^{el}_{LS} = 1$ and $g^{el}_{HS} = 3 \times 5$. When the orbital degeneracy is removed on symmetry lowering, the high-spin state ${}^5A_{1g}$ will possess $g^{el}_{HS} = 5$ giving rise to the lowest estimate of $\Delta S = R \ln 5 = 13.6\,\mathrm{JK^{-1}\,mol^{-1}}$.

The spin crossover can be monitored by several experimental techniques as listed in Table 1. However the monitored properties can be transformed to a common basis that is the high-spin mole fraction x_{HS}. For more deep information the reader should consult the literature [1–9].

Hamiltonian and Kets

The Hamiltonian appropriate for the spin crossover system is a two-level *Ising*-like Hamiltonian of the form

$$\hat{H} = (\Delta_0/2)\hat{\sigma} - J\langle\sigma\rangle\hat{\sigma} \tag{6}$$

where $\hat{\sigma}$ – operator of a fictitious spin that distinguishes between the LS and HS, Δ_0 – site formation energy (energy difference LS–HS), and $J>0$ – "ferromagnetic"-like or "cooperative" interaction (the sign in front of J is a matter of the

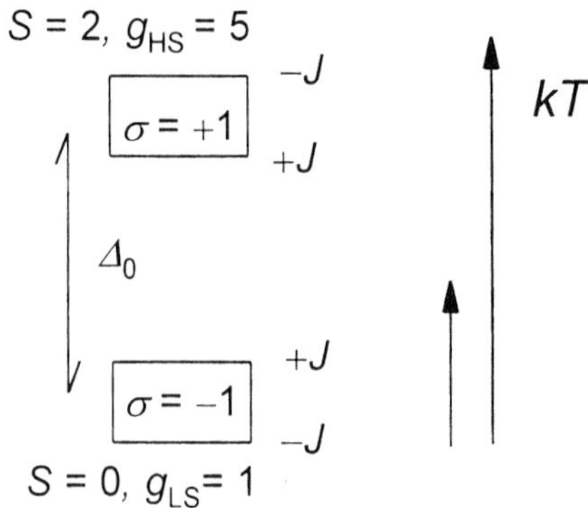

Fig. 3. Energy levels for the *Ising*-like Hamiltonian

convention; this parameter not to be confused with the exchange constant). The thermal average of the fictitious spin $\langle\sigma\rangle$ acts as a mean-field term and it is

$$\langle\sigma\rangle = \frac{\sum_i \sigma_i \exp(-E_i/kT)}{\sum_i \exp(-E_i/kT)} \tag{7}$$

It scales the high-spin mole fraction as follows

$$x_{\mathrm{HS}} = (1 + \langle\sigma\rangle)/2 \tag{8}$$

The kets act as follows

$$\hat{\sigma}|-1\rangle = -1 \tag{9}$$

$$\hat{\sigma}|+1\rangle = +1 \tag{10}$$

and they yield two energy levels (Fig. 3)

$$E_1(\sigma = -1) = -\Delta_0/2 + J\langle\sigma\rangle \tag{11}$$

$$E_2(\sigma = +1) = +\Delta_0/2 - J\langle\sigma\rangle \tag{12}$$

The key problem of the *Ising*-like model and its variants is the determination of the thermal average of the fictitious spin. There are two routes for such a purpose.

1) The equilibrium constant of a unimolecular reaction LS ↔ HS is expressed through the reaction *Gibbs* energy

$$K = \exp(-\Delta_{\mathrm{r}}G/RT) = \exp[-(G^0_{T,\mathrm{HS}} - G^0_{T,\mathrm{LS}})/RT] \tag{13}$$

On ignoring the difference between ΔF and ΔG for a solid sample (equivalent but distinguishable particles are assumed)

$$G \approx F = -RT \ln z \tag{14}$$

where

$$z_{\mathrm{LS}} = z^{\mathrm{el}}_{\mathrm{LS}} \cdot z^{\mathrm{vib}}_{\mathrm{LS}} \tag{15}$$

stands for the partition function of the given reactant (LS) and product (HS), respectively. Then

$$K = \frac{z^{\mathrm{el}}_{\mathrm{HS}} \cdot z^{\mathrm{vib}}_{\mathrm{HS}}}{z^{\mathrm{el}}_{\mathrm{LS}} \cdot z^{\mathrm{vib}}_{\mathrm{LS}}} \tag{16}$$

On the other hand there is

$$K = \frac{x_{\mathrm{HS}}}{x_{\mathrm{LS}}} = \frac{x_{\mathrm{HS}}}{1 - x_{\mathrm{HS}}} = \frac{1 + \langle\sigma\rangle}{1 - \langle\sigma\rangle} \tag{17}$$

from where one gets the equation

$$\langle\sigma\rangle = \frac{-1 + K}{+1 + K} \tag{18}$$

2) A direct application of the *Boltzmann* statistics for two possible values of the fictitious spin yields (the total partition function is applied)

$$\langle\sigma\rangle = \frac{(-1)z_{\mathrm{LS}}^{\mathrm{el}} \cdot z_{\mathrm{LS}}^{\mathrm{vib}} + (+1)z_{\mathrm{HS}}^{\mathrm{el}} \cdot z_{\mathrm{HS}}^{\mathrm{vib}}}{z_{\mathrm{LS}}^{\mathrm{el}} \cdot z_{\mathrm{LS}}^{\mathrm{vib}} + z_{\mathrm{HS}}^{\mathrm{el}} \cdot z_{\mathrm{HS}}^{\mathrm{vib}}} = \frac{-1 + K}{+1 + K} \tag{19}$$

where the expression for K is identical as above.

For the two-level *Ising*-like model the electronic partition functions are

$$z_{\mathrm{LS}}^{\mathrm{el}} = g_{\mathrm{LS}}^{\mathrm{el}} \exp[-(-\Delta_0/2 + J\langle\sigma\rangle)/kT] \tag{20}$$

$$z_{\mathrm{HS}}^{\mathrm{el}} = g_{\mathrm{HS}}^{\mathrm{el}} \exp[-(+\Delta_0/2 - J\langle\sigma\rangle)/kT] \tag{21}$$

and thus the equilibrium constant becomes

$$K = \left(\frac{z_{\mathrm{HS}}^{\mathrm{vib}} \cdot g_{\mathrm{HS}}^{\mathrm{el}}}{z_{\mathrm{LS}}^{\mathrm{vib}} \cdot g_{\mathrm{LS}}^{\mathrm{el}}}\right) \exp[-(\Delta_0 - 2J\langle\sigma\rangle)/kT] \tag{22}$$

In the simplest case – the *model A* – a constant preexponential factor is assumed and termed the effective degeneracy ratio

$$r_{\mathrm{eff}} = \left(\frac{z_{\mathrm{HS}}^{\mathrm{vib}} \cdot g_{\mathrm{HS}}^{\mathrm{el}}}{z_{\mathrm{LS}}^{\mathrm{vib}} \cdot g_{\mathrm{LS}}^{\mathrm{el}}}\right) \tag{23}$$

Then the implicit equation is to be obeyed

$$\langle\sigma\rangle = \frac{-1 + r_{\mathrm{eff}} \exp[-(\Delta_0 - 2J\langle\sigma\rangle)/kT]}{+1 + r_{\mathrm{eff}} \exp[-(\Delta_0 - 2J\langle\sigma\rangle)/kT]} \tag{24}$$

and this needs to be solved through an iterative procedure. The free parameters of the model are: r_{eff}, Δ_0 and J. These microscopic parameters are related to the thermodynamic quantities through

$$\Delta S = R \ln r_{\mathrm{eff}} \tag{25}$$

$$\Delta H = N_{\mathrm{A}} \Delta_0 \tag{26}$$

The involvement of the molecular vibrations – the *model B* – proceeds through the vibration partition function

$$\begin{aligned} z_{\mathrm{LS}}^{\mathrm{vib}} &= \prod_{i=1}^{3n-6} \frac{\exp(-h\nu_{\mathrm{LS},i}/2kT)}{1 - \exp(-h\nu_{\mathrm{LS},i}/kT)} \\ &= \exp\left[\sum_{i=1}^{3n-6} (h\nu_{\mathrm{LS},i}/2kT)\right] \prod_{i=1}^{3n-6} \frac{1}{1 - \exp(-h\nu_{\mathrm{LS},i}/kT)} \end{aligned} \tag{27}$$

and analogously for the HS molecules. Then the equilibrium constant becomes expressed as

$$K = \left(\frac{g_{\mathrm{HS}}^{\mathrm{el}}}{g_{\mathrm{LS}}^{\mathrm{el}}}\right)\left[\prod_{i=1}^{3n-6}\frac{1-\exp(-h\upsilon_{\mathrm{LS},i}/kT)}{1-\exp(-h\upsilon_{\mathrm{HS},i}/kT)}\right]$$
$$\exp\{-[\Delta_0+(\varepsilon_{\mathrm{HS}}-\varepsilon_{\mathrm{LS}})-2J\langle\sigma\rangle]/kT\} \quad (28)$$

where the energy of the zero-point vibration was summed up over all vibration modes

$$\varepsilon_{\mathrm{LS}} = \frac{1}{2}\sum_{i=1}^{3n-6} h\upsilon_{\mathrm{LS},i} \quad (29)$$

and analogously for the HS.

There are two approximations to the model:

1. The relevant (low-energy) modes are averaged to give $h\bar{\upsilon}_{\mathrm{LS}}$ and $h\bar{\upsilon}_{\mathrm{HS}}$; then

$$K = \left(\frac{g_{\mathrm{HS}}^{\mathrm{el}}}{g_{\mathrm{LS}}^{\mathrm{el}}}\right)\left[\frac{1-\exp(-h\bar{\upsilon}_{\mathrm{LS}}/kT)}{1-\exp(-h\bar{\upsilon}_{\mathrm{HS}}/kT)}\right]^{3n-6}$$
$$\exp\{-[\Delta_0+(3n-6)/2(h\bar{\upsilon}_{\mathrm{HS}}-h\bar{\upsilon}_{\mathrm{LS}})-2J\langle\sigma\rangle]/kT\} \quad (30)$$

2. In the limit of low vibration frequencies $h\upsilon \ll kT$ is fulfilled. Then the exponentials can be expanded into a *Taylor* series and their truncation after the second term yields

$$K = \left(\frac{g_{\mathrm{HS}}^{\mathrm{el}}}{g_{\mathrm{LS}}^{\mathrm{el}}}\right)\left[\frac{h\bar{\upsilon}_{\mathrm{LS}}}{h\bar{\upsilon}_{\mathrm{HS}}}\right]^{3n-6}\exp\{-[\Delta_0+(3n-6)/2(h\bar{\upsilon}_{\mathrm{HS}}-h\bar{\upsilon}_{\mathrm{LS}})-2J\langle\sigma\rangle]/kT\} \quad (31)$$

and now

$$r_{\mathrm{eff}} = \left(\frac{g_{\mathrm{HS}}^{\mathrm{el}}}{g_{\mathrm{LS}}^{\mathrm{el}}}\right)\left[\frac{h\bar{\upsilon}_{\mathrm{LS}}}{h\bar{\upsilon}_{\mathrm{HS}}}\right]^{3n-6} \quad (32)$$

$$\Delta_{\mathrm{eff}} = \Delta_0 + (3n-6)/2(h\bar{\upsilon}_{\mathrm{HS}}-h\bar{\upsilon}_{\mathrm{LS}}) \quad (33)$$

For hexacoordinate Fe(II) complexes 15 vibrational modes of the chromophore are relevant and the experimental data show that $(h\bar{\upsilon}_{\mathrm{LS}}) \approx 1.5(h\bar{\upsilon}_{\mathrm{HS}})$. Then the rough estimate is $r_{\mathrm{eff}} = 5(1.5)^{15} = 2189$ and consequently $\Delta S = R\ln r_{\mathrm{eff}} = 8.3 \times \ln(2189) = 64\,\mathrm{J\,K^{-1}\,mol^{-1}}$.

Final Formulae and Modeling

The final formulae of the *Ising*-like model of the spin crossover are collected in Table 2. A modeling is given by Fig. 4 and the following important findings become evident.

1. An increase of ΔH (at constant ΔS) raises the transition temperature.
2. An increase of ΔH, and simultaneous accommodation of ΔS to keep the transition temperature $T_c = \Delta H/\Delta S$ constant, causes an increased abruptness of the conversion curve $x_{\mathrm{HS}} = f(T)$.

Table 2. Formulae of the *Ising*-like models for mononuclear spin crossover systems

Derivation	Hamiltonian $\hat{H} = (\Delta_0/2)\hat{\sigma} - J\langle\sigma\rangle\hat{\sigma}$ in mean-field approximation Δ_0 – site formation energy (energy difference $E_{HS} - E_{LS}$) $J > 0$ – "ferromagnetic"-like or "cooperative" interaction
Implicit equation to be iterated	$\langle\sigma\rangle_T = \dfrac{-1 + f(\langle\sigma\rangle_T)}{1 + f(\langle\sigma\rangle_T)}$
High-spin mole fraction	$x_{HS} = (1 + \langle\sigma\rangle)/2$
Model A (*Ising*-like)	$f^{(A)} = r_{eff}\exp[-(\Delta_0 - 2J\langle\sigma\rangle_T)/kT]$ $r_{eff} = r_{el}r_{vib} > 5$ – effective degeneracy ratio; $r_{el} = g^{el}_{HS}/g^{el}_{LS}$ $\Delta S = R\ln r_{eff}$; $\Delta H = N_A\Delta_0$
Model B (*Ising*-like & vibrations)	$f^{(B)} = r_{eff}(T)\times\exp\{-[\Delta_{eff} - 2J\langle\sigma\rangle_T]/kT\}$ $r_{eff}(T) = \dfrac{g^{el}_{HS}}{g^{el}_{HS}}\left[\dfrac{1-\exp(h\bar{\upsilon}_{LS}/kT)}{1-\exp(h\bar{\upsilon}_{HS}/kT)}\right]^m$ $\Delta_{eff} = \Delta_0 + m(h\bar{\upsilon}_{HS} - h\bar{\upsilon}_{LS})/2$ m – active modes ($m = 15$ for a hexacoordinate complex) $h\bar{\upsilon}_{HS}$ and $h\bar{\upsilon}_{LS}$ – averaged vibration energies
Model C (*Ising*-like & domains)	$f^{(C)} = \exp\{-[\Delta H - T\Delta S - \gamma(2x_{HS} - 1)]n/RT\}$ $= \exp\{-[\Delta_0 - kT\ln r_{eff} - 2J\langle\sigma\rangle_T]n/kT\}$ n – optimum domain size
Model D (*Ising*-like & parameter distribution)	$f_i^{(D)} = \exp\{-[\Delta_0 - kT\ln r_{eff} - 2n_iJ\langle\sigma_i\rangle_T]/kT\}$ $x_i = (1 + \langle\sigma_i\rangle_T)/2$ $w_i \approx \exp[-(n_i - n_{opt})^2/\delta]$ – Gaussian distribution $x_{HS} = \left[\sum_{i=1}^{Mesh} w_i\cdot x_i\right] \Big/ \left[\sum_{i=1}^{Mesh} w_i\right]$ – a statistical average
Equilibrium constant	$\ln K = \ln\dfrac{x_{HS}}{1-x_{HS}} = -[\Delta H - T\Delta S + \gamma(1 - 2x_{HS})]n/RT$

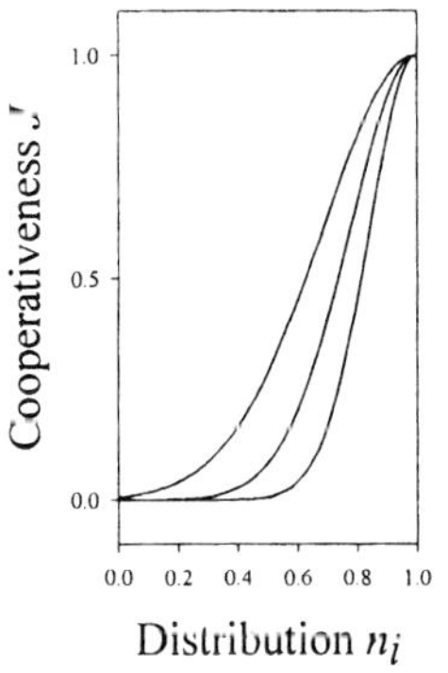

3. The increased domain size raises the abruptness of the conversion curve.
4. The increased cooperativeness causes a non-linearity near T_c (inverse S-shaped curvature) in the *van't Hoff* plot. above the critical value of $J > kT_c$ a hysteresis is obtained: the conversion curves on the heating and the cooling directions have different profiles.
5. The role of the vibrations makes a non-linearity of the *van't Hoff* line at low temperature; in an extreme case the system can return to the high-spin state on cooling.

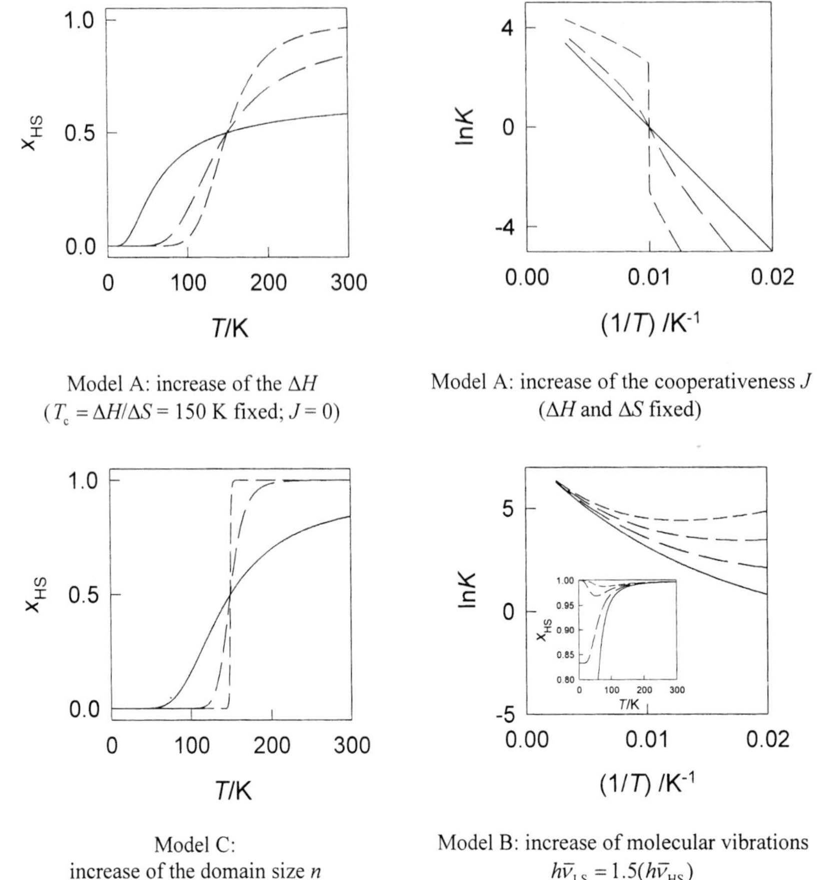

Fig. 4. Modeling of the spin crossover with the *Ising*-like model. Individual lines (full, long dashed, short dashed) correspond to the increase of a varied parameter

Extensions of the *Ising*-Like Model

Two basic derivations for the spin crossover models were presented in literature:

1) former thermodynamic approach starting with the mixing entropy, postulated interaction terms, the *Gibbs* energy and its stationery with respect to x_{HS};
2) a novel microscopic approach starting with the postulated Hamiltonian, derived energy levels, partition function, and the thermodynamic properties.

Thermodynamic Approach

a) The mixing entropy is expressed through the statistical probability of the state. In an assembly of N molecules there is a number of $x_{HS}N$ molecules in the HS state and the residual number of $(1 - x_{HS})N$ molecules in the LS state. The mixing entropy S_{mix} accounts for the fact that there are several ways of distributing the LS and HS molecules within the assembly of N molecules

$$S_{\text{mix}} = k \ln W = k \ln \frac{N!}{(xN)![(1-x)N]!}$$
$$= k\{N \ln N - xN \ln(xN) - (1-x)N \ln[(1-x)N]\}$$
$$= -kN\{x \ln x + (1-x)\ln(1-x)\} \tag{34}$$

where we applied the *Stirling* formula for the factorials ($x \equiv x_{\text{HS}}$ for the sake of clarity).

b) When the formation of domains of like spin is assumed then the number of molecules per domain is $n = N/D$ if the domains have a uniform size. The mixing entropy alters to

$$S_{\text{mix}} = k \ln \frac{D!}{(xD)![(1-x)D]!} = -kD\{x \ln x + (1-x)\ln(1-x)\} \tag{35}$$

For one mole ($N = N_{\text{A}}$) of molecules the molar mixing entropy can be rewritten as

$$S_{\text{mix}} = -(R/n)\{x \ln x + (1-x)\ln(1-x)\} \tag{36}$$

where $R = N_{\text{A}}k$ holds true.

c) The intermolecular interaction term I_x can be expressed in the general form

$$I_x = I_{\text{LL}}(1-x)^2 + 2I_{\text{LH}}x(1-x) + I_{\text{HH}}x^2 = J_0 + J_1 x + J_2 x^2 \tag{37}$$

where I_{LL}, I_{LH}, I_{HH} refer to interactions between LS–LS, LS–HS and HS–HS pairs, respectively. Then $J_0 = I_{\text{LL}}$, $J_1 = 2(I_{\text{LH}} - I_{\text{LL}})$, and $J_2 = I_{\text{LL}} + I_{\text{HH}} - 2I_{\text{LH}}$.

d) The molar *Gibbs* energy can be constructed from the following contributions

$$G_x = xG_{\text{HS}} + (1-x)G_{\text{LS}} - TS_{\text{mix}} + I_x \tag{38}$$

where G_{HS} (G_{LS}) is the molar *Gibbs* energy for the HS (LS) molecules.

e) The condition for the equilibrium demands

$$\left(\frac{\partial G_x}{\partial x}\right)_{T,p} = G_{\text{HS}} - G_{\text{LS}} + (R/n)T \ln\left(\frac{x}{1-x}\right) + J_1 + 2J_2 x = 0 \tag{39}$$

from where one gets the final, implicit equation for the high-spin mole fraction

$$x = \frac{1}{1 + \exp[n(\Delta H - T\Delta S + J_1 + 2J_2 x)/RT]} \tag{40}$$

The last, very general equation – *model C* has many links to more approximate models as reviewed by Table 3.

The basic assumption of the *solution model* is that the interaction term involves the cooperativeness γ through the form

$$I_x = \gamma x(1-x) \tag{41}$$

This formula has its origin in the intercentre interaction

$$I_x = J_0 + J_1 x + J_2 x^2 = I_{\text{LL}} + 2(I_{\text{LH}} - I_{\text{LL}})x + (I_{\text{LL}} + I_{\text{HH}} - 2I_{\text{LH}})x^2 \tag{42}$$

Table 3. Review of the spin crossover models

Model	Parameters $I_x = J_0 + J_1x + J_2x^2$	$x = f(T)$
1. Thermodynamic models		
Domain model *Sorai & Seki* [10]	$J_1 = J_2 = 0$	$x = 1/\{1 + \exp(n\Delta G/RT)\}$
Solution model *Drickamer* [11]	$J_1 = -J_2 = \gamma$; $I_x = \gamma x(1-x)$	$x = 1/\{1 + \exp[(\Delta G + \gamma - 2\gamma x)/RT]\}$
Interaction model *McGarvey et al.* [12]	$J_1 \neq 0$, $J_2 \neq 0$	$x = 1/\{1 + \exp[(\Delta G + J_1 + 2J_2x)/RT]\}$
Zimmermann & König [13]	$J_1 = 0$, $J_2 = -J$	$x = 1/\{1 + \exp[(\Delta - RT\ln Z - 2Jx)/RT]\}$
Spiering et al. [14, 15]	$J_1 = \Delta_x$, $J_2 = -\Gamma_x$	$x = 1/\{1 + \exp[(\Delta G + \Delta_x - 2\Gamma_x x)/RT]\}$
Interaction & domain [16]	$J_1 \neq 0$, $J_2 \neq 0$	$x = 1/\{1 + \exp[n(\Delta H - T\Delta S + J_1 + 2J_2x)/RT]\}$
Interaction model for two-step and binuclear systems [35, 28]		
2. Microscopic models		
Ising-like [17, 18]; model A	$\hat{I}_\sigma = -J\langle\sigma\rangle\hat{\sigma}$ $\langle\sigma\rangle = 2x - 1$	$x = 1/\{1 + \exp[(\Delta_0 - kT\ln r_{\text{eff}} - 2J(2x-1))/kT]\}$
Ising-like with vibrations [19]; model B	$\hat{I}_\sigma = -J\langle\sigma\rangle\hat{\sigma}$ $\langle\sigma\rangle = 2x - 1$	$x = 1/\{1 + \exp[(\Delta_{\text{eff}} - kT\ln r_{\text{eff},T} - 2J(2x-1))/kT]\}$ $r_{\text{eff},T} = \frac{g_{\text{HS}}^{\text{el}}}{g_{\text{HS}}^{\text{el}}}\left[\frac{1-\exp(h\bar{v}_{\text{LS}}/kT)}{1-\exp(h\bar{v}_{\text{HS}}/kT)}\right]^m$ $\Delta_{\text{eff}} = \Delta_0 + m(h\bar{v}_{\text{HS}} - h\bar{v}_{\text{LS}})/2$
Ising-like & domain model; model C	$\hat{I}_\sigma = -J\langle\sigma\rangle\hat{\sigma}$ $\langle\sigma\rangle = 2x - 1$	$x = 1/\{1 + \exp[n(\Delta_{\text{eff}} - kT\ln r_{\text{eff},T} - 2J(2x-1))/kT]\}$
Parameter distribution [20]; model D	$x_{\text{HS}} = \left[\sum_{i=1}^{\text{Mesh}} w_i \cdot x_i\right] \Big/ \left[\sum_{i=1}^{\text{Mesh}} w_i\right]$	$x_i = 1/\{1 + \exp[(\Delta_{\text{eff}} - kT\ln r_{\text{eff},T} - 2n_iJ(2x_i-1))/kT]\}$
Two-step *Ising*-like [21]		$\langle\sigma_A\rangle = \frac{-1 + r_{\text{eff}}\exp\{-[\Delta_0 - 2(J_A\langle\sigma_A\rangle + J_{AB}\langle\sigma_B\rangle)]/kT\}}{1 + r_{\text{eff}}\exp\{-[\Delta_0 - 2(J_A\langle\sigma_A\rangle + J_{AB}\langle\sigma_B\rangle)]/kT\}}$ $\langle\sigma_B\rangle = \frac{-1 + r_{\text{eff}}\exp\{-[\Delta_0 - 2(J_B\langle\sigma_B\rangle + J_{AB}\langle\sigma_A\rangle)]/kT\}}{1 + r_{\text{eff}}\exp\{-[\Delta_0 - 2(J_B\langle\sigma_B\rangle + J_{AB}\langle\sigma_A\rangle)]/kT\}}$
1D-*Ising*-like [29]		
Ising-like for binuclear compounds [22]	$x = (2 + \langle\sigma_A\rangle + \langle\sigma_B\rangle)/4$	$\langle\sigma_A\rangle = [-\exp(-E_1/kT) - r_{\text{eff}}\exp(-E_2/kT) + r_{\text{eff}}\exp(-E_3/kT) + r_{\text{eff}}^2\exp(-E_4/kT)]/Z_0$ $\langle\sigma_B\rangle = [-\exp(-E_1/kT) + r_{\text{eff}}\exp(-E_2/kT) - r_{\text{eff}}\exp(-E_3/kT) + r_{\text{eff}}^2\exp(-E_4/kT)]/Z_0$ $Z_0 = [\exp(-E_1/kT) + r_{\text{eff}}\exp(-E_2/kT) + r_{\text{eff}}\exp(-E_3/kT) + r_{\text{eff}}^2\exp(-E_4/kT)]$

(continued)

Table 3 (*continued*)

Model	Parameters $I_x = J_0 + J_1 x + J_2 x^2$	$x = f(T)$
		$E_1 = -\Delta_0 + (\langle\sigma_A\rangle + \langle\sigma_B\rangle)(J + J') - J_{AB}$ $E_2 = (\langle\sigma_A\rangle - \langle\sigma_B\rangle)(J - J') + J_{AB}$ $E_3 = -(\langle\sigma_A\rangle - \langle\sigma_B\rangle)(J - J') + J_{AB}$ $E_4 = \Delta_0 - (\langle\sigma_A\rangle + \langle\sigma_B\rangle)(J + J') - J_{AB}$
3. Other models		
Vibronic models (electron–phonon coupling) [13, 23]		

yielding the relationship within the solution and/or solution & domain model as follows

$$\frac{\partial I_x}{\partial x} = J_1 + 2J_2 x = (J_1 + J_2) - J_2(1 - 2x) = \gamma(1 - 2x) \tag{43}$$

The remainder

$$J_1 + J_2 = I_{\mathrm{HH}} - I_{\mathrm{LL}} \tag{44}$$

is thought either to vanish or to be absorbed to the effective parameter Δ_{eff}. Then the cooperativeness becomes

$$\gamma = -J_2 = 2I_{\mathrm{LH}} - I_{\mathrm{LL}} - I_{\mathrm{HH}} \tag{45}$$

This means an excess of the interaction energy between the molecules of the different spin relative to the interaction energy of the molecules of the like spin. It is a measure of the tendency for molecules of one type to interact effectively (to be surrounded) by molecules of the like spin.

The solution model, in fact, is fully equivalent to the two-level *Ising*-like model through the correspondence $\gamma/R = 2J$.

Model of a Parameter Distribution

This model has been motivated by the fact that the solid state samples are far from their ideal behavior and some drop in the cooperativeness could be described through a statistical distribution. As the sizable cooperativeness is responsible for the eventual hysteresis, the above effect will manifest itself in the profile of the conversion curve.

The key idea of this *model D* is that the optimum cooperativeness drops as

$$J_i = n_i J \tag{46}$$

Here i is the mesh point, say 1/100 of the value of $n_{\mathrm{opt}} = 1$. Then the factor entering the implicit equation for $\langle\sigma_i\rangle$ becomes

$$f_i^{(\mathrm{D})} = \exp\{-[\Delta_0 - kT \ln r_{\mathrm{eff}} - 2n_i J\langle\sigma_i\rangle]/kT\} \tag{47}$$

The equation is to be iterated for the given trial set of parameters (Δ_0, r_{eff}, J), for a given temperature, and for the given mesh point. Moreover, the iteration should start differently for the data point taken in the heating direction ($\sigma_i^{(0),\uparrow} = -1$ is used as an initial trial) and the cooling direction ($\sigma_i^{(0),\downarrow} = +1$). The statistical average is provided by the formula

$$x_{\mathrm{HS}} = \left[\sum_{i=1}^{\mathrm{Mesh}} w_i \cdot x_i\right] \Big/ \left[\sum_{i=1}^{\mathrm{Mesh}} w_i\right] \tag{48}$$

where the weights can be determined from the postulated distribution, e.g. the Gaussian distribution in the form of

$$w_i \approx \exp[-(n_i - n_{\mathrm{opt}})^2/\delta] \tag{49}$$

Additional parameter δ determines the width of the distribution (Fig. 5):

1) for $\delta \approx 0$ a sharp distribution exists and the model D collapses to the model C (or A) with fixed parameters. The hysteresis loop of the conversion curve possesses the rectangular walls.
2) The increase of δ manifests itself in angled walls of the hysteresis loop and decreased hysteresis width.
3) At the same time the completeness of the spin crossover is lowered and the conversion curve becomes smoother, resembling suppress of the cooperativeness.

The existence of the hysteresis originates in the fact that the *Gibbs* energy possesses two minima at different temperature; the system falls into one of them depending on the history of the heating/cooling regime.

Model for Two-Step Spin Crossover

Some compounds exhibit a spin crossover of the form that the fraction x_{HS} of molecules in the HS state increases with temperature in two steps; a plateau of a few K exists between these steps. This behavior can be explained by considering two sublattices (A and B) containing the same number of molecules [21]. The *Ising*-like Hamiltonians corresponding to the respective lattices, in the mean field approach, are defined as follows

$$\hat{H}_A = (\Delta_0/2)\hat{\sigma}_A - (J_A\langle\sigma_A\rangle + J_{AB}\langle\sigma_B\rangle)\hat{\sigma}_A \tag{50}$$

$$\hat{H}_B = (\Delta_0/2)\hat{\sigma}_B - (J_B\langle\sigma_B\rangle + J_{AB}\langle\sigma_A\rangle)\hat{\sigma}_B \tag{51}$$

where J_A and J_B are the intra-sublattice interaction parameters for $\mathrm{A}\cdots\mathrm{A}$ and $\mathrm{B}\cdots\mathrm{B}$ pairs; J_{AB} is the inter-sublattice interaction parameter for $\mathrm{A}\cdots\mathrm{B}$ pairs. Their positive values mean a "ferromagnetic-like" or cooperative interaction. The mole fractions of the HS state are interrelated through

$$\langle\sigma_A\rangle = 2x_A - 1 \tag{52}$$

$$\langle\sigma_B\rangle = 2x_B - 1 \tag{53}$$

The corresponding eigenvalues are

$$E_{A2} = (\Delta_0/2) - (J_A\langle\sigma_A\rangle + J_{AB}\langle\sigma_B\rangle) \tag{54}$$

$$E_{A1} = -(\Delta_0/2) + (J_A\langle\sigma_A\rangle + J_{AB}\langle\sigma_B\rangle) \tag{55}$$

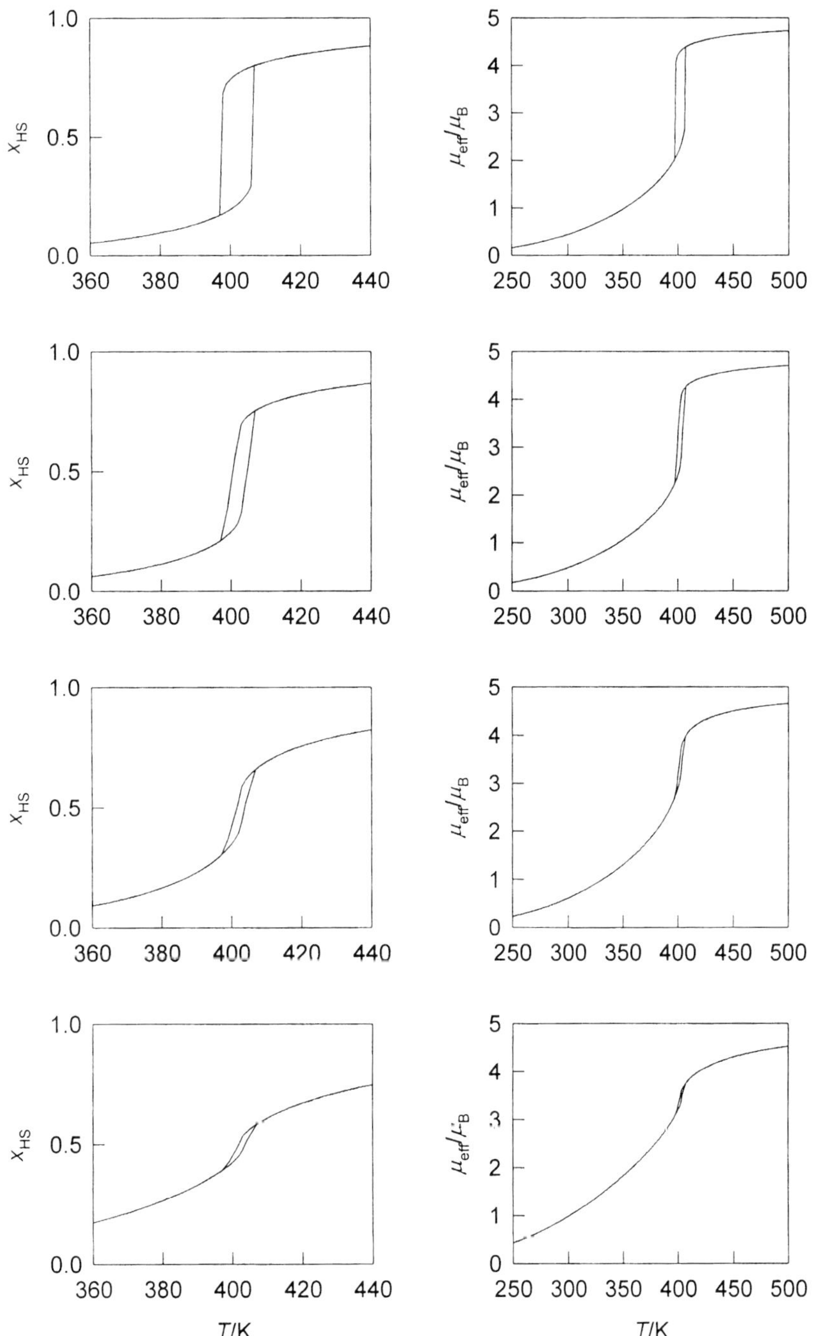

Fig. 5. Modeling of the spin crossover with a Gaussian distribution of the cooperativeness for $\delta = 0.00001$, 0.01, 0.1, and 1.0. The other spin-crossover parameters were fixed ($\Delta_0/k = 2144$ K, $J/k = 452$ K, $r_{eff} = 205$; $g_{HS} = 2.0$); $0 \leq n_i \leq n_{opt} = n_{max} = 1$

$$E_{B2} = (\Delta_0/2) - (J_B\langle\sigma_B\rangle + J_{AB}\langle\sigma_A\rangle) \tag{56}$$

$$E_{B1} = -(\Delta_0/2) + (J_B\langle\sigma_B\rangle + J_{AB}\langle\sigma_A\rangle) \tag{57}$$

Thermal population within the *Boltzmann* statistics yields

$$\langle \sigma_A \rangle = \frac{-1 + K_A}{1 + K_A} \tag{58}$$

where

$$K_A = \frac{z_{A,\mathrm{HS}}}{z_{A,\mathrm{LS}}} = \frac{g_{\mathrm{HS}}^{\mathrm{el}} \exp(-E_{A2}/kT)}{g_{\mathrm{LS}}^{\mathrm{el}} \exp(-E_{A1}/kT)} = r_{\mathrm{eff}} \exp[-(E_{A2} - E_{A1})/kT] \tag{59}$$

$$r_{\mathrm{eff}} = g_{\mathrm{HS}}^{\mathrm{el}} / g_{\mathrm{LS}}^{\mathrm{el}} \tag{60}$$

and analogously for the sublattice B. Then two coupled equations should be fulfilled simultaneously

$$\langle \sigma_A \rangle = \frac{-1 + r_{\mathrm{eff}} \exp\{-[\Delta_0 - 2(J_A\langle \sigma_A \rangle + J_{AB}\langle \sigma_B \rangle)]/kT\}}{1 + r_{\mathrm{eff}} \exp\{-[\Delta_0 - 2(J_A\langle \sigma_A \rangle + J_{AB}\langle \sigma_B \rangle)]/kT\}} \tag{61}$$

$$\langle \sigma_B \rangle = \frac{-1 + r_{\mathrm{eff}} \exp\{-[\Delta_0 - 2(J_B\langle \sigma_B \rangle + J_{AB}\langle \sigma_A \rangle)]/kT\}}{1 + r_{\mathrm{eff}} \exp\{-[\Delta_0 - 2(J_B\langle \sigma_B \rangle + J_{AB}\langle \sigma_A \rangle)]/kT\}} \tag{62}$$

These equations can be solved by an iterative procedure. The free parameters of the model cover J_A, J_B, J_{AB}, Δ_0, and r_{eff}. For modeling see Fig. 6.

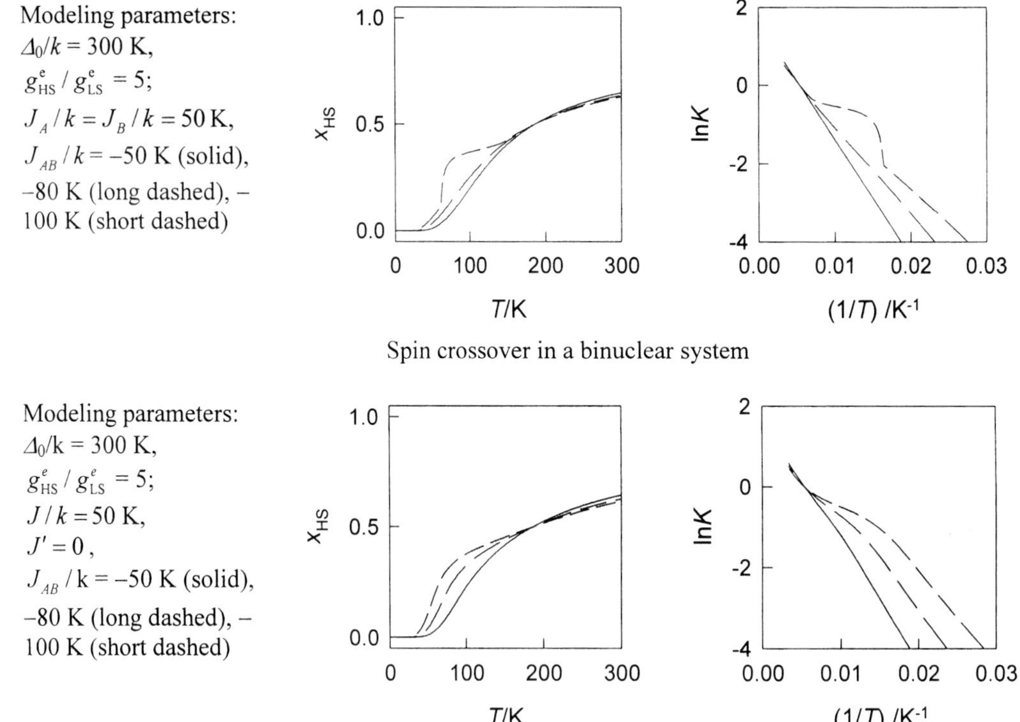

Fig. 6. Modelling of the spin crossover for a two-step case and a binuclear complex

Model for Binuclear Compounds

The case of the binuclear compounds formed of a symmetric pair of metallic centers the *Ising*-like Hamiltonian is

$$\hat{H} = -J_{AB}(\hat{\sigma}_A \cdot \hat{\sigma}_B) + \frac{\Delta_0}{2}(\hat{\sigma}_A + \hat{\sigma}_B) - J(\langle\sigma_A\rangle\hat{\sigma}_A + \langle\sigma_B\rangle\hat{\sigma}_B) - J'(\langle\sigma_A\rangle\hat{\sigma}_B + \langle\sigma_B\rangle\hat{\sigma}_A)$$

The intracomplex A–B interaction parameter J_{AB} adopts positive values for a "ferromagnetic-like" (cooperative) interaction of the fictitious spins. This has nothing to do with the isotropic exchange coupling constant J_{ex} that determines the energy levels in a binuclear system and has some connection to the site-formation energy. The intermolecular interaction parameters are J for A $\cdots$ A and B $\cdots$ B pairs, and J' for A $\cdots$ B and B $\cdots$ A, respectively (these are eventually neglected). The energy levels results in the form (see Fig. 7)

$$E_1(\sigma_A = -1, \sigma_B = -1) = E_{LL} = -\Delta_0 + (\langle\sigma_A\rangle + \langle\sigma_B\rangle)(J + J') - J_{AB} \quad (63)$$

$$E_2(\sigma_A = -1, \sigma_B = +1) = E_{LH} = (\langle\sigma_A\rangle - \langle\sigma_B\rangle)(J - J') + J_{AB} \quad (64)$$

$$E_3(\sigma_A = +1, \sigma_B = -1) = E_{HL} = -(\langle\sigma_A\rangle - \langle\sigma_B\rangle)(J - J') + J_{AB} \quad (65)$$

$$E_4(\sigma_A = +1, \sigma_B = +1) = E_{HH} = \Delta_0 - (\langle\sigma_A\rangle + \langle\sigma_B\rangle)(J + J') - J_{AB} \quad (66)$$

The partition function of the system is constructed as follows

$$Z = \sum_{i=1}^{4} g_i \exp(-E_i/kT)$$
$$= g_{LS}^2[\exp(-E_1/kT) + r_{eff}\exp(-E_2/kT) + r_{eff}\exp(-E_3/kT) + r_{eff}^2\exp(-E_4/kT)]$$

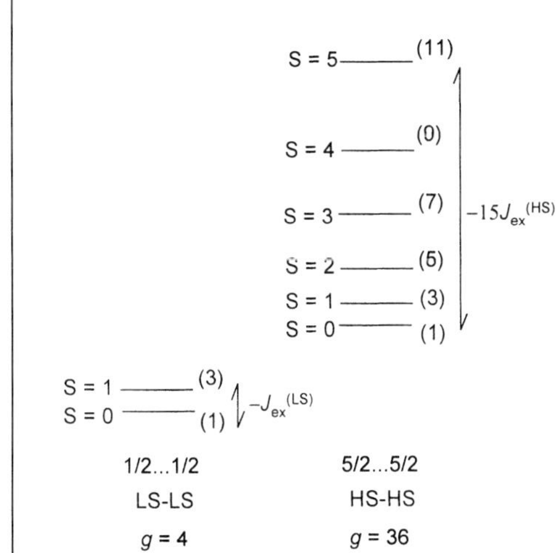

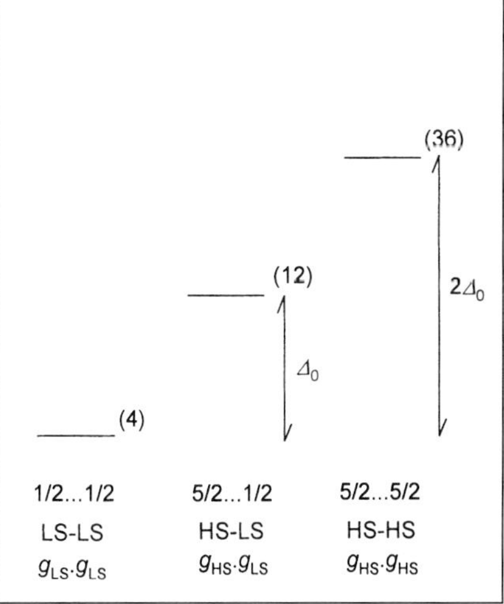

Fig. 7. A relationship between the energy levels of the isotropic (*Heisenberg*) exchange (left) and the *Ising*-like model of the spin crossover (right) for a binuclear Fe(III) complex. Degeneracies of the respective energy levels are given in parenthesis

where the effective degeneracy ratio, r_{eff}, occurs. The thermal average of the individual (formal) spin values are calculated as follows

$$\begin{aligned}\langle\sigma_A\rangle &= \sum_{i=1}^{4}\sigma_{A,i}g_i\exp(-E_i/kT)\\ &= g_{\text{LS}}^2[-\exp(-E_1/kT) - r_{\text{eff}}\exp(-E_2/kT)\\ &\qquad + r_{\text{eff}}\exp(-E_3/kT) + r_{\text{eff}}^2\exp(-E_4/kT)]/Z\end{aligned} \tag{67}$$

and

$$\begin{aligned}\langle\sigma_B\rangle = g_{\text{LS}}^2[&-\exp(-E_1/kT) + r_{\text{eff}}\exp(-E_2/kT) - r_{\text{eff}}\exp(-E_3/kT)\\ &+ r_{\text{eff}}^2\exp(-E_4/kT)]/Z\end{aligned} \tag{68}$$

Such a pair of the coupled equations can be solved by an iterative procedure. When $J = J'$ are assumed, the values of $\langle\sigma_A\rangle$ and $\langle\sigma_B\rangle$ are necessarily equal. Finally, the high-spin mole fraction is

$$x = (2 + \langle\sigma_A\rangle + \langle\sigma_B\rangle)/4 \tag{69}$$

The modeling is shown in Fig. 6 and some recent applications were presented elsewhere [24].

Application of the Distribution Model

A need of the new spin crossover model has been motivated by some experimental facts that could not be explained by previous models. The molecular complex $[Fe(bzimpy)_2](ClO_4)_2\cdot 0.25H_2O$ (hereafter **1**) was identified as a spin crossover system with high transition temperature of $T_c = 403$ K and a hysteresis width of 12 K. This is true for a freshly prepared microcrystalline sample [20]. On the sample aging and its treatment like graining a marked loss of the cooperativeness is observed which manifests in these features of the conversion curve: 1) the hysteresis width becomes lower; 2) the wall of the hysteresis loop become more angled; 3) a back-ground signal increases; 4) the conversion seem be incomplete. The distribution model D is quite successful in reproducing of all these features, as shown in Fig. 8.

One could expect that the above model can be applied to some other cases. A loss of the cooperativeness has been identified for $[Fe(PM\text{-}BiA)_2(NCS)_2]$ when passing from the crystalline sample 1 to its powder counterpart 2 prepared by a fast precipitation [25]. The dilution of the $[Fe(ptz)_6](BF_4)_2$ complex in an analogous Zn-matrix led to a systematic decrease of the abruptness (cooperativeness) of the conversion curves as well as a decrease of T_c that correlates with the dilution degree [26]. A decrease of the transition temperature along with a change of the profile of the conversion curve has been observed in $[Fe(pap)_2]ClO_4$ system as a time effect [27]: one weak after preparation gave a substantial effect.

Remember that the cooperativeness has its origin in the intercentre interaction (irrespective of its nature). Thus any break of such an interaction (point defects, dislocations, surfaces, hetero-atoms, degradation and oxidation products) will lower cooperativeness in a statistical manner.

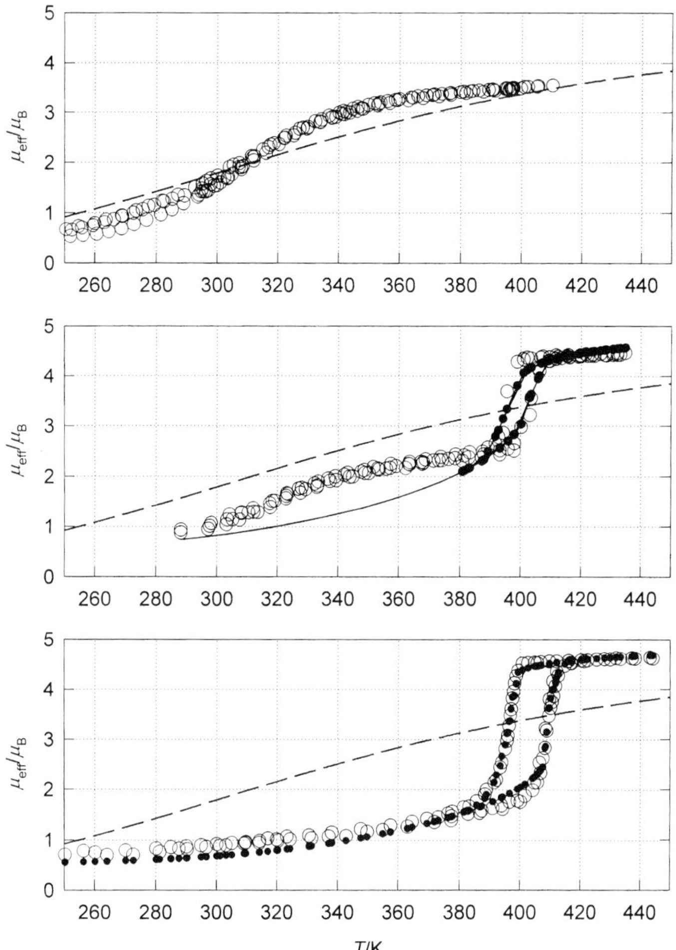

Fig. 8. Temperature variation of the effective magnetic moment (open symbols) for samples of **1** with different history (top – a one year-old sample, center – three-month old and powdered sample, bottom – a freshly prepared microcrystalline sample). Full points – fitted data using the theoretical model D. Solid line (center) – predicted. Dashed line – a theoretical curve in the absence of the cooperativeness

A Need of New Models

Not all experimental data were reproduced satisfactorily by the existing models of the spin crossover. The first problem occurs when the hysteresis loop possesses a strong asymmetry as found experimentally many times [30–33].

Second, a general, appropriate model should recover not only the conversion curve (usually constructed of the magnetic susceptibility data or the *Mössbauer* spectra data) but also the whole profile of the heat capacity [34].

Third, there are three-nuclear and polynuclear complexes exhibiting the spin crossover. A successful model for them is absent so far.

Fourth, a more complete inclusion of the interactions in the solid state is feasible, as outlined elsewhere [28, 35].

References

[1] Goodwin HA (1976) Coord Chem Rev **18**: 293
[2] Gütlich P (1981) Struct Bonding **44**: 83
[3] Beattie JK (1988) Advan Inorg Chem **32**: 1
[4] Toftlund H (1989) Coord Chem Rev **94**: 67
[5] König E (1991) Struct Bonding **76**: 51
[6] Kahn O, Krober J, Jay C (1992) Adv Mater **11**: 718
[7] Gütlich P, Hauser A, Spiering H (1994) Angew Chem **106**: 2109
[8] Kahn O (1993) Molecular Magnetism. VCH, New York
[9] Boča R (1999) Theoretical Foundations of Molecular Magnetism. Elsevier, Amsterdam
[10] Sorai M, Seki S (1974) J Phys Chem Solids **35**: 555
[11] Slichter CP, Drickamer HG (1972) J Chem Phys **56**: 2142
[12] Rao PS, Ganguli P, McGarvey BR (1981) Inorg Chem **20**: 3682
[13] Zimmermann R, König E (1977) J Phys Chem Solids **38**: 779
[14] Adler P, Wiehl L, Meissner E, Köhler CP, Spiering H, Gütlich P (1987) J Phys Chem Solids **48**: 517
[15] Spiering H, Meissner E, Köppen H, Müller EW, Gütlich P (1982) Chem Phys **68**: 65
[16] Cantin C, Kliava J, Marbeuf A, Mikailitchenko S (1999) Eur Phys J **B12**: 525
[17] Bari RA, Sivardiére J (1972) Phys Rev **B5**: 4466
[18] Wajnflasz J (1970) J Phys Stat Sol **40**: 537
[19] Bousseksou A, Constant-Machado H, Varret F (1995) J Phys **15**: 747
[20] Boča R, Boča M, Dlháň L', Fakl K, Fuess H, Haase W, Jaroščiak R, Papánková B, Renz F, Vrbová M, Werner R (2001) Inorg Chem **40**: 3025
[21] Bousseksou A, Nasser J, Linares J, Boukheddaden K, Varret F (1992) J Phys **12**: 1381
[22] Bousseksou A, Varret F, Nasser J (1993) J Phys **13**: 1463
[23] Kambara T (1979) J Chem Phys **70**: 4199
[24] Boča R, Fukuda Y, Gembický M, Herchel R, Jaroščiak R, Linert W, Renz F, Yuzurihara J (2000) Chem Phys Lett **325**: 411
[25] Letard J-F, Guionneau P, Rabardel L, Howard JAK, Goeta AE, Chasseau D, Kahn O (1998) Inorg Chem **37**: 4432
[26] Jung J, Schmitt G, Wiehl L, Hauser A, Knorr K, Spiering H, Gutlich P (1996) Z Phys **B100**: 523
[27] Hayami S, Maeda Y (1997) Inorg Chim Acta **255**: 181
[28] Real J-A, Bolvin H, Bousseksou A, Dworkin A, Kahn O, Varret F, Zarembowitch J (1992) J Am Chem Soc **114**: 4650
[29] Linares J, Spiering H, Varret F (1999) Eur Phys J **B10**: 271
[30] Wiehl L, Kiel G, Köhler CP, Spiering H, Gütlich P (1986) Inorg Chem **25**: 1565
[31] Müller EW, Ensling J, Spiering H, Gütlich P (1983) Inorg Chem **22**: 2074
[32] Grasjean F, Long GL, Hutchinson BB, Ohlhausen LN, Neill P, Holcomb JD (1989) Inorg Chem **28**: 4406
[33] Niel V, Martinez-Agudo JM, Munoz MC, Gaspar AB, Real JA (2001) Inorg Chem **40**: 3838
[34] Nakamoto T, Tan Z-C, Sorai M (2001) Inorg Chem **40**: 3805
[35] Koudriavtsev AB (1999) Chem Phys **241**: 109

Invited Review

Quantum Spin Dynamics in Molecular Magnets

Michael N. Leuenberger, Florian Meier, and **Daniel Loss***

Department of Physics and Astronomy, University of Basel, CH-4056 Basel, Switzerland

Received May 7, 2002; accepted May 22, 2002
Published online September 19, 2002

Summary. The detailed theoretical understanding of quantum spin dynamics in various molecular magnets is an important step on the roadway to technological applications of these systems. Quantum effects in both ferromagnetic and antiferromagnetic molecular clusters are, by now, theoretically well understood. Ferromagnetic molecular clusters allow one to study the interplay of incoherent quantum tunneling and thermally activated transitions between states with different spin orientation. The *Berry* phase oscillations found in Fe_8 are signatures of the quantum mechanical interference of different tunneling paths. Antiferromagnetic molecular clusters are promising candidates for the observation of coherent quantum tunneling on the mesoscopic scale. Although challenging, application of molecular magnetic clusters for data storage and quantum data processing are within experimental reach already with present day technology.

Keywords: Molecular Magnets; Spin quantum tunneling; Quantum computing.

Introduction

Molecular magnets have attracted considerable interest recently because of their potential for data storage and data processing [1]. In addition to possible future technological applications, molecular magnets are also interesting from an academic point of view because they show quantum effects on the mesoscopic scale [2] in the form of tunneling of magnetization. In the following, we review some of our theoretical work on quantum spin dynamics in molecular magnets.

Ferromagnetic molecular magnets such as Mn_{12} and Fe_8 show incoherent tunneling of the magnetization [3–6] and allow one to study the interplay of thermally activated processes and quantum tunneling. The spin tunneling leads to two effects. Firstly, the magnetization relaxation is accelerated whenever spin states of opposite direction become degenerate due to the variation of the external longitudinal

* Corresponding author. E-mail: Daniel.Loss@unibas.ch

magnetic field [7–11]. Secondly, the spin acquires a *Berry* phase during the tunneling process, which leads to oscillations of the tunnel splitting as a function of the external transverse magnetic field [12–15].

Due to the strong quantum spin dynamics induced by antiferromagnetic exchange interaction [16–19], antiferromagnetic molecular magnets such as ferric wheels belong to the most promising candidates for the observation of coherent quantum tunneling on the mesoscopic scale [20–23]. In contrast to incoherent tunneling, in quantum coherent tunneling spins tunnel back and forth between energetically degenerate configurations at a tunneling rate which is *large* compared to the decoherence rate. The detection of coherent quantum tunneling is more challenging in antiferromagnetic molecular magnets than in ferromagnetic systems, but is feasible with present day experimental techniques.

Understanding the properties of molecular magnets is only a first step on the roadway to technological applications. A possible next step will be the preparation and control of a well defined single-spin quantum state of a molecular cluster. Although challenging, this task appears feasible with present day experiments and would allow one to carry out quantum computing with molecular magnets [1]. The idea is to use the *Grover* quantum search algorithm [24] to read-in and decode information stored in the phases of a single-spin state.

Spin Tunneling in Mn_{12}-Acetate

The magnetization relaxation of crystals and powders made of molecular magnets Mn_{12} has attracted much recent interest since several experiments [25–29] have indicated unusually long relaxation times as well as increased rates [7, 8, 30] whenever two spin states become degenerate in response to a varying longitudinal magnetic field H_z. According to earlier suggestions [31, 27] this phenomenon has been interpreted as a manifestation of incoherent macroscopic quantum tunneling (MQT) of the spin.

As long as the external magnetic field H_z is much smaller than the internal exchange interactions between the Mn ions of the Mn_{12} cluster, the Mn_{12} cluster behaves like a large single spin $\mathbf{S}$ of length $s = 10$. For temperatures $T \gtrsim 1$ K its spin dynamics can be described by a spin Hamiltonian of form $\mathcal{H} = \mathcal{H}_a + \mathcal{H}_Z + \mathcal{H}_{sp} + \mathcal{H}_T$ including the coupling between this large spin and the phonons in the crystal [9–11, 32–37]. In particular,

$$\mathcal{H}_a = -AS_z^2 - BS_z^4 \tag{1}$$

represents the magnetic anisotropy where $A \gg B > 0$. The *Zeeman* term through which the external magnetic field H_z couples to the spin $\mathbf{S}$ is given by $\mathcal{H}_Z = g\mu_B H_z S_z$, while the tunneling between S_z-states is governed by

$$\mathcal{H}_T = -\frac{1}{2}B_4(S_+^4 + S_-^4) + g\mu_B H_x S_x, \tag{2}$$

where $H_x = |\mathbf{H}| \sin\theta$ ($\ll H_z$) is the transverse field, with θ being the misalignment angle. The values of the anisotropy constants A, B, B_4 have been determined by

ESR experiments [38, 39]. Finally, the most general spin-phonon coupling reads

$$\begin{aligned}\mathcal{H}_{\mathrm{sp}} &= g_1(\epsilon_{xx} - \epsilon_{yy}) \otimes (S_x^2 - S_y^2) + \frac{1}{2} g_2 \epsilon_{xy} \otimes \{S_x, S_y\} \\ &+ \frac{1}{2} g_3(\epsilon_{xz} \otimes \{S_x, S_z\} + \epsilon_{yz} \otimes \{S_y, S_z\}) \\ &+ \frac{1}{2} g_4(\omega_{xz} \otimes \{S_x, S_z\} + \omega_{yz} \otimes \{S_y, S_z\}), \end{aligned} \tag{3}$$

where g_i are the spin-phonon coupling constants, and $\epsilon_{\alpha\beta}(\omega_{\alpha\beta})$ is the (anti-)symmetric part of the strain tensor. From the comparison between experimental data [7, 8] and calculation it turns out that the constants $g_i \approx A \; \forall i$ [9–11].

We denote by $|m\rangle$, $-s \leq m \leq s$, the eigenstate of the unperturbed Hamiltonian $\mathcal{H}_{\mathrm{a}} + \mathcal{H}_{\mathrm{Z}}$ with eigenvalue $\varepsilon_m = -Am^2 - Bm^4 + g\mu_B H_z m$. If the external magnetic field H_z is increased one gets doubly degenerate spin states whenever a level m coincides with a level m' on the opposite side of the potential barrier. The resonance condition for double degeneracy, *i.e.* $\varepsilon_m = \varepsilon_{m'}$, leads to the resonance field

$$H_z^{mm'} = \frac{n}{g\mu_B}[A + B(m^2 + m'^2)]. \tag{4}$$

As usual, we refer to $n = m + m' =$ even (odd) as even (odd) resonances.

The relaxation of the magnetization is described in terms of a generalized master equation for the reduced density matrix $\rho(t)$ which includes off-diagonal terms due to resonances [9, 10]. We use the notation $\rho_{mm'} = \langle m|\rho|m'\rangle$, $\rho_m = \langle m|\rho|m\rangle$. In the stationary limit $\dot{\rho}_{mm'} \approx 0$ a complete master equation

$$\dot{\rho}_m = -W_m \rho_m + \sum_{n \neq m, m'} W_{mn} \rho_n + \Gamma_m^{m'}(\rho_{m'} - \rho_m) \tag{5}$$

can be derived, where

$$\Gamma_m^{m'} = E_{mm'}^2 \frac{W_m + W_{m'}}{4\xi_{mm'}^2 + \hbar^2 (W_m + W_{m'})^2} \tag{6}$$

is the incoherent tunneling rate from m to m' in the presence of phonon-damping [10]. The spin-phonon rates $W_{m \pm 1,m}$ and $W_{m \pm 2,m}$ are evaluated by means of *Fermi*'s golden rule [10].

In Refs. [9, 10] the master equation is solved exactly to find the largest relaxation time. The result is plotted in Fig. 1. The even resonances are induced by the quartic B_4-anisotropy, whereas the odd resonances are induced by product-combinations of $B_4 S_\pm^4$- and $H_x S_x$-terms. In Fig. 2 the peaks of the resonance at $H_z = 0$ (induced only by the B_4-term) are displayed for four different temperatures. All the peaks are of single Lorentzian shape as a result of the 2-state transition rate $\Gamma_m^{m'}$ given in (6), which agrees well with the measurements [7].

It is instructive to determine the dominant transition paths via which the spin can relax. For this an approximate analytic expression for the relaxation time can be derived by means of conservation laws that resemble *Kirchhoff*'s rules for electrical circuits [9, 10].

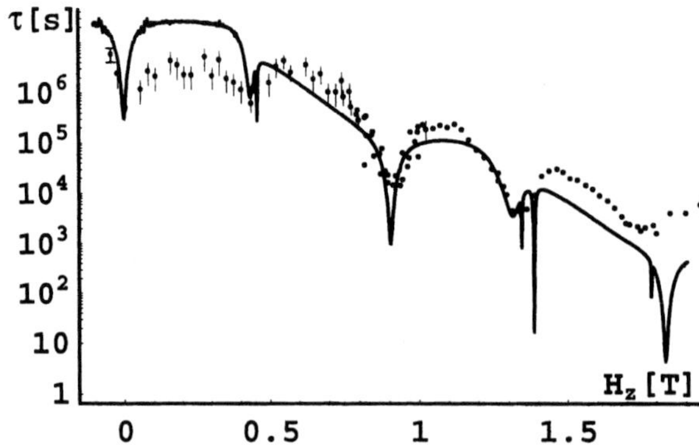

Fig. 1. Full line: semilogarithmic plot of calculated relaxation time τ as function of magnetic field H_z at $T = 1.9\,\mathrm{K}$. Dots and error bars: data taken from Ref. [8]

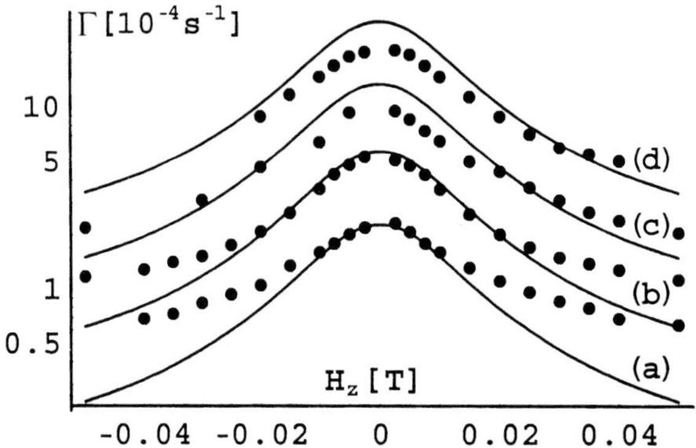

Fig. 2. Full lines: semilogarithmic plots of calculated relaxation rate $\Gamma = 1/\tau$ as function of H_z for the first resonance peak at (a) $T = 2.5\,\mathrm{K}$, (b) $T = 2.6\,\mathrm{K}$, (c) $T = 2.7\,\mathrm{K}$, and (d) $T = 2.8\,\mathrm{K}$. All peaks are of single *Lorentzian* shape. Dots: data taken from Ref. [7]

Incoherent *Zener* Tunneling in Fe_8

Besides Mn_{12} there have been several experiments on the molecular magnet Fe_8 that revealed macroscopic quantum tunneling of the spin [40–42, 12, 13]. In particular, recent measurements on Fe_8 [12, 13] lead to the development of the concept of the incoherent *Zener* tunneling [14]. The resulting *Zener* tunneling probability P_{inc} exhibits *Berry* phase oscillations as a function of the external transverse field H_x.

For many physical systems the *Landau–Zener* model [43] has become an important tool for studying tunneling transitions [44–47]. It must be noted that all quantum systems to which the *Zener* model [43] is applicable can be described by *pure* states and their *coherent* time evolution. Reference [14] generalizes the *Zener* theory in the sense that also the *incoherent* evolution of *mixed* states is taken into account (see also Refs. [44] and [48–52] for a comparison). In particular, the theory presented in Ref. [14] agrees well with recent measurements of $P_{\mathrm{inc}}(H_x)$ for various temperatures in Fe_8 [12, 13].

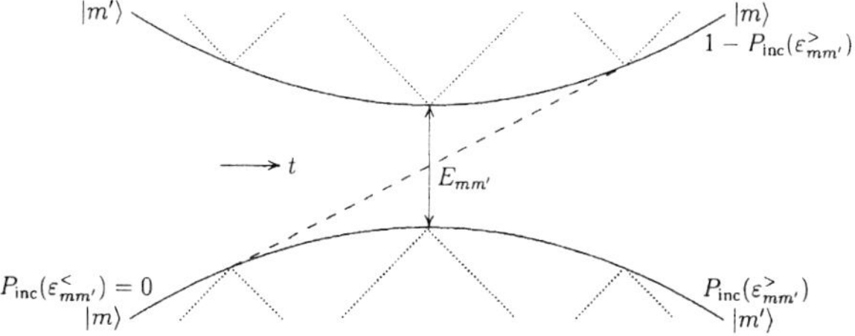

Fig. 3. Energy level crossing diagram for incoherent *Zener* transitions. Dotted lines: transitions due to interaction with environment, leading to a linewidth $\gamma_{mm'}$

For the *Zener* transition usually only the asymptotic limit is of interest. Therefore it is required that the range over which $\varepsilon_{mm'}(t) = \varepsilon_m - \varepsilon_{m'}$ is swept is much larger than the tunnel splitting $E_{mm'}$ and the decoherence rate $\hbar\gamma_{mm'}$ (see below and Fig. 3). In addition, the evolution of the spin system is restricted to times t that are much longer than the decoherence time $\tau_{\text{d}} = 1/\gamma_{mm'}$. In this case, tunneling transitions between pairs of degenerate excited states are incoherent. This tunneling is only observable if the temperature T is kept well below the activation energy of the potential barrier. Accordingly, one is interested only in times t that are larger than the relaxation times of the excited states. Thus, the formalism presented in Refs. [9, 10] can be applied. It was shown in Ref. [14] that the *Zener* tunneling can be described by Eq. (5), where

$$\Gamma_m^{m'}(t) = \frac{E_{mm'}^2}{2} \frac{\gamma_{mm'}}{\varepsilon_{mm'}^2(t) + \hbar^2\gamma_{mm'}^2} \tag{7}$$

is time-dependent, in contrast to Eq. (6). As usual, the abbreviations $\gamma_{mm'} = (W_m + W_{m'})/2$ and $W_m = \sum_n W_{nm}$ are used, where W_{nm} denotes the approximately time-independent transition rate from $|m\rangle$ to $|n\rangle$, which can be obtained via *Fermi*'s golden rule [9, 10]. The tunnel splitting [9, 10] is given by

$$E_{mm'} = 2\left| \sum_{\substack{m_1,\ldots,m_N \\ m_i \neq m,m'}} \frac{V_{m,m_1}}{\varepsilon_m - \varepsilon_{m_1}} \prod_{i=1}^{N-1} \frac{V_{m_i,m_{i+1}}}{\varepsilon_m - \varepsilon_{m_{i+1}}} V_{m_N,m'} \right|. \tag{8}$$

V_{m_i,m_j} denote off-diagonal matrix elements of the total Hamiltonian $\mathcal{H}_{\text{tot}}$.

Since all resonances n lead to similar results, Eq. (5) is solved only in the unbiased case – corresponding to $n=0$ (see below) – where the ground states $|s\rangle$, $|-s\rangle$ and the excited states $|m\rangle$, $|-m\rangle$, $m \in [[s] - s + 1, s - 1]$ of the spin system with spin s are pairwise degenerate. In addition, it is assumed that the excited states are already in their stationary state, i.e., $\dot{\rho}_m = 0\,\forall m \neq s, -s$. Eq. (5) leads then to

$$1 - P_{\text{inc}} \equiv \Delta\rho(t) = \exp\left\{ -\int_{t_0}^{t} dt' \Gamma_{\text{tot}}(t') \right\}, \tag{9}$$

where $\Delta\rho(t) = \rho_s - \rho_{-s}$, which satisfies the initial condition $\Delta\rho(t = t_0) = 1$, and thus $P_{\text{inc}}(t = t_0) = 0$. The total time-dependent relaxation rate is given by $\Gamma_{\text{tot}} = 2[\Gamma_s^{-s} + \Gamma_{\text{th}}]$, where the thermal rate Γ_{th}, which determines the incoherent relaxation via the excited states, is evaluated by means of relaxation diagrams [9, 10].

Assuming linear time dependence, i.e., $\varepsilon_{mm'}(t) = \alpha_m^{m'} t$, in the transition region [43], and with $\left|\varepsilon_{mm'}^{\langle,\rangle}\right| \gg \hbar\gamma_{mm'}$ one obtains from Eq. (9)

$$\Delta\rho = \exp\left\{-\frac{2E_{s,-s}^2}{\hbar\alpha_s^{-s}} \arctan\left(\frac{\alpha_s^{-s}}{\hbar\gamma_{s,-s}} t\right) - \int_{-t}^{t} dt'\Gamma_{\text{th}}\right\}$$
$$\approx \exp\left\{-\frac{\pi E_{s,-s}^2}{\hbar\alpha_s^{-s}} - \int_{-t}^{t} dt'\Gamma_{\text{th}}\right\}, \tag{10}$$

where $t_0 = -t$. In the low-temperature limit $T \to 0$ the excited states are not populated anymore and thus Γ_{th}, which consists of intermediate rates that are weighted by *Boltzmann* factors b_m [9, 10], vanishes. Consequently, Eq. (10) simplifies to

$$\Delta\rho = \exp\left\{-\frac{\pi E_{s,-s}^2}{\hbar\alpha_s^{-s}}\right\} = \exp\left\{-\frac{\pi E_{s,-s}^2}{\hbar|\dot{\varepsilon}_{s,-s}(0)|}\right\}. \tag{11}$$

The exponent in Eq. (11) differs by a factor of 2 from the *Zener* exponent [43]. This is not surprising since Γ_{tot} is the relaxation rate of $\Delta\rho$, where both ρ_s and ρ_{-s} are changed in time by the same amount, and *not* an escape rate like in the case of coherent *Zener* transition, where only the population of the initial state is changed in time. Equation (11) implies $P_{\text{inc}} = 1$ for $|\dot{\varepsilon}_{s,-s}(0)| \to 0$ (adiabatic limit) and $P_{\text{inc}} = 0$ for $|\dot{\varepsilon}_{s,-s}(0)| \to \infty$ (sudden limit).

In accordance with earlier work [12, 13, 40–42, 53] Ref. [14] uses a single-spin Hamiltonian $\mathcal{H} = \mathcal{H}_{\text{a}} + \mathcal{H}_{\text{T}} + \mathcal{H}_{\text{Z}} + \mathcal{H}_{\text{sp}}$ that describes sufficiently well the behavior of the large spin $\mathbf{S}$ with $s = 10$ of a Fe_8 cluster. After fitting the parameters the incoherent *Zener* theory is in excellent agreement with experiments [12, 13] for the temperature range $0.05\,\text{K} \leq T \leq 0.7\,\text{K}$ if the states $|\pm 10\rangle$, $|\pm 9\rangle$, and $|\pm 8\rangle$ are taken into account. In particular, the path leading through $|\pm 8\rangle$ gives a non-negligible contribution for $T \gtrsim 0.6\,\text{K}$. Solving the relaxation diagram shown in Fig. 4 one obtains from Eq. (10) for Fe_8 in the case $n = 0$

$$\Gamma_{\text{tot}} = 2\left(\Gamma_{10}^{-10} + \sum_{n=9}^{8} \frac{b_n}{\frac{2}{W_{10,n}} + \frac{1}{\Gamma_n^{-n}}}\right),$$
$$\Delta\rho = \exp\left\{-\frac{\pi E_{10,-10}^2}{\hbar\alpha_{10}^{-10}} - \sum_{n=9}^{8} \frac{\pi E_{n,-n}^2 W_{10,n} b_n}{\alpha_n^{-n}\sqrt{E_{n,-n}^2 + \hbar^2 W_{10,n}^2}}\right\}, \tag{12}$$

where the approximation $\gamma_{n,-n} \approx W_{10,n}$ and $\left|\varepsilon_{mm'}^{\langle,\rangle}\right| \gg E_{n,-n}, \gamma_{n,-n}$ is used. $P_{\text{inc}} = 1 - \Delta\rho$, which is plotted in Fig. 5, is in good agreement with the measurements [13].

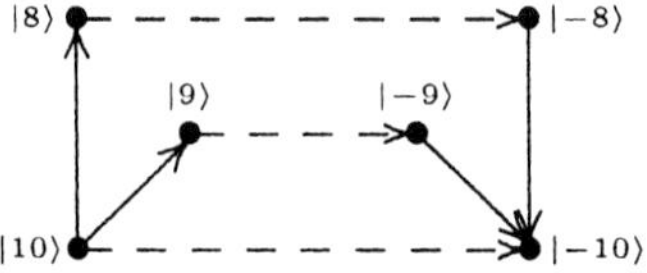

Fig. 4. Unbiased ($n = 0$) relaxation diagram for Fe_8. Full (dashed) lines: thermal (tunneling) transitions

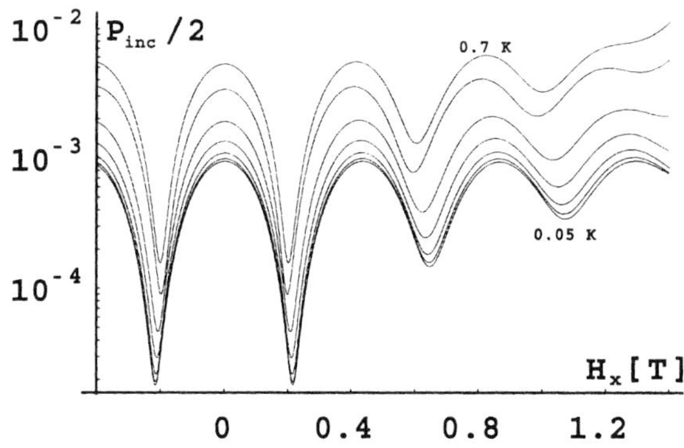

Fig. 5. *Zener* transition probability $P_{\text{inc}}(H_x)$ for temperatures $T=0.7\,\text{K}$, 0.65 K, 0.6 K, 0.55 K, 0.5 K, 0.45 K, and 0.05 K. The fit agrees well with data (Ref. [13]). Note that P_{inc} is equal to $2P$ in Ref. [13]

Coherent *Néel* Vector Tunneling in Antiferromagnetic Molecular Wheels

Antiferromagnetic molecular clusters belong to the most promising candidates for the observation of coherent quantum tunneling on the mesoscopic scale currently available [20]. Several systems in which an even number N of antiferromagnetically coupled ions is arranged on a ring have been synthesized to date [54–57]. These systems are well described by the spin Hamiltonian

$$\hat{H} = J\sum_{i=1}^{N} \hat{\mathbf{s}}_i \cdot \hat{\mathbf{s}}_{i+1} + g\mu_B \mathbf{B} \cdot \sum_{i=1}^{N} \hat{\mathbf{s}}_i - k_z \sum_{i=1}^{N} \hat{s}_{i,z}^2, \tag{13}$$

where $\hat{\mathbf{s}}_i$ is the spin operator at site i with spin quantum number s, $\hat{\mathbf{s}}_{N+1} \equiv \hat{\mathbf{s}}_1$, J is the nearest-neighbor exchange, $\mathbf{B}$ the magnetic field, and $k_z{>}0$ the single-ion anisotropy directed along the ring axis. The parameters J and k_z have been well established both for various ferric wheels [54–56, 58–62] with $N = 6, 8, 10$, and, more recently, also for a Cr wheel [57]. For $\mathbf{B} = 0$, the classical ground-state spin configuration of the wheel shows alternating (*Néel*) order with the spins pointing along $\pm\mathbf{e}_z$. The two states with the *Néel* vector $\mathbf{n}$ along $\pm\mathbf{e}_z$ (Fig. 6), labeled $|\uparrow\rangle$ and $|\downarrow\rangle$, are energetically degenerate and separated by an energy barrier of height Nk_zs^2. Because antiferromagnetic exchange induces dynamics of *Néel* ordered spins, the states $|\uparrow\rangle$ and $|\downarrow\rangle$ are not energy eigenstates. Rather, a molecule prepared in spin state $|\uparrow\rangle$ would tunnel coherently between $|\uparrow\rangle$ and $|\downarrow\rangle$ at a rate Δ/h, where Δ is the tunnel splitting [16, 17]. This tunneling of the *Néel* vector corresponds to a simultaneous tunneling of all N spins within the wheel through a potential barrier governed by the easy-axis anisotropy. Within the framework of coherent state spin path integrals, an explicit expression for the tunnel splitting Δ as a function of the magnetic field $\mathbf{B}$ has been derived [20]. A magnetic field applied in the ring plane, B_x, gives rise to a *Berry* phase acquired by the spins during tunneling [15, 63]. The resulting interference of different tunneling paths leads to a sinusoidal dependence of Δ on B_x, which allows one to continuously tune the tunnel splitting from 0 to a maximum value which is of order of some Kelvin for the antiferromagnetic wheels synthesized to date.

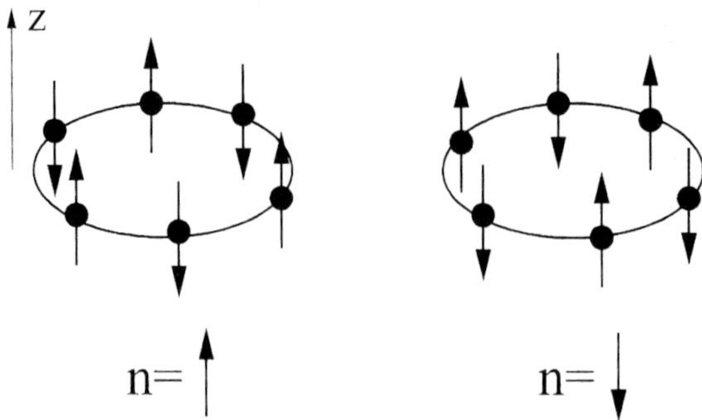

Fig. 6. The two degenerate classical ground state spin configurations of an antiferromagnetic molecular wheel with easy axis anisotropy

The tunnel splitting Δ also enters the energy spectrum of the antiferromagnetic wheel as level spacing between the ground and first excited state. Thus, Δ can be experimentally determined from various quantities such as magnetization, static susceptibility, and specific heat. Even more information on the physical properties of antiferromagnetic wheels [Eq. (13)] can be obtained from a theoretical and experimental investigation of dynamical quantities, such as the correlation functions of the total spin $\hat{\mathbf{S}} = \sum_{i=1}^{N} \hat{\mathbf{s}}_i$ or of single spins within the wheel [21–23]. By symmetry arguments, it follows that the correlation function of total spin, $\langle \hat{S}_\alpha(t)\hat{S}_\alpha(0)\rangle$, which is experimentally accessible via measurement of the alternating current (AC) susceptibility does not contain a component which oscillates with the tunnel frequency Δ/h [21–23]. Hence, neither the tunnel splitting nor the decoherence rate of *Néel* vector tunneling can be obtained by experimental techniques which couple to the *total* spin of the wheel. In contrast, the correlation function of a single spin

$$\langle \hat{s}_{i,z}(t)\hat{s}_{i,z}(0)\rangle \simeq s^2\left(\frac{e^{-\beta\Delta/2}}{2\cosh(\beta\Delta/2)}e^{i\Delta t/\hbar} + \frac{e^{\beta\Delta/2}}{2\cosh(\beta\Delta/2)}e^{-i\Delta t/\hbar}\right) \tag{14}$$

exhibits the time dependence characteristic of coherent tunneling of the quantity $\hat{s}_{i,z}$ with a tunneling rate Δ/h [21]. We conclude that *local* spin probes are required for the observation of the *Néel* vector dynamics. Nuclear spins which couple (predominantly) to a given single spin $\hat{\mathbf{s}}_i$ are ideal candidates for such probes [21] and have already been used to study spin cross-relaxation between electron and nuclear spins in ferric wheels [64, 65].

For simplicity, we consider a single nuclear spin $\hat{\mathbf{I}}$, $I = 1/2$, coupled to one electron spin by a hyperfine contact interaction $\hat{H}' = A\hat{\mathbf{s}}_1 \cdot \hat{\mathbf{I}}$. According to Eq. (14), the tunneling electron spin $\hat{\mathbf{s}}_1$ produces a rapidly oscillating hyperfine field $As\cos(\Delta t/\hbar)$ at the site of the nucleus. Signatures of the coherent electron spin tunneling can thus also be found in the nuclear susceptibility. For a static magnetic field applied in the plane of the ring, B_x, it can be shown that the nuclear susceptibility

$$\chi''_{I,yy}(\omega) \simeq \frac{\pi}{4}\left[\tanh\left(\frac{\beta\gamma_I B_x}{2}\right)\delta(\omega - \gamma_I B_x/\hbar) + \left(\frac{As}{\Delta}\right)^2 \tanh\left(\frac{\beta\Delta}{2}\right)\delta(\omega - \Delta/\hbar)\right] - [\omega \to -\omega] \tag{15}$$

exhibits a *satellite resonance at the tunnel splitting* Δ *of the electron spin system* [21]. Here, $\gamma_I B_x$ is the *Larmor* frequency of the nuclear spin and the first term in Eq. (15) corresponds to the transition between the *Zeeman*-split energy levels of I. Because typically $As \simeq 1\,\text{mK}$ and $\Delta \lesssim 2\,\text{K}$ in Fe_{10}, the spectral weight of the satellite peak is small compared to the one of the first term in Eq. (15) unless the magnetic field is tuned such that Δ is significantly reduced compared to its maximum value. The observation of the satellite peak in $\chi''_{I,yy}(\omega)$ is challenging, but possible with current experimental techniques [21]. The experiment must be conducted with single crystals of an antiferromagnetic molecular wheel with sufficiently large anisotropy $k_z > 2J/(Ns)^2$ at high, tunable fields (10 T) and low temperatures (2 K). Moreover, because the tunnel splitting $\Delta(\mathbf{B})$ depends sensitively on the relative orientation of $\mathbf{B}$ and the easy axis [59, 60], careful field sweeps are necessary to ensure that the satellite peak in Eq. (15) has a large spectral weight.

The need for local spin probes such as NMR or inelastic neutron scattering to detect coherent *Néel* vector tunneling can be traced back to the translation symmetry of the spin Hamiltonian $\hat{H}$ [22, 23]. If this symmetry is broken, e.g. by doping of the wheel, ESR also provides an adequate technique for the detection of coherent *Néel* vector tunneling. If one of the original Fe or Cr ions of the wheel with spin $s = 5/2$ or $s = 3/2$, respectively, is replaced by an ion with different spin $s' \neq s$, this will in general also result in a different exchange constant J' and single ion anisotropy k'_z at the dopand site, i.e.,

$$\hat{H} = J\sum_{i=2}^{N-1} \hat{\mathbf{s}}_i \cdot \hat{\mathbf{s}}_{i+1} + J'(\hat{\mathbf{s}}_1 \cdot \hat{\mathbf{s}}_2 + \hat{\mathbf{s}}_1 \cdot \hat{\mathbf{s}}_N) + g\mu_B \mathbf{B} \cdot \sum_{i=1}^{N} \hat{\mathbf{s}}_i - \left(k'_z \hat{s}_{1,z}^2 + k_z \sum_{i=2}^{N} \hat{s}_{i,z}^2 \right). \quad (16)$$

Although thermodynamic quantities, such as magnetization, of the doped wheel may differ significantly from the ones of the undoped wheel, the picture of spin tunneling in antiferromagnetic molecular systems [16, 17] remains valid [22]. However, due to unequal sublattice spins, a net total spin remains even in the *Néel* ordered state of the doped wheel (Fig. 7) which allows one to distinguish the configurations sketched in Fig. 6 according to their total spin. The dynamics of the total spin $\hat{\mathbf{S}}$ is coupled to the one of the *Néel* vector, and coherent tunneling of the *Néel* vector results in a coherent oscillation of the total spin such that the

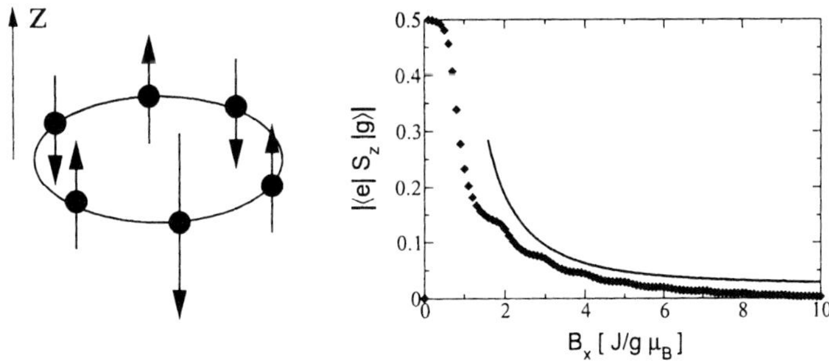

Fig. 7. The doped antiferromagnetic molecular wheel acquires a tracer spin which follows the *Néel* vector dynamics (left panel). Comparison of results obtained for the matrix element $|\langle e|\hat{S}_z|g\rangle|$ with a coherent state spin path integral formalism (solid line) and by numerical exact diagonalization (symbols) for $N = 4$, $s = 5/2$, $s' = 2$, $J' = J$, $k_z = k'_z = 0.055\,\text{J}$ (right panel)

tunneling dynamics can also be probed by ESR. The AC susceptibility shows a resonance peak at the tunnel splitting Δ,

$$\chi''_{zz}(\omega \simeq \Delta/\hbar) = \pi(g\mu_B)^2|\langle e|\hat{S}_z|g\rangle|^2 \tanh\left(\frac{\beta\Delta}{2}\right)\delta(\omega - \Delta/\hbar). \quad (17)$$

with a transition matrix element between the ground state $|g\rangle$ and first excited state $|e\rangle$,

$$|\langle e|\hat{S}_z|g\rangle| \simeq |s' - s|\frac{8Jk_z s^2}{(g\mu_B B_x)^2} \quad (18)$$

for $g\mu_B B_x \gg s\sqrt{8Jk_z}$. The matrix element in Eq. (18) determines the spectral weight of the absorption peak in the ESR spectrum. The analytical dependence has been determined within a semiclassical framework and is in good agreement with numerical results obtained from exact diagonalization of small systems (Fig. 7).

In conclusion, several antiferromagnetic molecular wheels synthesized recently are promising candidates for the observation of coherent *Néel* vector tunneling. Although the observation of this phenomenon is experimentally challenging, nuclear magnetic resonance, inelastic neutron scattering, and ESR on doped wheels are adequate experimental techniques. The theory of coherent spin quantum tunneling as presented above applies to zero-dimensional systems, such as small ferric wheels. With increasing wheel size, the possibility of different magnetic domains arises and new exciting quantum effects in the dynamics of domain walls come into play [66–72]. Molecular wheels will also allow one to study the transition from zero- to one-dimensional quantum behavior with increasing system size.

Quantum Computing with Molecular Magnets

Shor and *Grover* demonstrated that a quantum computer can outperform any classical computer in factoring numbers [73] and in searching a database [24] by exploiting the parallelism of quantum mechanics. Recently, the latter has been successfully implemented [74] using *Rydberg* atoms. In Ref. [1] an implementation of *Grover*'s algorithm was proposed that uses molecular magnets [7, 8, 13, 41, 75]. It was shown theoretically that molecular magnets can be used to build dense and efficient memory devices based on the *Grover* algorithm. In particular, one single crystal can serve as a storage unit of a dynamic random access memory device. Fast electron spin resonance pulses can be used to decode and read out stored numbers of up to 10^5, with access times as short as 10^{-10} seconds. This proposal should be feasible using the molecular magnets Fe_8 and Mn_{12}.

Suppose we want to find a phone number in a phone book consisting of $N = 2^n$ entries. Usually it takes $N/2$ queries on average to be successful. Even if the N entries were encoded binary, a classical computer would need approximately $\log_2 N$ queries to find the desired phone number [24]. But the computational parallelism provided by the superposition and interference of quantum states enables the *Grover* algorithm to reduce the search to one single query [24]. This query can be implemented in terms of a unitary transformation applied to the single spin of a molecular magnet. Such molecular magnets, forming identical and largely independent units, are embedded in a single crystal so that the ensemble nature of such a crystal provides a natural amplification of the magnetic moment of a single spin. However, for the *Grover* algorithm to succeed, it is necessary to find ways to generate arbitrary superpositions of spin eigenstates. For spins larger than $1/2$ this turns out to be a highly non-trivial task as spin excitations induced by magnetic dipole transitions in conventional electron spin resonance (ESR) can change the magnetic quantum number m by only ± 1. To circumvent such physical limitations it was proposed to use multifrequency coherent magnetic radiation that allows the controlled generation of arbitrary spin superpositions. In particular, it was shown that by means of advanced ESR techniques it is possible to coherently populate and manipulate many spin states simultaneously by applying one single pulse of a magnetic a.c. field containing an appropriate number of matched frequencies. This a.c. field creates a nonlinear response of the magnet via multiphoton absorption processes involving particular sequences of σ and π photons which allows the encoding and, similarly, the decoding of states. Finally, the subsequent read-out of the decoded quantum state can be achieved by means of pulsed ESR techniques. These exploit the non-equidistance of energy levels which is typical of molecular magnets.

Molecular magnets have the important advantage that they can be grown naturally as single crystals of up to 10–100 μm length containing about 10^{12} to 10^{15} (largely) independent units so that only minimal sample preparation is required. The molecular magnets are described by a single-spin Hamiltonian of the form $\mathcal{H}_{\text{spin}} = \mathcal{H}_{\text{a}} + V + \mathcal{H}_{\text{sp}} + \mathcal{H}_{\text{T}}$ [9, 10, 14], where $\mathcal{H}_{\text{a}} = -AS_z^2 - BS_z^4$ represents the magnetic anisotropy ($A \gg B > 0$). The *Zeeman* term $V = g\mu_B \mathbf{H} \cdot \mathbf{S}$ describes the coupling between the external magnetic field $\mathbf{H}$ and the spin $\mathbf{S}$ of length s. The calculational states are given by the $2s + 1$ eigenstates of $\mathcal{H}_{\text{a}} + g\mu_B H_z S_z$ with eigenenergies $\varepsilon_m = -Am^2 - Bm^4 + g\mu_B H_z m, -s \leq m \leq s$. The corresponding classical anisotropy potential energy $E(\theta) = -As^2 \cos^2\theta - Bs^4 \cos^4\theta + g\mu_B H_z s \cos\theta$, is obtained by the substitution $S_z = s\cos\theta$, where θ is the polar spherical angle. We have introduced the notation $m, m' = m - m'$. By applying a bias field H_z such that $g\mu_B H_z > E_{mm'}$, tunneling can be completely suppressed and thus $\mathcal{H}_{\text{T}}$ can be neglected [9, 10, 14]. For temperatures of below 1 K transitions due to spin-phonon interactions ($\mathcal{H}_{\text{sp}}$) can also be neglected. In this regime, the level lifetime in Fe_8 and Mn_{12} is estimated to be about $\tau_{\text{d}} = 10^{-7}$s, limited mainly by hyperfine and/or dipolar interactions [1].

Since the *Grover* algorithm requires that all the transition probabilities are almost the same, Ref. [1] proposes that all the transition amplitudes between the states $|s\rangle$ and $|m\rangle$, $m = 1, 2, \ldots, s - 1$, are of the same order in perturbation V. This allows us to use perturbation theory. A different approach uses the magnetic field amplitudes to adjust the appropriate transition amplitudes [76]. Both methods work only if the energy levels are not equidistant, which is typically the case in molecular magnets owing to anisotropies. In general, if we choose to work with the states $m = m_0, m_0 + 1, \ldots, s - 1$, where $m_0 = 1, 2, \ldots, s - 1$, we have to go up to nth order in perturbation, where $n = s - m_0$ is the number of computational states used for the *Grover* search algorithm (see below), to obtain the first non-vanishing contribution. Figure 9 shows the transitions for $s = 10$ and $m_0 = 5$. The nth-order transitions correspond to the nonlinear response of the spin system to strong magnetic fields. Thus, a coherent magnetic pulse of duration T is needed with a discrete frequency spectrum $\{\omega_m\}$, say, for Mn_{12} between 20 and 300 GHz and a single low-frequency 0 around 100 MHz. The low-frequency field $\mathbf{H}_z(t) = H_0(t)\cos(\omega_0 t)\mathbf{e}_z$, applied along the easy-axis, couples to the spin of the molecular magnet through the Hamiltonian

$$V_{\text{low}} = g\mu_B H_0(t)\cos(\omega_0 t)S_z, \tag{19}$$

where $\hbar\omega_0 \ll \varepsilon_{m_0} - \varepsilon_{m_0+1}$ and $\mathbf{e}_z$ is the unit vector pointing along the z axis. The π photons of V_{low} supply the necessary energy for the resonance condition (see below). They give rise to virtual transitions with $\Delta m = 0$, that is, they do not transfer any angular momentum, see Fig. 9. The perturbation Hamiltonian for the high-frequency transitions from $|s\rangle$ to virtual states that are just below $|m\rangle$, $m = m_0, \ldots, s-1$, given by the transverse fields $\mathbf{H}_\perp^-(t) = \sum_{m=m_0}^{s-1} H_m(t)[\cos(\omega_m t + \Phi_m)\mathbf{e}_x - \sin(\omega_m t + \Phi_m)\mathbf{e}_y]$, reads

$$\begin{aligned} V_{\text{high}}(t) &= \sum_{m=m_0}^{s-1} g\mu_B H_m(t)[\cos(\omega_m t + \Phi_m)S_x - \sin(\omega_m t + \Phi_m)S_y] \\ &= \sum_{m=m_0}^{s-1} \frac{g\mu_B H_m(t)}{2}[e^{i(\omega_m t+\Phi_m)}S_+ + e^{-i(\omega_m t+\Phi_m)}S_-], \end{aligned} \tag{20}$$

with phases Φ_m (see below), where we have introduced the unit vectors $\mathbf{e}_x$ and $\mathbf{e}_y$ pointing along the x and y axis, respectively. These transverse fields rotate clockwise and thus produce left circularly polarized σ^- photons which induce only transitions in the left well (see Fig. 8). In general, absorption (emission) of σ^- photons gives rise to $\Delta m = -1$ $(\Delta m = +1)$ transitions, and vice versa in the case of σ^+ photons. Anti-clockwise rotating magnetic fields of the $\mathbf{H}_\perp^+(t) = \sum_{m=m_0}^{s-1} H_m(t)[\cos(\omega_m t + \Phi_m)\mathbf{e}_x + \sin(\omega_m t + \Phi_m)\mathbf{e}_y]$ can be used to induce spin transitions only in the right well (Fig. 8). In this way, both wells can be accessed independently.

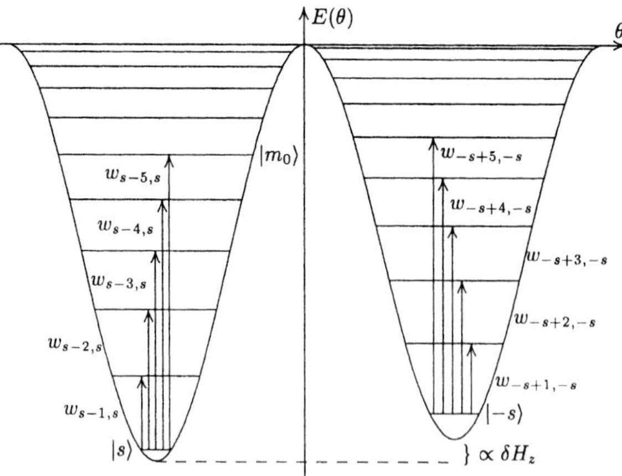

Fig. 8. Double well potential seen by the spin due to magnetic anisotropies in Mn_{12}. Arrows depict transitions between spin eigenstates driven by the external magnetic field **H**

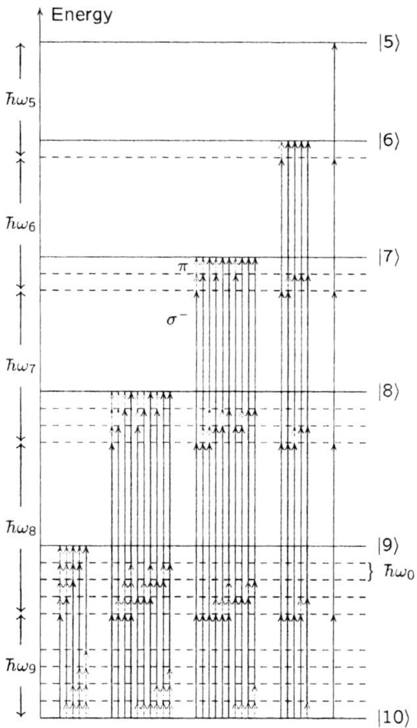

Fig. 9. *Feynman* diagrams $\mathcal{F}$ that contribute to $S_{m,s}^{(5)}$ for $s=10$ and $m_0=5$ describing transitions (of 5th order in V) in the left well of the spin system (see Fig. 8). The solid and dotted arrows indicate σ^- and π transitions governed by Eq. (20) and Eq. (19), respectively. We note that $S_{m,s}^{(j)}=0$ for $j<n$, and $S_{m,s}^{(j)} \ll S_{m,s}^{(n)}$ for $j>n$

Next we calculate the quantum amplitudes for the transitions induced by the magnetic a.c. fields (see Fig. 9) by evaluating the S-matrix perturbatively. The jth-order term of the perturbation series of the S-matrix in powers of the total perturbation Hamiltonian $V(t)=V_{\text{low}}(t)+V_{\text{high}}(t)$ is expressed by

$$S_{m,s}^{(j)} = \left(\frac{1}{i\hbar}\right)^j \prod_{k=1}^{j-1} \int_{-\infty}^{\infty} dt_k \int_{-\infty}^{\infty} dt_j \Theta(t_k - t_{k+1}) \\ \times U(\infty, t_1)V(t_1)U(t_1, t_2)V(t_2)\cdots V(t_j)U(t_j, -\infty), \tag{21}$$

which corresponds to the sum over all *Feynman* diagrams $\mathcal{F}$ of order j, and where $U(t,t_0)=e^{-i(\mathcal{H}_a+g\mu_B\delta H_z)(t-t_0)/\hbar}$ is the free propagator, $\Theta(t)$ is the *Heavyside* function. The total S-matrix is then given by $S=\sum_{j=0}^{\infty} S^{(j)}$. The high-frequency virtual transition changing m from s to $s-1$ is induced by the frequency $\omega_{s-1}=\omega_{s-1,s}-(n-1)\omega_0$. The other high frequencies ω_m, $m=m_0,\ldots,s-2$, of the high-frequency fields H_m mismatch the level separations by ω_0, that is, $\hbar\omega_m=\varepsilon_m-\varepsilon_{m+1}+\hbar\omega_0$ (Fig. 9). As the levels are not equidistant, it is possible to choose the low and high frequencies in such a way that $S_{m,s}^{(j)}=0$ for $j<n$, in which case the resonance condition is not satisfied, that is, energy is not conserved. In addition, the higher-order amplitudes $|S_{m,s}^{(j)}|$ are negligible compared to $|S_{m,s}^{(j)}|$ for $j>n$. Using rectangular pulse shapes, $H_k(t)=H_k$, if $-T/2<t<T/2$, and 0 otherwise, for $k=0$ and

$k \geqslant m_0$, one obtains ($m \geqslant m_0$)

$$S_{m,s}^{(n)} = \sum_{\mathcal{F}} \Omega_m \frac{2\pi}{i} \left(\frac{g\mu_B}{2\hbar} \right)^n \frac{\prod_{k=m}^{s-1} H_k e^{-i\Phi_k} H_0^{m-m_0} p_{m,s}(\mathcal{F})}{(-1)^{q_{\mathcal{F}}} q_{\mathcal{F}}! r_s(\mathcal{F})! \omega_0^{n-1}}$$
$$\times \delta^{(T)} \left(\omega_{m,s} - \sum_{k=m}^{s-1} \omega_k - (m - m_0)\omega_0 \right), \tag{22}$$

where $\Omega_m = (m - m_0)!$ is the symmetry factor of the *Feynman* diagrams $\mathcal{F}$ (see Fig. 9), $q_{\mathcal{F}} = m - m_0 - r_s(\mathcal{F})$, $p_{m,s}(\mathcal{F}) = \prod_{k=m}^{s} \langle k|S_z|k\rangle^{r_k(\mathcal{F})}$, $r_k(\mathcal{F}) = 0, 1, 2, \ldots, \leq m - m_0$ is the number of π transitions directly above or below the state $|k\rangle$, depending on the particular *Feynman* diagram $\mathcal{F}$, and $\delta^{(T)}(\omega) = \frac{1}{2\pi}\int_{-T/2}^{+T/2} e^{i\omega t} dt = \sin(\omega T/2)/\pi\omega$ is the delta-function of width $1/T$, ensuring overall energy conservation for $\omega T \gg 1$. The duration T of the magnetic pulses must be shorter than the lifetimes τ_d of the states $|m\rangle$ (see Fig. 8).

In general the magnetic field amplitudes H_k must be chosen in such a way that perturbation theory is still valid and the transition probabilities are almost equal, which is required by the *Grover* algorithm. In Ref. [1] the amplitudes H_k do not differ too much from each other due to the partial destructive interference of the different transition diagrams shown in Fig. 8. Reference [76] shows that the transition probabilities can be increased by increasing both the magnetic field amplitudes and the detuning energies under the condition that the magnetic field amplitudes remain smaller than the detuning energies. In this way, both high multiphoton *Rabi* oscillation frequencies and small quantum computation times can be attained. This makes both methods [1, 76] very robust against decoherence sources.

In order to perform the *Grover* algorithm, one needs the relative phases φ_m between the transition amplitudes $S_{m,s}^{(n)}$, which is determined by $\Phi_m = \sum_{k=s-1}^{m+1} \Phi_k + \varphi_m$, where Φ_m are the relative phases between the magnetic fields $H_m(t)$. In this way, it is possible to read-in and decode the desired phases Φ_m for each state $|m\rangle$. The read-out is performed by standard spectroscopy with pulsed ESR, where the circularly polarized radiation can now be incoherent because only the absorption intensity of only one pulse is needed. Figure 10 shows an example where the state $|7\rangle$ is populated within the computational basis (excluding the ground state). Then, we would observe only transitions from $|7\rangle$ to $|8\rangle$ at the frequency $\omega = \omega_{8,7}$ and transitions from $|7\rangle$ to $|6\rangle$ at $\omega = \omega_{6,7}$, which uniquely identifies the populated level, because the levels are not equidistant. This spectrum identifies all the populated states unambiguously. Alternatively, one could measure the magnetization of the sample with a high precision (see e.g. Ref. [77]). We emphasize that the entire *Grover* algorithm (read-in, decoding, read-out) requires three subsequent pulses each of duration T with $\tau_d > T > \omega_0^{-1} > \omega_m^{-1} > \omega_{m,m\pm 1}^{-1}$. This gives a 'clock-speed' of about 10 GHz for Mn_{12}, that is, the entire process of read-in, decoding, and read-out can be performed within about 10^{-10} s.

The proposal for implementing *Grover*'s algorithm works not only for molecular magnets but for any electron or nuclear spin system with non-equidistant energy levels, as is shown in Ref. [76] for nuclear spins in GaAs semiconductors. Instead of storing information in the phases of the eigenstates $|m\rangle$ [1], in Ref. [76] the eigenenergies of $|m\rangle$ in the generalized rotating frame are used for encoding

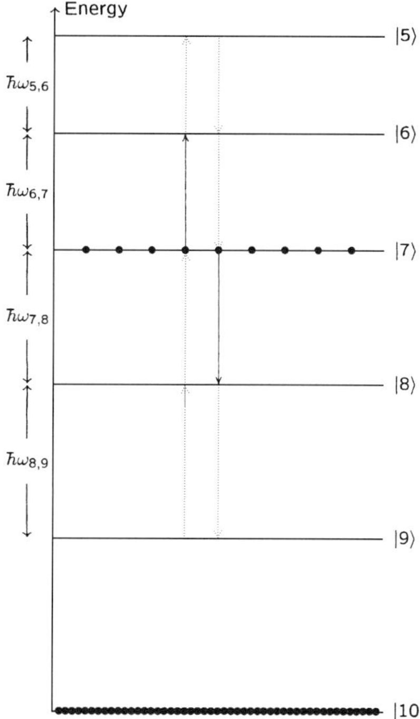

Fig. 10. Read-out of the information by ESR in the linear regime. In this example only the state $|7\rangle$ of the computational basis is populated. Thus only the transitions marked by solid arrows induce a response that can be observed in a ESR spectrum

information. The decoding is performed by bringing the delocalized state $(1/\sqrt{n})\sum_m |m\rangle$ into resonance with $|m\rangle$ in the generalized rotating frame. Although such spin systems cannot be scaled to arbitrarily large spin s – the larger a spin becomes, the faster it decoheres and the more classical its behavior will be – we can use such spin systems of given s to great advantage in building dense and highly efficient memory devices.

For a first test of the nonlinear response, one can irradiate the molecular magnet with an a.c. field of frequency $\omega_{s-2,s}/2$, which gives rise to a two-photon absorption and thus to a *Rabi* oscillation between the states $|s\rangle$ and $|s-2\rangle$. For stronger magnetic fields it is in principle possible to generate superpositions of *Rabi* oscillations between the states $|s\rangle$ and $|s-1\rangle$, $|s\rangle$ and $|s-2\rangle$, $|s\rangle$ and $|s-3\rangle$, and so on (see also Ref. [76]).

Acknowledgements

This work has been supported by the EU network Molnanomag, the BBW Bern, the Swiss National Science Foundation, DARPA, and ARO.

References

[1] Leuenberger MN, Loss D (2001) Nature **410**: 789

[2] Leggett AJ (1995) In: Gunther L, Barbara B (eds) Quantum Tunneling of Magnetization. Kluwer, Dordrecht, p 1

[3] Korenblit IY, Shender EF (1978) Sov Phys JETP **48**: 937
[4] van Hemmen JL, Süto A (1986) Europhys Lett **1**: 481; (1986) Physica B **141**: 37
[5] Enz M, Schilling R (1986) J Phys C Solid State Phys **19**: L711; (1986) J Phys C Solid State Phys **19**: 1765
[6] Chudnovsky EM, Gunther L (1988) Phys Rev Lett **60**: 661
[7] Friedman JR, Sarachik MP, Tejada J, Ziolo R (1996) Phys Rev Lett **76**: 3830
[8] Thomas L, Lionti F, Ballou R, Gatteschi D, Sessoli R, Barbara B (1996) Nature (London) **383**: 145
[9] Leuenberger MN, Loss D (1999) Europhys Lett **46**: 692
[10] Leuenberger MN, Loss D (2000) Phys Rev B **61**: 1286
[11] Leuenberger MN, Loss D (2000) Europhys Lett **52**: 247
[12] Wernsdorfer W, Sessoli R (1996) Science **284**: 133
[13] Wernsdorfer W, Sessoli R, Caneschi A, Gatteschi D, Cornia A (2000) Europhys Lett **50**: 552
[14] Leuenberger MN, Loss D (2000) Phys Rev B **61**: 12200
[15] Loss D, DiVincenzo DP, Grinstein G (1992) Phys Rev Lett **69**: 3232; (1992) von Delft J, Henley CL, *ibid* 3236; Leuenberger MN, Loss D (2001) Phys Rev B **63**: 054414
[16] Barbara B, Chudnovsky EM (1990) Phys Lett A **145**: 205
[17] Krive IV, Zaslavskii OB (1990) J Phys Condens Matter **2**: 9457
[18] Awschalom DD, Smyth JF, Grinstein G, DiVincenzo DP, Loss D (1992) Phys Rev Lett **68**: 3092; (1993) *ibid* **71**: 4279(E)
[19] Gider S, Awschalom DD, Douglas T, Mann S (1995) Science **268**: 77
[20] Chiolero A, Loss D (1998) Phys Rev Lett **80**: 169
[21] Meier F, Loss D (2001) Phys Rev Lett **86**: 5373
[22] Meier F, Loss D (2001) Phys Rev B **64**: 224411
[23] Honecker A, Meier F, Loss D, Normand B (2002) Europhys J **B27**: 487
[24] Grover LK (1997) Phys Rev Lett **79**: 325; *ibid* 4709; (1998) Phys Rev Lett **80**: 4329
[25] Paulsen C, Park JG (1995) In: Gunther L, Barbara B (eds) Quantum Tunneling of Magnetization. Kluwer, Dordrecht, p 189
[26] Paulsen C, Park JG, Barbara B, Sessoli R, Caneschi A (1995) J Magn Magn Mater **140–144**: 379
[27] Novak MA, Sessoli R (1995) In: Gunther L, Barbara B (eds) Quantum Tunneling of Magnetization. Kluwer, Dordrecht, p 171
[28] Sessoli R, Gatteschi D, Caneschi A, Novak MA (1993) Nature (London) **365**: 141
[29] Novak MA, Sessoli R, Caneschi A, Gatteschi D (1995) J Magn Magn Mater **146**: 211
[30] Hernández JM, Zhang XX, Luis F, Bartolomé J, Tejada J, Friedman JR, Sarachik MP, Ziolo R (1996) Europhys Lett **35**: 301
[31] Barbara B, Wernsdorfer W, Sampaio LC, Park JG, Paulsen C, Novak MA, Ferré R, Mailly D, Sessoli R, Caneschi A, Hasselbach K, Benoit A, Thomas L (1995) J Magn Magn Mater **140–144**: 1825
[32] Villain J, Hartmann-Boutron F, Sessoli R, Rettori A (1994) Europhys Lett **27**: 159
[33] Fort A, Rettori A, Villain J, Gatteschi D, Sessoli R (1998) Phys Rev Lett **80**: 612
[34] Luis F, Bartolomé J, Fernández F (1998) Phys Rev B **57**: 505
[35] Hartmann-Boutron F, Politi P, Villain J (1996) Int J Mod Phys B **10**: 2577
[36] Garanin DA, Chudnovsky EM (1997) Phys Rev B **56**: 11102
[37] Hernández JM, Zhang XX, Luis F, Tejada J, Ziolo R (1997) Phys Rev B **55**: 5858
[38] Barra AL, Gatteschi D, Sessoli R (1997) Phys Rev B **56**: 8192
[39] Zhong Y, Sarachik MP, Friedman JR, Robinson RA, Kelley TM, Nakotte H, Christianson AC, Trouw F, Aubin SMJ, Hendrickson DN (1999) J Appl Phys **85**: 5636
[40] Barra A-L, Debrunner P, Gatteschi D, Schulz CE, Sessoli R (1996) Europhys Lett **35**: 133
[41] Sangregorio C, Ohm T, Paulsen C, Sessoli R, Gatteschi D (1997) Phys Rev Lett **78**: 4645

[42] Ohm T, Sangregorio C, Paulsen C (1998) Europhys J **B6**: 595; (1998) J Low Temp Phys **113**: 1141
[43] Landau LD (1932) Phys Z Sowjetunion **2**: 46; Zener C (1932) Proc R Soc London A **137**: 696
[44] Shimshoni E, Gefen Y (1991) Ann Phys **210**: 16
[45] Averin D, Bardas A (1995) Phys Rev Lett **75**: 1831
[46] Crothers DS, Huges JG (1977) J Phys **B10**: L557; Nikitin EE, Umanskii SYa (1984) Theory of Slow Atomic Collisions (Springer, Berlin)
[47] Garg A, Onuchi NJ, Ambegaokar V (1985) J Chem Phys **83**: 4491
[48] Gefen Y, Ben Jacob E, Caldeira AO (1987) Phys Rev B **36**: 2770
[49] Kayanuma Y (1987) Phys Rev Lett **58**: 1934
[50] Ao P, Rammer J (1989) Phys Rev Lett **62**: 3004; (1991) Phys Rev B **43**: 5397
[51] Shimshoni E, Stern A (1993) Phys Rev B **47**: 9523
[52] Leggett AJ, Chakravarty S, Dorsey AT, Fisher MPA, Garg A, Zwerger W (1987) Rev Mod Phys **59**: 1
[53] Caciuffo R, Amoretti G, Murani A, Sessoli R, Caneschi A, Gatteschi D (1998) Phys Rev Lett **81**: 4744
[54] Taft KL, Delfs CD, Papaefthymiou GC, Foner S, Gatteschi D, Lippard SJ (1994) J Am Chem Soc **116**: 823
[55] Caneschi A, Cornia A, Fabretti AC, Foner S, Gatteschi D, Grandi R, Schenetti L (1996) Chem Eur J **2**: 1379
[56] Waldmann O, Koch R, Schromm S, Schülein J, Müller P, Bernt I, Saalfrank RW, Hampel F (2001) Inorg Chem **40**: 2986
[57] van Slageren J, Sessoli R, Gatteschi D, Smith AA, Helliwell M, Winpenny REP, Cornia A, Barra A-L, Jansen AGM, Rentschler E, Timco GA (2002) Chem Eur J **8**: 277
[58] Affronte M, Lasjaunias JC, Cornia A, Caneschi A (1999) Phys Rev B **60**: 1161
[59] Cornia A, Jansen AGM, Affronte M (1999) Phys Rev B **60**: 12177
[60] Normand B, Wang X, Zotos X, Loss D (2001) Phys Rev B **63**: 184409
[61] Pilawa B, Desquiotz R, Kelemen MT, Weickenmeier M, Geisselmann A (1998) J Magn Magn Mat **177–181**: 748
[62] Waldmann O, Schülein J, Koch R, Müller P, Bernt I, Saalfrank RW, Andres HP, Güdel HU, Allenspach P (1999) Inorg Chem **38**: 5879
[63] Garg A (1993) Europhys Lett **22**: 205
[64] Julien M-H, Jang ZH, Lascialfari A, Borsa F, Horvatić M, Caneschi A, Gatteschi D (1999) Phys Rev Lett **83**: 227
[65] Cornia A, Fort A, Pini MG, Rettori A (2000) Europhys Lett **50**: 88
[66] Loss D (1998) In: Skjeltrop AT, Sherrington D (eds) Dynamical Properties of Unconventional Magnetic Systems. Kluwer, Dordrecht, p 29
[67] Braun H-B, Loss D (1994) J Appl Phys **76**: 6177
[68] Braun H-B, Loss D (1995) Europhys Lett **31**: 555
[69] Braun H-B, Loss D (1996) Phys Rev B **53**: 3237
[70] Braun H-B, Loss D (1996) J Appl Phys **79**: 6107
[71] Braun H-B, Kyriakidis J, Loss D (1997) Phys Rev B **56**: 8129
[72] Kyriakidis J, Loss D (1998) Phys Rev B **58**: 5568
[73] Shor P (1994) Proceedings of the 35th Annual Symposium on the Foundations of Computer Science. IEEE Press, Los Alamitos, p 124
[74] Ahn J, Weinacht TC, Bucksbaum PH (2000) Science **287**: 463
[75] Thiaville A, Miltat J (1999) Science **284**: 1939
[76] Leuenberger MN, Loss D, Poggio M, Awschalom DD (2002) cond-mat/0204355
[77] Salis G, Fuchs DT, Kikkawa JM, Awschalom DD, Ohno Y, Ohno H (2001) Phys Rev Lett **86**: 2677; Salis G, Awschalom DD, Ohno Y, Ohno H (2001) Phys Rev B **64**: 195304

Invited Review

Ab Initio Calculations Versus Polarized Neutron Diffraction for the Spin Density of Free Radicals

E. Ressouche* and **J. Schweizer**

CEA/Grenoble, DSM – DRFMC/SPSMS-MDN, 38054 Grenoble Cedex 9, France

Received October 15, 2002; accepted October 28, 2002
Published online January 7, 2003

Summary. The determination of the magnetization distribution using polarized neutron diffraction has played a key role during the last twenty years in the field of molecular magnetism. This distribution can also be obtained by first principle *ab initio* calculations. Such calculations always rely on approximations and the question that arises is to know whether the obtained results are reliable enough to represent accurately the properties of these molecules. The comparison between polarized neutron experimental results and *ab initio* calculations has turned to provide stringent tests for these methods. In the present article a comparison between experimental and theoretical results is made and is illustrated by examples based on magnetic free radicals.

Keywords. Polarized neutron diffraction; Magnetization distribution; First principle *ab initio* calculations; *Hartree-Fock* approximation; Local density approximation; Magnetic molecular compounds; Free radicals.

Introduction

Molecular magnetism is a recently emerging field in science dealing with the conception, the design, the study, and the use of molecule based magnetic compounds with new but predictable properties. Such a class of compounds is supposed to fill a gap in the range of existing materials, and several new applications have already been invented. Due to the extraordinary flexible chemistry of organic and organometallic compounds, an infinite number of possibilities exists, and a goal for the future is to play with different molecules to finely tune the sought properties. But such a goal requires a perfect understanding of all the phenomena involved in these compounds.

* Corresponding author. E-mail: ressouche@cea.fr

The molecular wave function is of fundamental importance, off course: if we were able to obtain an exact solution for the *Schrödinger* equation, then we would be able to predict any property for any system, and, as a result, experimentalists would loose their job. Unfortunately (not for experimentalists!), we are never able to obtain an exact solution, even in the simplest possible cases. In order to attempt a calculation of the wave function in a solid, some approximations must always be made. Once the idea to get an exact solution is abandoned, then the question to know whether the approximations made are valid or not remains. From an exact solution, any particular property could be predicted with confidence, but an approximate solution may predict the result of one type of experiment accurately whilst giving a widely incorrect result for another.

The charge density, as measured in an X-ray scattering experiment, or the spin density, as obtained from polarized neutron diffraction (PND) are two quantities closely related to the wave function. The charge density is the sum of the charge distribution of the two spin states $\rho^+(\mathrm{r})$ and $\rho^-(\mathrm{r})$, whereas the spin density simply corresponds to the difference between those two quantities. The charge density is nothing but the probability to find an electron in a volume element around one particular point, and is the integral of the squared wave function over all space coordinates. The spin density is just the probability to find an unpaired electron in the same volume. Techniques other than PND can be used as measures of spin densities, such as magnetization measurements, resonance technique (NMR, EPR or *Mössbauer* spectroscopy). But, with these techniques, the relationship between the quantities actually measured and the spin density is often indirect, or they represent only a small part of the information (e.g. density at the nuclei...).

In this context, it is rather natural to use the comparison between the experimental densities maps obtained from X-ray or neutrons to first principle *ab initio* calculations as stringent tests for the validity of the approximations they imply. And because it involves only the highest lying molecular orbitals, the spin distribution, from the very first polarized neutron experiments performed on molecular systems, turned out to be much more a severe test than the charge density distribution, as we will see in the next sections.

Results and Discussion

UHF Versus Experiment on Isolated Molecules: A Failure of the Calculations

There are two main families of *ab initio* calculations. The first family, based on the *Hartree Fock* approximation, calculates the multicoordinate molecular wave functions, expressed as *Slater* determinants of single particle atomic wave functions. The Hamiltonian is a sum of single particle Hamiltonians including kinetic energy, *Coulomb* attraction by the nucleus and *Coulomb* repulsion by the other electrons. To express this last term, one assumes that the electrons are distributed over their wave function. This is in fact the main approximation. Such an assumption does not take into account the correlation between electrons, which results in an electron which does not stay unperturbed over his wave function when another one passes by. The calculation is processed self consistently. Each molecular wave function is occupied by two electrons, one with spin up (+) and one with spin down (−). The unpaired electron occupies the SOMO (Singly Occupied Molecular Orbital). In the Restricted *Hartree Fock* scheme (RHF), the orbital part of the wave functions of electrons of spin (+) and (−) are the same. Therefore, the spin density obtained by this method is automatically positive. As we will see, some regions of space may carry negative spin densities (spin polarization effect), and it is impossible for RHF methods to correctly account for this effect. Such methods are thus automatically discarded. In the Unrestricted *Hartree Fock* scheme (UHF), the spatial part of the wave functions of electrons of spin (+) and (−) may be different. The spin density thus obtained may be negative in some parts of the molecule.

To go beyond the *Hartree Fock* approximation, that is to take into account the correlation energy which exists between the electrons, one may admix excited states and optimize the mixing coefficients. The method is time consuming. According to the complexity of the development, one has the *Moller-Plesset* methods (MP2, MP3) [1] or the full Configuration Interaction (full CI) [2]. Another trick to reinject artificially the correlation term is the use of polarized basis sets. This approach consists in including in the calculations orbitals which are in principle empty, e.g. 2p/3d for an hydrogen or 3d/4f for an oxygen atom.

The first radical studied by polarized neutron diffraction was a nitroxide biradical, the tanol suberate [3], which was thought at the time to be the first ferromagnetic organic compound, with a very low T_C. Actually, this compound orders antiferromagnetically, but its behavior becomes ferromagnetic as soon as a small external field is applied [4]. The molecule consists in two rings $C_5H_5(CH_3)_4NO$, linked by a long chain $-OOC(CH_2)_6COO-$. Therefore, the two NO groups belonging to the same biradical are separated by more than 10 Å, and the shortest distance between two NO in the crystal is 6 Å. Each nitroxide group can therefore be considered as a well isolated radical. Polarized neutron experiments were made in the paramagnetic state at 4.2 K, with the magnetization aligned by an external magnetic field. The main results of these experiments were that the spin density in tanol suberate is mainly localized around the N and O atoms of the nitroxide groups, and that the spin is roughly equally shared by these two atoms [3]. This result strongly renewed the interest in the nature of the spin distribution in nitroxide free radicals. These radicals were extensively studied experimentally since they were synthesized, especially by ESR techniques, which give access to the spin density at nuclei positions. From the local values of the density at the nuclei semi-empirical relations exist that can be used to deduce more global information in terms of spin populations. However, such an estimation is indirect and rather ambiguous.

From the theoretical point of view, a large number of calculations on nitroxide free radicals were already performed, prior to the neutron results, by using semi empirical methods. Mainly because of computational problems (the reader should not forget that the laptop (s)he is using now everyday is a few order of magnitudes more powerful than computers at this time), *ab initio* calculations were only performed on the simplest H_2NO model. Elaborate computations using double perturbation theory and configuration interaction, leading to a good matching with ESR results, indicated a strong localization of the unpaired electron around the oxygen atom [5]. And this qualitatively well established result turned out to be in

Table 1. Theoretical *Mulliken* spin populations for the tanol suberate; (a) UHF calculations performed in Ref. [6]; (b) LSD calculations of Ref. [15], polarized basis set; (c) from Ref. [16]

Molecule	Atoms	UHF (a)	LSD (b)	SDCI/6-31G (c)	Iterative CI/6-31G (c)
H_2NO	N	0.22	0.44	0.31	0.37
	O	0.82	0.57	0.72	0.66
$(CH_3)_2NO$	N	0.35	0.47	0.34	0.43
	O	0.71	0.46	0.66	0.56
$(C_2H_5)_2NO$	N	0.37	–	–	–
	O	0.69	–	–	–

contradiction with the almost equidistribution between nitrogen and oxygen found with neutrons. As a result, new calculations were performed by *Gillon et al.* [6], in particular to see whether it was meaningful to compare H_2NO and the true molecule. In other words, the question was to know if the difference between theory and experiment was only due to substitution of H by larger alkyl groups or not. *Ab initio* UHF calculations were then performed on a series of radicals ranging from H_2NO to $C_5H_{10}NO$. Comparison of spin populations and spin density maps in the series showed a net spin migration from oxygen to nitrogen when H was replaced by larger groups (Table 1). The predominant effect was the substitution of the first two methyl groups adjacent to the nitrogen. However, the final result on the most substituted model was a distribution of 35%/65% on N and O, compared to 21%/79% for H_2NO and 50%/50% experimentally. This result was important, since it showed the influence of the substitution, and that H_2NO was not a reference model for the theoretical spin density in nitroxides as it was considered so far. But only 50% of the difference between experiment in tanol suberate and calculations in H_2NO was explained. 50% was still to be explained.

DFT Versus Experiment on Isolated Molecules: A Relevant Comparison

The second family of electronic calculations, called the Density Functional Theory (DFT), is based on the *Hohenberg-Kohn* theorem [7], which states that the energy of an ensemble of electrons is a functional of the charge density only and that the fundamental state is the state which minimizes this functional. The method calculates directly the electron density instead of the wave functions, which simplifies the problem very much. The functional contains a kinematic term, a *Coulomb* term and an exchange-correlation term. However, the analytical expression of this last term cannot be derived in the general case, and approximations have to be made. A first approximation is the Local Density Approximation (LDA), which assumes that the exchange-correlation term for electrons in a crystal has the same expression as for an homogeneous interacting electron gas [8]. This approximation has been extended to magnetic systems (Local Spin Density Approximation or LSD) [9, 10] by introducing a functional of two electronic densities $\rho^+(r)$ and $\rho^-(r)$, where the exchange-correlation term is the same as for an homogeneous interacting and partly polarized electron gas (*Vosko-Wilk-Nusair* functional [11]).

Other approaches have been proposed to go beyond this local approximation. One of them is the gradient method [12] (sometimes also referred to as nonlocal).

A parenthesis should be opened here, concerning the basis sets used by the different DFT programs. There are two main programs for performing such calculations: DGAUSS [13] and DMOL [14]. Whereas DGAUSS uses gaussian basis sets, DMOL can also use atomic-like basis sets. This difference is of some importance, and will be discussed further in the text.

At the beginning of the eighties, LSD calculations were mainly applied to solid state electronic structure calculations, and very few to molecular systems. *Delley et al.* performed exactly the same calculations than *Gillon et al.* on the series of molecules H_2NO to $C_5H_{10}NO$ [15]. In marked contrast to what was found with UHF, that is a strong spin localization around the oxygen atom, LSD calculations yield a one to one distribution between N/O for $C_5H_{10}NO$, as found experimentally (Table 1). Another important result of this paper was a simulation of an hydrogen bond as it occurs in another nitroxide free radical: crystalline tanol. Hydrogen bridges provide another mechanism to affect the spin balance between nitrogen and oxygen, resulting in an additional transfer from O to N. This was, thus, the first time that the active role of an hydrogen bond in propagating magnetic exchange interactions was evidenced.

This result had a very strong impact on the community of theoreticians performing *ab initio* calculations, provoking in particular skepticism for scientists using UHF methods. Following these two pioneering results, and owing to the increase of power of computers, such calculations were repeated, to confirm these results. *Wang et al.* studied the same series of radicals using iterative CI methods [16]. Calculations at different geometries with various basis sets were performed. They confirmed the substitution effect, leading to the conclusion that H_2NO was not a suitable theoretical prototype for nitroxide radicals, and they observed further shifts from oxygen to nitrogen due to inclusion of higher excitations into the CI expansion and the use of better (*i.e.* much larger) basis sets. Those calculations, however, were not able to reproduce the observed spin balance, leading to a 43 %/57 % distribution in the most favorable case. Roughly at the same time, the same kind of comparison between polarized neutron experiment and *ab initio* calculations was repeated for another nitroxide derivative: the tano [17]. The influence of the alkyl substitution was studied for a series of molecules starting from H_2NO in the experimental geometry of tano up to the complete (real) molecule using UHF methods [17, 18]. The influence of the basis set was also checked. Here again, the conclusions were still the same: UHF was not able to reproduce the experimental equidistribution observed in tano (Table 2 and Fig. 1), even using large basis sets and the true molecule (which means very long calculations). On the contrary, LSD calculations performed by the same authors turned out to be very stable, with roughly no influence of the basis set used, and were able to give, in one order of magnitude less computing time, much more reliable (*i.e.* closer to the experiment) results than very sophisticated UHF calculations using large polarized basis sets.

A few years later, the radical-based cyano-acceptor tetracyanoethylene $(TCNE)^{\bullet -}$ received considerable interest when the compound $[Fe(C_5Me_5)_2]^{\bullet +}(TCNE)^{\bullet -}$ was discovered to be a ferromagnet with $T_C = 4.8$ K [19]. Another successful synthesis involving $(TCNE)^{\bullet -}$ was that of $V(TCNE)_x{\bullet}yCH_2Cl_2$ [20], which magnetically order

Table 2. Theoretical *Mulliken* spin populations for the tano [17, 18]; UHF calculations have been performed using the basis set of Ref. [6], LSD calculations according to [15]

Molecule	Atoms	UHF	LSD
H_2NO	N	0.24	–
	O	0.80	–
$(CH_3)_2NO$	N	0.38	–
	O	0.69	–
$(C_2H_5)_2NO$	N	0.41	–
	O	0.66	–
$(C_3H_7)_2NO$	N	0.43	–
	O	0.64	–
Tano: $C_9H_{16}NO_2$	N	0.42	0.48
	O	0.65	0.48

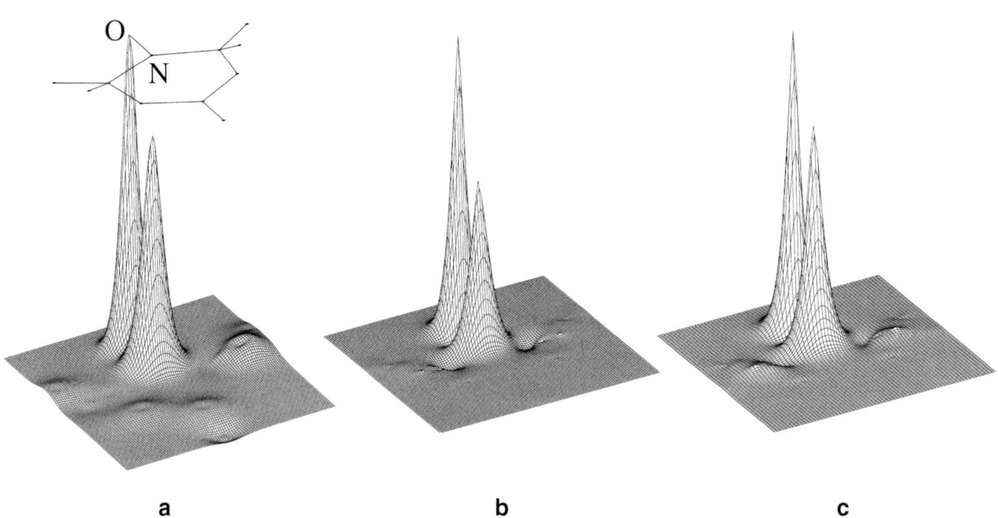

Fig. 1. Spin density projections along the π-direction for the free radical tano; (a) Experimental map; (b) UHF calculation, performed using the polarized basis set of Ref. [6]; (c) LSD calculation, using the polarized basis set described in Ref. [15]

above room temperature. The spin density in $(TCNE)^{\bullet -}$ has been determined with polarized neutrons, and compared with *ab initio* calculations (Table 3) [21]. The singly occupied orbital of $(TCNE)^{\bullet -}$ is the π^* antibonding molecular orbital consistent with molecular orbital predictions. The main part of the density is localized on the central sp^2 carbon atoms, but a combination of both the spin delocalization effect and the spin polarization effect yields a distribution of 33 %, -5 % and 13 % of the total spin on the sp^2 carbons, the sp carbons and nitrogen atoms, respectively (Fig. 2). In Table 3 are also presented the different results of *ab initio* calculations. As it can be clearly seen, none of the UHF methods is able to reproduce the experimental distribution, since these methods overestimate the negative contribution on the intermediate sp carbon atoms. On the contrary, LSD based methods predict with a rather good accuracy what has been found experimentally.

Table 3. Experimental and theoretical atomic spin populations for the radical $(TCNE)^{\bullet -}$; UHF calculations were performed using GAUSSIAN 92 [22], DFT calculations using DGAUSS [13]

Atoms	UHF 3-21G	MP2 3-21G	UHF 6-311G*	LDFT TZVP	NLDFT TZVP	Experiment
C_1 (sp^2)	0.51	0.52	0.48	0.28	0.30	0.33
C_2 (sp^2)	0.56	0.55	0.50	0.28	0.29	0.33
C_3 (sp)	− 0.50	− 0.62	− 0.40	− 0.02	− 0.04	− 0.05
C_4 (sp)	− 0.70	− 0.67	− 0.60	− 0.03	− 0.05	− 0.04
C_5 (sp)	− 0.55	− 0.53	− 0.46	− 0.03	− 0.04	− 0.03
C_6 (sp)	− 0.61	− 0.63	− 0.54	− 0.03	− 0.05	− 0.08
N_3	0.49	0.50	0.42	0.13	0.14	0.12
N_4	0.69	0.70	0.60	0.15	0.16	0.12
N_5	0.53	0.52	0.44	0.14	0.14	0.13
N_6	0.60	0.51	0.53	0.13	0.15	0.16

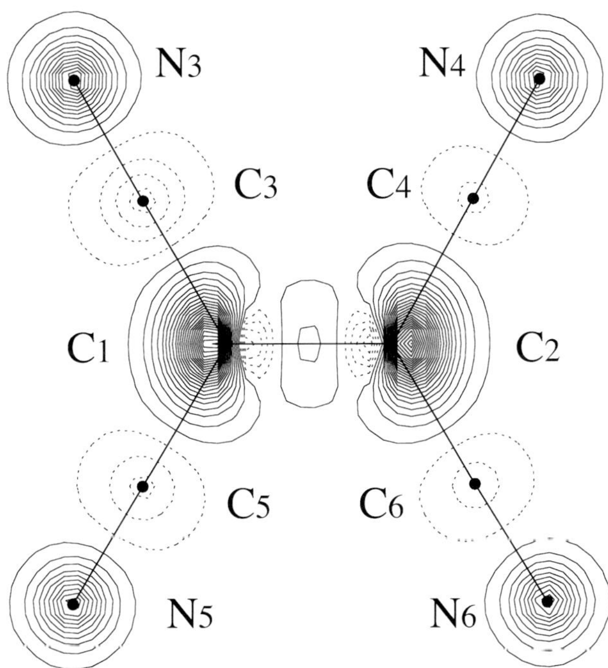

Fig. 2. Experimental spin density in the tetracyanoethylene $(TCNE)^{\bullet -}$ anion-radical, determined by PND

Another very important result on the ability of these two kinds of methods to account for the spin polarization effect was met during the experimental study of the phenyl nitronyl nitroxide by *Zheludev et al.* [23]. In recent years, nitronyl nitroxide free radicals have played a key role in the engineering of molecule-based magnetic materials: they are stable, capable of being handled under ordinary conditions, and they carry a spin 1/2. The phenyl derivative (*NitPh*) crystallizes in the monoclinic space group $P2_1/c$ with two molecules per asymmetric unit, and remains paramagnetic down to very low temperature, with very little interactions between neighboring molecules. It was, thus, a good example of a well isolated member of the nitronyl nitroxide family. Polarized neutron experiments were performed on this compound in the paramagnetic state, with the magnetization aligned by an external magnetic field. The experimental spin density map is represented in Fig. 3 as a projection onto the O–N–C–N–O plane. As expected from simple molecular orbital arguments, the unpaired electron occupies a π^* antibonding molecular

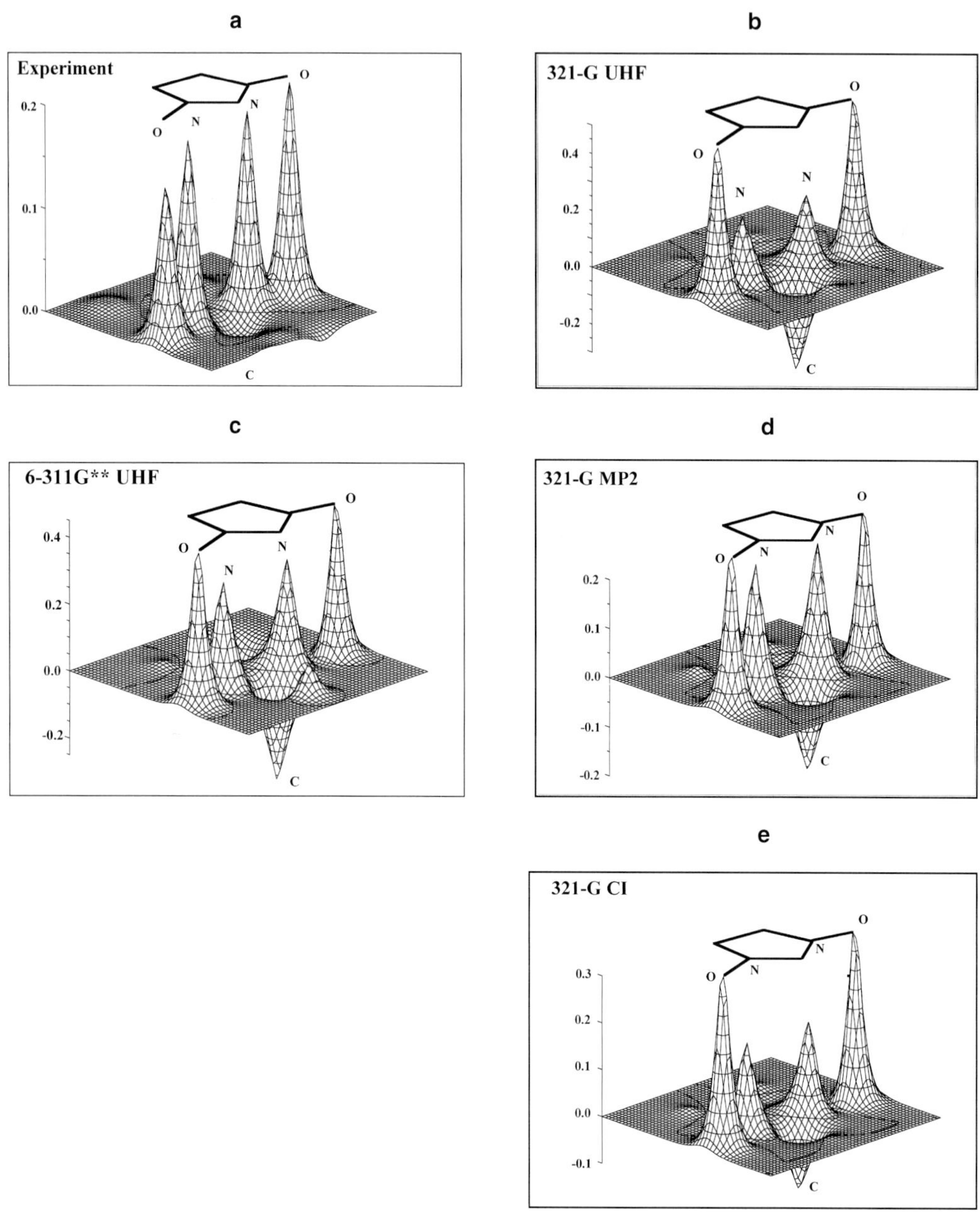

Fig. 3. Spin density projections along the π-direction for the phenyl nitronyl nitroxide free radical; (a) Experimental map; (b) UHF calculation using the small 3-21G basis set; (c) UHF calculation using the large 6-311G** basis set; (d) MP2 calculation in 3-21G; (e) Full CI in 3-21G

orbital mainly built from the $2p_z$ atomic orbitals of the two oxygen and the two nitrogen atoms. The spin density is equally shared between these four atoms. The fraction of spin delocalized on the rest of the molecule is rather small. One particular feature deserves a special attention: there is a large negative contribution (roughly 1/3 of the other contributions) on the sp^2 bridging carbon atom. Such an effect, or spin polarization effect, is conveniently explained in terms of a triplet HOMO–LUMO excitation induced by the unpaired spin of the SOMO. These frontier orbitals have a large contribution from the atomic orbital $2p_z$ on this carbon, which becomes polarized. The resulting density is negative, as pointed out by *Anderson* [24], since the positive spins on the frontier molecular orbitals are attracted by the positive spin on the SOMO, leaving behind the negative spins on the node of the SOMO, that is on the carbon site. The sign alternating spin densities found on the carbon atoms of the phenyl ring are also manifestations of this spin polarization effect.

A comparison between these experimental results and the spin densities calculated by the different *ab initio* methods has been made in Ref. [23]. The theoretical spin populations on the oxygen, nitrogen and central carbon are reported in Table 4, together with the experimental values averaged and normalized to $1\mu_B/NitPh$ molecule. We report here UHF results with two basis sets: the rather small 3–21G and the huge polarized 6–311G**, the results of the MP2 method and the full CI. The resulting theoretical spin density maps from UHF methods are also presented in Fig. 3, where they are compared with the experimental map. Due to the large computing time, those UHF calculations have been done for an optimized, planar and truncated molecule, where the phenyl and methyl groups were replaced by hydrogen atoms. As it is clear from the maps and from Table 4, all these methods overestimate considerably the spin polarization effect, that is the negative spin density on the central carbon. Another discrepancy, for all but MP2, is the ratio O/N. Surprisingly, full CI, which in principle is more elaborate than MP2, is farther from reality. We can then conclude that the methods based on *Hartree Fock* approximation not only are unable to account correctly for the spin density in nitronyl nitroxides, but also that they are unstable: the results depend very much on which method is used and which basis set has been taken for the calculation. Such a conclusion could also be drawn from the studies on simple nitroxides and on the $(TCNE)^{\bullet -}$ anion radical previously described.

In Table 4 are also displayed the results obtained by the LSD family calculations: at the local approximation level (VWN functional), with DZVP (double zeta) and TZVP (triple zeta) basis sets, and at the non local level referred as B88VWN. Those calculations were performed on the real molecule, in its experimental

Table 4. Experimental and theoretical atomic spin populations for the ONCNO fragment of *NitPh*; DFT calculations were performed using (a) the DGAUSS [13] and (b) the DMOL [14] programs

Atoms	UHF 3-21G	UHF 6-311G**	MP2 3-21G	CI 3-21G	VWN[(a)] DZVP	VWN[(a)] TZVP	VWN[(b)] DNPP	B88VWN[(b)] DNPP	Experiment
O	0.50	0.40	0.32	0.39	0.32	0.30	0.29	0.29	0.27
N	0.27	0.36	0.34	0.23	0.21	0.22	0.20	0.21	0.27
C(sp^2)	−0.55	−0.52	−0.31	−0.24	−0.08	−0.07	−0.02	−0.04	−0.11

Table 5. Theoretical atomic spin populations calculated by DGAUSS (TZVP) for one of the two *NitPh* molecules and compared to experimental PND populations

Atoms	Experiment	DGAUSS (TZVP)
O_4	0.277 (13)	0.319
N_4	0.278 (15)	0.238
C_{20}	−0.121 (17)	−0.065
N_3	0.278 (16)	0.210
O_3	0.247 (13)	0.288
C_{14}	0.024 (16)	0.004
C_{15}	0.000 (13)	−0.009
C_{16}	0.025 (13)	0.002
C_{17}	−0.016 (12)	−0.009
C_{18}	0.011 (12)	0.002
C_{19}	−0.037 (13)	−0.009
C_{21}	−0.025 (16)	−0.004
C_{22}	0.009 (13)	−0.005
C_{23}	0.055 (12)	0.021
C_{24}	0.009 (13)	0.000
C_{25}	−0.008 (12)	0.001
C_{26}	−0.005 (13)	0.023

geometry. Two points must be emphasized: on the one hand these calculations are stable: changing approximation or basis set changes very little the calculated spin populations. On the other hand, even if the observed ratio O/N is not exactly reproduced, the theoretical results are not very far from the experimental ones.

As the DFT calculations have been made on the complete *NitPh* molecule, we can look at the amount of spin which is localized on the ONCNO fragment. PND has found 0.97 for the first molecule and 0.96 for the second one. The DFT predictions vary a little (Table 4): 0.98 and 0.97 with the program DGAUSS and the basis set DZVP and TZVP, respectively, and 0.96 with the program DMOL.

Table 5 compares more closely the spin density found by DGAUSS to the experimental one for one of the two molecules (Fig. 4). On the phenyl rings, DGAUSS reproduces very well the spin alternation which has been found experimentally, but it reduces the values of the positive and negative spin populations. If one compares the sum of the absolute values of the populations on the carbon atoms, DGAUSS puts forward 0.035 while it was found experimentally 0.113. It is the same for the sum of the absolute values on the $C(CH_3)_2$ groups: 0.054 for DFT and 0.112 found experimentally.

To conclude this paragraph, we can say that DFT methods account rather well for the spin densities of the central part of the free radical, but we can already anticipate that there will be problems to predict the delocalized part of the spin density.

DFT Versus Experiment on Interacting Molecules: A Difficult Challenge

When molecules are interacting, the pathways of the exchange interactions imply, most of the times, atoms which do not belong to the localized part of the spin

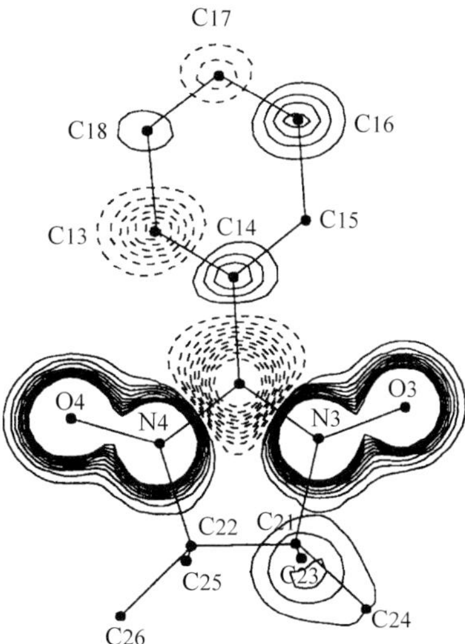

Fig. 4. Experimental spin density (low contours) of one of the two *NitPh* molecules of the asymmetric unit, determined by PND

density distribution. These trails are generally revealed in PND experiments by a wake of spin density on the concerned atoms. As first principle calculations are not so reliable to account for the delocalized spin distribution, we can expect that DFT will have some difficulties in visualizing such effects.

The effects of an hydrogen bond mediating the magnetic exchange interactions on the spin density have been evidenced in the *NitPy* radical (6-HC≡C*Py*NN). In this compound the molecules form ferromagnetic chains, the pyridine of one molecule being connected to the oxygen of the next molecule by a weak C≡C–H ⋯ O hydrogen bond. The spin density determined by PND [25] is represented in Fig. 5 and the corresponding spin populations are reported in Table 6 and Table 7. Several points deserve attention in this distribution:

- on the one hand, the part of the spin population localized on the ONCNO fragment is 0.877 only, compared to 0.971 and 0.959 for the two "isolated" molecules of the *NitPh* radical. This reflects that in *NitPy* the exchange interactions between molecules imply a delocalization of the spin.
- the spin population on oxygen O_1, the oxygen atom which participates to the hydrogen bond, is depleted compared to that of O_2, the other oxygen atom of the group ONCNO.
- carbon C_6 and hydrogen H_{16}, on the C≡C–H ⋯ O hydrogen bond, carry a substantial spin population: 0.030 and 0.045.

DFT calculations are presented in Tables 6 and 7. For an isolated molecule (Table 6), two calculations have been performed with the program DGAUSS with different basis sets, and one with DMOL. Molecules interacting in the crystal have been studied with DMOL [26]. As expected, the main features of the spin distribution on the central fragment ONCNO, particularly the negative value on the

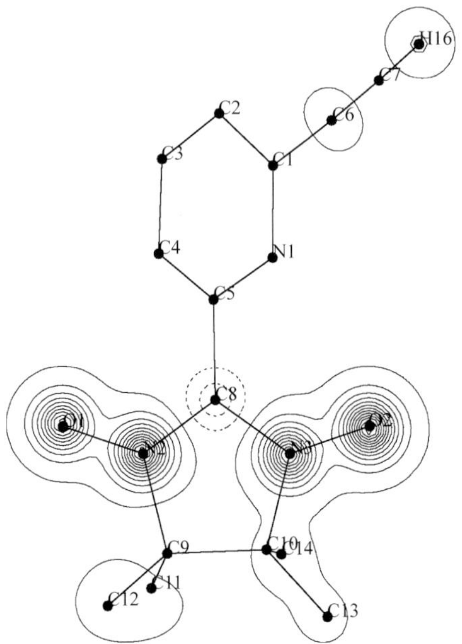

Fig. 5. Experimental spin density of the *NitPy* radical, determined by PND

Table 6. Theoretical atomic spin populations calculated by DGAUSS (DZVP), DGAUSS (TZVP) and DMOL for one isolated molecule of the *NitPy* radical and compared to experimental PND populations

Atoms	Experiment	Isolated molecule DGAUSS (DZVP)	Isolated molecule DGAUSS (TZVP)	Isolated molecule DMOL
O_1	0.203 (10)	0.292	0.270	0.274
N_2	0.242 (12)	0.208	0.212	0.187
C_8	− 0.071 (11)	− 0.069	− 0.086	− 0.076
N_3	0.225 (12)	0.217	0.230	0.198
O_2	0.278 (7)	0.335	0.312	0.312
Sum ONCNO	0.877 (24)	0.983	0.938	0.895
H_{16}	0.045 (10)	0.000	0.000	0.000

Table 7. Theoretical atomic spin populations calculated by DGAUSS (DZVP), DGAUSS (TZVP) and DMOL for interacting *NitPy* molecules and compared to experimental PND populations

Atoms	Experiment	Two interacting molecules DGAUSS (DZVP)	Two interacting molecules DGAUSS (TZVP)	Crystal DMOL
O_1	0.203 (10)	0.273	0.243	0.239
N_2	0.242 (12)	0.216	0.224	0.188
C_8	− 0.071 (11)	− 0.066	− 0.084	− 0.075
N_3	0.225 (12)	0.216	0.230	0.212
O_2	0.278 (7)	0.341	0.318	0.299
Sum ONCNO	0.877 (24)	0.980	0.931	0.863
H_{16}	0.045 (10)	0.001	0.002	0.004

central carbon, are encountered. However, if we refer to the points underlined above:

- the total amount of spin population localized on the central fragment is not stable: DMOL and DGAUSS programs do not give the same localized/delocalized partition of the spin density. For the isolated molecule, this part varies from 0.983 for DGAUSS (DZVP) to 0.938 for DGAUSS (TZVP) and 0.895 for DMOL. Compared to experiment, DGAUSS localizes too much. Program DMOL has the possibility to make also spin density calculations when the molecule is inserted in the crystal. With the same computational conditions, this sum decreases only from 0.895 for the isolated molecule to 0.863 for the molecules interacting in the crystal.
- the effect of the hydrogen bond has been looked at with DGAUSS by calculations on two interacting molecules, having the geometry of the crystal, and with DMOL on the whole crystal. The results are reported in Table 7. As expected, the spin population on O_1, the oxygen atom connected to the hydrogen bond, decreases, compared to isolated molecules, for molecules in interaction. This reduction is 0.019 for DGAUSS (DZVP), 0.027 for DGAUSS (TZVP) and 0.035 for DMOL. However, even in the isolated molecule, the spin population is smaller on O_1 than on O_2, the atom which does not participate to the bond. It has been shown that this difference in the isolated molecule comes from the presence of the nitrogen atom which dissymetrizes the piperidine ring [27].
- the spin population on H_{16} was zero for all calculations of the isolated molecules. For molecules in interaction DFT found some spin density on the hydrogen atom of the bond: 0.001 for DGAUSS (DZVP), 0.002 for DGAUSS (TZVP), and 0.004 for DMOL in the crystal. These values are one order of magnitude less than that observed with the neutrons (0.045). However, all these calculations underline the pathway of the magnetic exchange interactions which goes through the hydrogen bond. DFT sees the magnetic interactions but underevaluates their influence on the spin density and, as expected, DGAUSS produces effects which are weaker than DMOL.

Another example of perturbation of the spin density due to magnetic exchange interaction and revealed by PND in nitronyl nitroxide free radicals has been encountered in *Nit(SMe)Ph*. This compound is also ferromagnetic ($T_C = 0.2\,K$) and the packing of molecules in the crystal reveals a weak intermolecular contact (3.72 Å) between oxygen O_2 of the nitronyl nitroxide and carbon C_{14} of the methylthio group of the next molecule. The experimental spin density is reported in Table 8 and displayed in Fig. 6. The striking point in this figure is the presence of some spin density on the atom C_{14}, which supports the existence of a magnetic pathway along the molecule and through the C_{14}–O_2 contact [28].

The spin density has been calculated by the DGAUSS program, with a DZVP basis set for an isolated molecule and for two molecules in interaction, with a geometry corresponding to the crystal packing [26, 27]. The results of this calculation are also presented in Table 8. As expected, the localization of the spin density on ONCNO is much too strong. Nevertheless, for the bond DGAUSS finds a depletion of spin density on the oxygen atom concerned by the contact, but this depletion has not been detected by the neutrons. Furthermore, on C_{14} the magnetic

Table 8. Theoretical atomic spin populations calculated by DGAUSS (DZVP) for one isolated *Nit*(S*Me*)*Ph* molecule and for two interacting molecules, compared to experimental PND populations

Atoms	Experiment	One isolated molecule DGAUSS (DZVP)	Two molecules DGAUSS (DZVP)
O_1	0.226 (7)	0.304	0.307
N_1	0.272 (9)	0.221	0.221
C_1	− 0.099 (8)	− 0.063	− 0.063
N_2	0.247 (9)	0.225	0.229
O_2	0.226 (8)	0.308	0.298
Sum ONCNO	0.872 (18)	0.995	0.992
C_{14}	0.031 (7)	0.000	0.001

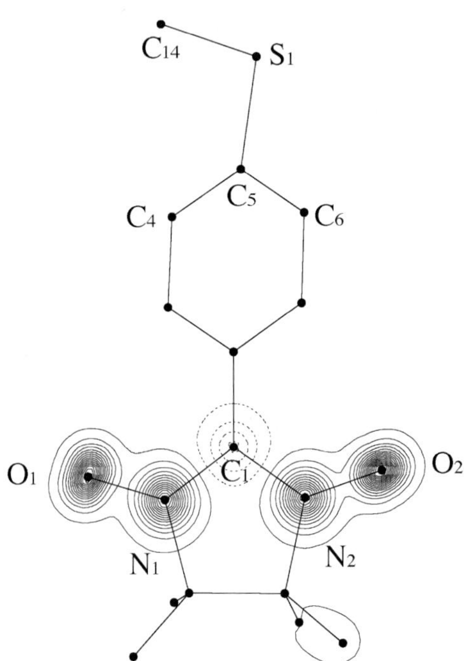

Fig. 6. Experimental spin density of the *Nit*(S*Me*)*Ph* radical, determined by PND

exchange interaction induces some spin density, only 0.001 according to DFT, when 0.031 were found by PND. Here again, *ab initio* calculations see qualitative effects, but quantitatively, these effects are much too small compared with the experimental ones.

Comparative DFT Calculations: A Possible Way to Trace the Magnetic Interactions

DFT calculations are reliable to describe the main part of the spin density in free radicals, but underestimate enormously the effects of small perturbations. However, it is possible to use such calculations in a comparative way. Having in parallel, with the same program and the same basis set, two computations differing

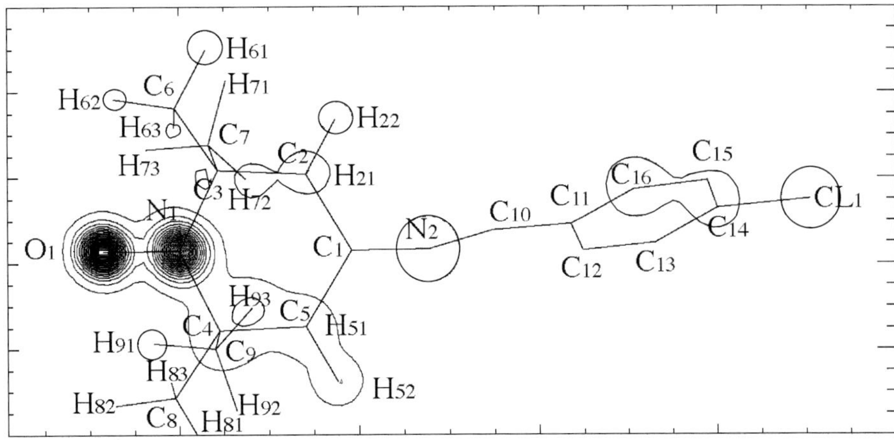

Fig. 7. Experimental spin density of the Cl-*TEMPO* radical, determined by PND

Table 9. Theoretical atomic spin populations calculated by DGAUSS for one isolated Cl-*TEMPO* molecule and for two interacting molecules, compared to experimental PND populations

Atoms	Experiment	One isolated molecule	Two interacting molecules	
			molecule 5	molecule 1
O_1	0.401 (14)	0.490	0.477	0.489
N_1	0.393 (19)	0.444	0.451	0.445
Sum NO	0.794 (24)	0.934	0.928	0.934
C_2	0.006 (16)	0.007	0.006	0.011
H_{22}	− 0.032 (13)	0.000	0.000	0.000
C_6	− 0.055 (16)	0.003	0.004	0.004
H_{61}	0.038 (12)	− 0.001	− 0.001	− 0.002

only by one given perturbation, it is possible to obtain not quantitative results, but, at least, a qualitative information concerning this perturbation.

The search of the magnetic pathways in the ferromagnetic ($T_C = 0.28$ K) compound Cl–*TEMPO* is an illustration of this approach [29]. The spin density, as determined by PND is represented in Fig. 7. Almost 80% of the spin is localized on the NO group, the rest being diluted on different atoms of the other parts of the molecule (Table 9). In particular, the presence of a significant spin density on H_{22} (− 0.032) and on H_{61} (0.038) suggests that along the *c* axis of the monoclinic cell (Fig. 8) the contacts between adjacent molecules (2.73 Å for O_1–H_{22} and 2.37 Å for O_1–H_{61}) propagate the ferromagnetic interaction.

This hypothesis has been put to the test by comparing DFT spin densities on one isolated molecule and on two interacting molecules: molecule 1 and molecule 5, in the geometry of the crystal packing. The spin populations, obtained with the DGAUSS program, are reported in Table 9. As expected, the localization obtained by DGAUSS on the NO group is much too strong. However, when molecule 5 interacts with molecule 1 through the two contacts explicited above, on the one hand, on molecule 5, this concentration of spin on the oxygen atom of the NO

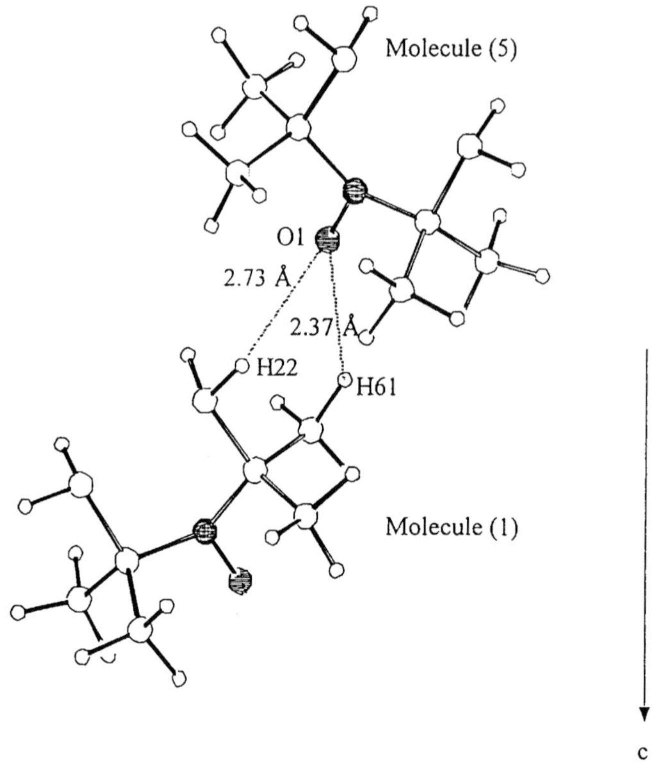

Fig. 8. Hydrogen bonds contacts between two molecules of Cl-*TEMPO*

group is reduced, and the balance between oxygen and nitrogen is modified in favour of nitrogen. On the other hand, on molecule 1, the spin density on the methylene atom H_{22} has not been modified, but the carbon atom C_2, which is attached to it, has now a spin population which is increased by 0.004. Furthermore, the negative spin on the methyl atom H_{61} changes from -0.001 to -0.002. From this comparison, we cannot conclude anything about the strength of the exchange interactions, but we can confirm that magnetic interactions propagate through such contacts.

Another example concerns the hydrogen bonds in the *PNNB* complex [30]. This complex consists of chains where *NitPh* radicals alternate with phenylboronic acid $B(OH)_2$*Ph* molecules. Along these chains, the ONCNO groups of the *NitPh* are connected together via the HOBOH atoms of the phenylboronic acid through hydrogen bonds NO $\cdots$ HOB. At room temperature the structure is monoclinic, but on cooling, due to the ordering of the methyl groups, the symmetry is reduced and the structure becomes triclinic. Therefore, at low temperature, there exist two crystallographically non equivalent *NitPh* molecules. The hydrogen bond distances, which were almost equivalent at room temperature, are now quite different: 1.96, 1.84, 1.96, and 1.91 Å for $H_{24} \cdots O_2$, $H_3 \cdots O_1$, $H_4 \cdots O_{21}$ and $H_{23} \cdots O_{22}$, respectively, each *NitPh* molecule being connected to two phenylboronic acids through a long and a short hydrogen bond. The spin density, determined by PND at low temperature, shows for each of the two *NitPh* radicals a strong asymmetry between the two NO groups, as shown in Fig. 9. For the two molecules the

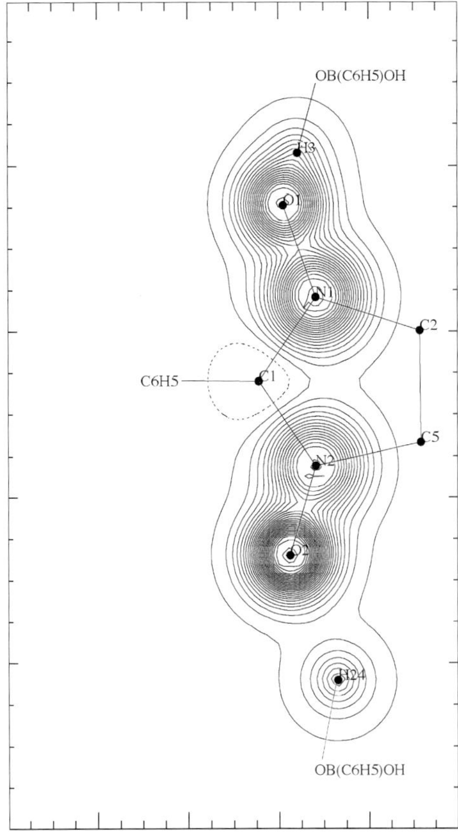

Fig. 9. Experimental spin density of the *PNNB* complex, determined by PND

Table 10. Theoretical atomic spin populations calculated by DGAUSS for one isolated *NitPh* molecule and for one *NitPh* radical connected to two phenylboronic acids, compared to experimental PND populations

Atoms	Experiment	Isolated *NitPh*	Connected molecules
H_3	0.024 (36)	0.000	−0.001
O_1	0.227 (25)	0.290	0.253
N_1	0.307 (25)	0.220	0.233
C_1	−0.025 (20)	−0.089	−0.101
N_2	0.246 (25)	0.224	0.252
O_2	0.328 (25)	0.285	0.268
H_{24}	0.056 (25)	0.000	0.002

spin density on oxygen is the weaker for the shorter length of the hydrogen bond, which is in accordance with former results.

To confirm this view, a DFT calculation (DGAUSS) was carried out for the geometry of one of the two *NitPh* molecules, first isolated and then surrounded by the two phenylboronic acids. The results are presented in Table 10. A bond distance of 1.96 Å ($H_{24} \cdots O_2$) induces a depletion of the spin population of 0.017 on the oxygen atom, while a bond distance of 1.84 Å ($H_3 \cdots O_1$) induces a depletion of 0.037. The shorter the distance is, the stronger the oxygen depletion is; the effect of the hydrogen bond length is firmly established by this comparative calculation.

Conclusions

The comparison of first principle predictions with the results of polarized neutron diffraction, experiments which measure the spin density everywhere in the molecules, has been and still is a very selective test to judge which approximation and which method is most reliable to account for spin densities in free radicals.

If it was obvious, as soon as negative spin densities appeared in some parts of the experimental distributions, that RHF was unable to yield a complete view of such maps, it took more time to convince people that UHF methods must be discarded as they are unstable and unable to account for the reality of the spin density.

Now, the superiority of DFT calculations to describe the spin density in free radicals is clear. On the one hand they are stable: calculations made with different programs give roughly the same results. On the other hand, they compare rather well with the spin densities measured by PND, at least for the parts of the molecules where the main part of the density is localized. For the delocalized part of the molecules for which the comparison with experiment is poor, DFT calculations produce results which depend on program and basis set. DGAUSS, with its gaussian basis sets, emphazises the localized part of the spin density much more than DMOL with an atomic like basis set. Both programs reduce enormously the influence on the spin density of the magnetic interactions between adjacent molecules, but DGAUSS does it more than DMOL. Nevertheless these programs may be useful in the investigation of magnetic exchange interactions, as long as one considers their predictions as qualitative and not quantitative.

Acknowledgments

We want to thank Dr. Paul Rey (CEA-Grenoble), Dr. Bernard Delley (PSI Zurich), Dr. Andrey Zheludev (Oak Ridge National Laboratory) and Dr. Yves Pontillon (CEA Grenoble) who have participated actively to most of the experiments and calculations reviewed in this paper.

References

[1] Moller C, Plessey MS (1934) Phys Rev **46**: 618
[2] Shavitt (1977) Methods of Configurational Interaction, In: Shaefer HF (ed) Modern Theoretical Chemistry, Plenum Press, New York, 189
[3] Brown PJ, Capiomont A, Gillon B, Schweizer J (1979) JMMM **14**: 289
[4] Veyret-Jeandey C (1981) J Physique (Paris) **42**: 875
[5] Ellinger Y, Subra R, Rassat A, Douady J, Berthier G (1975) J Am Chem Soc **97**: 476
[6] Gillon B, Becker P, Ellinger Y (1981) Mol Phys **48**: 763
[7] Hohenberg P, Kohn W (1964) Phys Rev **136**: 864
[8] Kohn W, Sham LJ (1965) Phys Rev **A140**: 1133
[9] von Bart U, Hedin L (1972) J Phys C **5**: 1629
[10] Rajagopal AK, Callaway J (1973) Phys Rev **B7**: 1912

[11] Vosko SH, Wilk L, Nausair M (1988) Can J Phys **58**: 1200
[12] Becke AD (1988) Phys Rev **A38**: 3098
[13] UniChem DGAUSS 1.1, Cray Research Inc, Cray Research Park, 655 Lone Oak Drive, Eagan, MN 55121
[14] DMOL, Biosym Technologies Inc, 9686 Scranton Road, San Diego, CA
[15] Delley B, Becker P, Gillon B (1984) J Chem Phys **80**: 4286
[16] Wang J, Smith Jr, VH (1993) Z Naturforsch **48A**: 109
[17] Bordeaux D, Boucherle JX, Delley B, Gillon B, Ressouche E, Schweizer J (1993) Z Naturforsch **48A**: 120
[18] Ressouche E (1991) PhD Thesis, Univ Grenoble
[19] Miller JS, Calabrese JC, Rommelmann H, Chittipeddi SR, Zhang JH, Reiff WM, Epstein AJ (1987) J Am Chem Soc **109**: 769
[20] Manriquez JM, Yee GT, McLean RS, Epstein AJ, Miller JS (1991) Science **252**: 1415
[21] Zheludev A, Grand A, Ressouche E, Schweizer J, Morin BG, Epstein AJ, Dixon DA, Miller JS (1994) J Am Chem Soc **116**: 7243
[22] Gaussian 92, Frisch MJ et al., Gaussian Inc, Pittsburgh, PA, 1992
[23] Zheludev A, Barone V, Bonnet M, Delley B, Grand A, Ressouche E, Rey P, Subra R, Schweizer J (1994) J Am Chem Soc **116**: 2019
[24] Anderson PW (1963) In: Rado GT, Suhl H (eds) 'Magnetism'. Academic Press, New York, vol 1, pp 25
[25] Romero FM, Ziessel R, Bonnet M, Pontillon Y, Ressouche E, Schweizer J, Delley B, Grand A, Paulsen C (2000) J Am Chem Soc **122**: 1298
[26] Delley B, Private communication
[27] Pontillon Y (1997) PhD Thesis, Univ Grenoble
[28] Pontillon Y, Caneschi A, Gatteschi D, Grand A, Ressouche E, Sessoli R, Schweizer J (1999) Chem Eur J **5**: 3616
[29] Pontillon Y, Grand A, Ishida T, Lelievre-Berna E, Nogami T, Ressouche E, Schweizer J (2000) J Mater Chem **10**: 1539
[30] Pontillon Y, Akita T, Grand A, Kobayashi K, Lelievre-Berna E, Pécaut J, Ressouche E, Schweizer J (1999) J Am Chem Soc **121**: 10126

A New Layered Compound Containing $[PMo_{12}O_{40}]^{3-}$ and Both 5- and 6-Coordinated Homoleptic (1-(2-Chloroethyl)tetrazole)Copper(II) Cations

Arno F. Stassen[1], **Eugenia Martínez Ferrero**[2], **Carlos Giménez-Saiz**[2], **Eugenio Coronado**[2], **Jaap G. Haasnoot**[1,*], and **Jan Reedijk**[1]

[1] Leiden Institute of Chemistry, Gorlaeus Laboratories, Leiden University, P.O. Box 9502, 2300 RA Leiden, The Netherlands

[2] Instituto de Ciencia Molecular, Universidad de Valencia, 46100 Burjassot, Spain

Received June 5, 2002; accepted June 12, 2002
Published online December 19, 2002

Summary. The synthesis, crystal structure and physical properties of the complex obtained from the reaction between the polyoxometalate anion $[PMo_{12}O_{40}]^{3-}$, copper(II) and the ligand 1-(2-chloroethyl)tetrazole (*teec*) are described. This compound has been synthesized as a model for designing materials containing both magnetic polyoxometalate anions and iron(II) spin-crossover cations.

The compound, with formula $[Cu(teec)_5]_2[Cu(teec)_6][PMo_{12}O_{40}]_2 \cdot 2H_2O$, consists of alternating layers of polyoxometalates and cationic complexes. Both, five and six coordinated Cu(II) ions are present, each positioned in different layers. Despite these layers having a similar width, the layer of pentacoordinated Cu(II) ions contains twice as many cationic complexes as the layer of hexacoordinated Cu(II) ions. This unusual coexistence of complexes with different coordination number is most likely caused by the steric hindrance induced by the bulky polyoxometalates in the layer of pentacoordinated Cu(II) which prevents the presence of a sixth *teec* ligand.

Keywords. Coordination chemistry; Heterocycles; Polyoxometalates; Structure elucidation; Tetrazole.

Introduction

The design of new hybrid materials combining two or more different properties not normally associated with a single material is a contemporary challenge in molecular chemistry. The various combinations have resulted in interesting magnetic [1–3], photophysical [4, 5] and electric [2] phenomena.

* Corresponding author. E-mail: haasnoot@chem.leidenuniv.nl

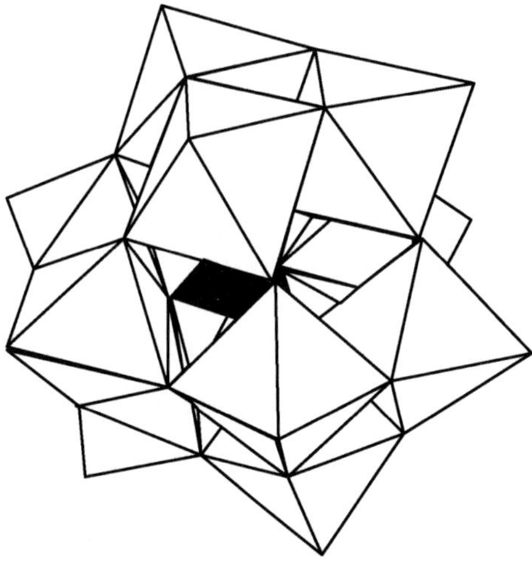

Fig. 1. Polyhedral representation of the *Keggin* polyoxometalate $[PMo_{12}O_{40}]^{3-}$

The aim of this research is to obtain model systems for coordination complexes, in which magnetic polyoxometalate anions are combined with cations undergoing a temperature-dependent spin transition. The 12-molybdophosphate anion has been chosen for this purpose. It has the well-known *Keggin* structure that consists of an arrangement of 12 MO_6 (M = W, Mo) octahedra sharing edges and corners surrounding a central XO_4 tetrahedron (see Fig. 1). Many different heteroatoms can occupy the central tetrahedral site (*e.g.* $X = S^{VI}$, P^{V}, As^{V}, Si^{IV}, Ge^{IV}, H_2^{2+}, B^{III}, Cr^{III}, Fe^{III}, Co^{III}, Co^{II}, Cu^{II}, Zn^{II}, ...) [6].

Polyoxometalates are widely used in chemistry [7], biology [8], physics [9] and material science [10]. In spin-crossover research, polyoxometalate anions can be of interest because of both their size and physical properties. The large size of the polyoxometalate anions can be used for synthesizing compounds, in which the spin-crossover iron(II) complexes are either positioned in two-dimensional planes, or in one-dimensional chains, i.e. by using polyoxometalates of different shape and/or charge. In this way, the elastic cooperativity required to obtain useful spin-crossover phenomena in the solid state is reduced from three dimensions to only two, or even to only one dimension. This feature should facilitate the study of cooperativity in these solids.

The combination of 1-(2-chloroethyl)tetrazole (*teec*) and copper(II) has been used, because this system [11–13] and the equivalent iron(II) spin crossover systems [14, 15] have been studied intensely over the last few years. The iron(II) complexes show a thermal spin transition at relatively high temperatures compared to other 1-alkyltetrazole compounds [16], but crystallize poorly. Therefore the copper(II) ion has been used, which has, with this ligand, proved to result in good quality single crystals.

Results and Discussion

Description of the structure

The asymmetric unit of $[Cu(teec)_5]_2[Cu(teec)_6][PMo_{12}O_{40}]_2 \cdot 2H_2O$ (**1**) is depicted in Fig. 2 (crystal data Table 1). The crystallographic unit cell contains two

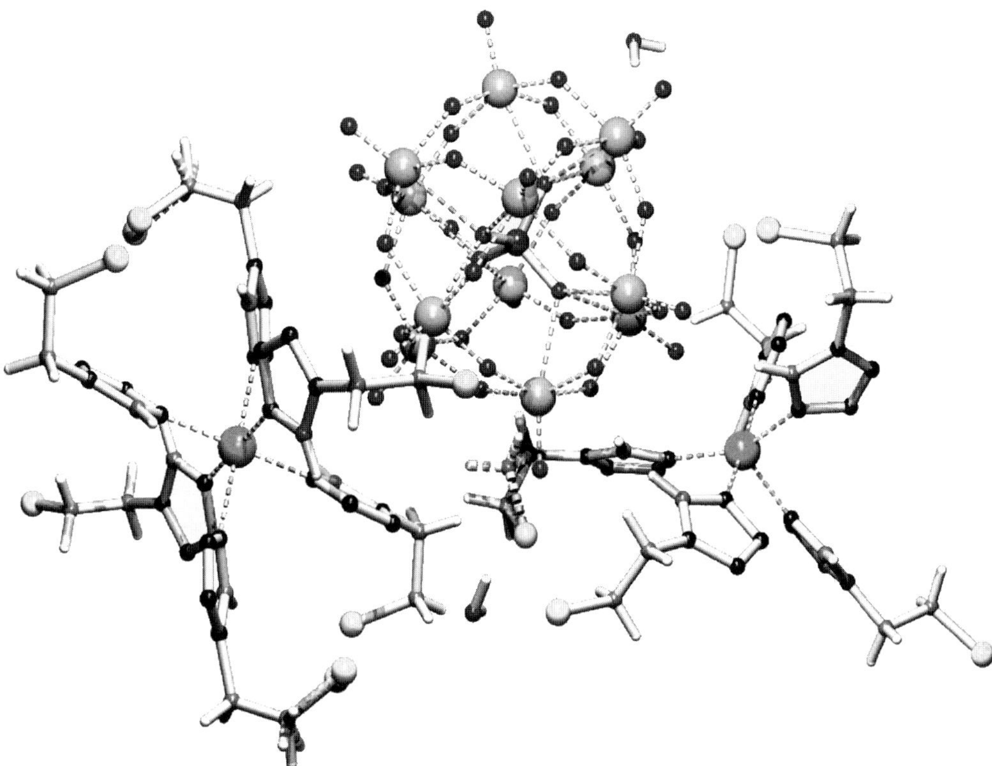

Fig. 2. Molecular structure [28, 29] of $[Cu(\textit{teec})_5]_2[Cu(\textit{teec})_6][PMo_{12}O_{40}]_2 \cdot 2H_2O$

$[PMo_{12}O_{40}]^{3-}$ anions, two copper(II) ions surrounded by five teec ligands, one copper(II) ion surrounded by six teec ligands and finally two water molecules. The $[Cu(\textit{teec})_6]^{2+}$ center is positioned on an inversion center (0, 1/2, 1/2), resulting in three teec ligands with unique positions, and three obtained by symmetry. One of the two $[PMo_{12}O_{40}]^{3-}$ anions and one of the five-coordinated copper(II) ions are obtained by symmetry ($-x$, $-y$, $-z$). The same occurs with the water molecules, which are equally distributed over four positions.

Compound **1** shows a layered structure with three different types of layers: layers of *Keggin* polyoxometalates (layers of type A), layers of complexes $[Cu(\textit{teec})_5]^{2+}$ (layers of type B) and layers of complexes $[Cu(\textit{teec})_6]^{2+}$ (layers of type C). These three layers are arranged in the structure following the pattern ...ABACABAC... (see Fig. 4).

The layer of $[Cu(\textit{teec})_5]^{2+}$ centers (B layer) contains twice as many copper centers as the layer of $[Cu(\textit{teec})_6]^{2+}$ (C layer), whereas the space available is not twice as large (the distance between the layers of $[PMo_{12}O_{40}]^{-3}$ anions is 14.55 Å for the C-type layers and 15.66 Å for the B-type layers).

In the layer of type B, the $[Cu(\textit{teec})_5]^{2+}$ centers (see Fig. 3, left) show one of the bond lengths (Cu(1)–N(9)) approximately 0.2 Å longer than the other four copper–nitrogen bonds. The two ligands coordinated via N(5) and N(17) are moved away from N(9), resulting in angles with N(9) of over 100°. The other two ligands make almost square angles with N(9). The angles within the distorted equatorial plane are between 87.7(5)° and 92.6(4)° (Table 2). The τ value for this

Table 1. Crystal data and structure refinement for $[Cu(teec)_5]_2[Cu(teec)_6][PMo_{12}O_{40}]_2 \cdot 2H_2O$

Compound	$[Cu(teec)_5]_2[Cu(teec)_6][PMo_{12}O_{40}]_2 \cdot 2H_2O$
Compound	$Cu_3(teec)_{16}[PMo_{12}O_{40}]_2 \cdot 2H_2O$
Chemical formula	$C_{48}H_{84}N_{64}Cu_3Cl_{16}P_2Mo_{24}O_{82}$
Molecular weight	5991.97
Crystal system	Triclinic
Space group	$P\bar{1}$
a, Å	11.717(4)
b, Å	12.010(17)
c, Å	30.21(3)
α, °	88.32(12)
β, °	80.93(5)
γ, °	71.40(5)
V, Å^3	3978(7)
Z	2
D_{calc}, g cm^{-3}	2.500
Absorption coefficient	2.608 mm^{-1}
Temperature	293(2) K
$F(000)$	2869.0
Crystal color, shape	Green, elongated plates
Crystal size	$0.48 \times 0.2 \times 0.05$ mm^3
Radiation, wavelength, Å	Mo Kα, 0.71069
Monochromator	Graphite
Diffractometer	Enraf Nonius CAD4
Theta range for data collection	1.37 to 24.98°
Index ranges	$0 \leq h \leq 13,\ -13 \leq k \leq 14,\ -35 \leq l \leq 35$
Reflections collected	14687
Independent reflections	13935 [R(int) = 0.0777]
Absorption correction	Psi-scan
Max. and min. transmission	0.9989 and 0.7661
Refinement method	Full-matrix-squares on F^2
Data/restrains/parameters	13935/5/1097
Goodness-of-fit on $F^2(S)$	1.058
Final R indices [I > 2sigma(I)]	$R1 = 0.0551$, $wR2 = 0.1534$
R indices (all data)	$R1 = 0.1581$, $wR2 = 0.19077$
$(\Delta/\sigma)_{av}$, $(\Delta/\sigma)_{max}$, e Å^{-3}	3.097, −1.719
R_{int}, R_σ	0.0777, 0.1064
Structure solution	SIR-97
Structure refinement	SHELXL-97

Calc $w = 1/[\sigma^2(F_o^2) + (0.1094P)^2 + 0.0000P]$ where $P = (F_o^2 + 2F_c^2)/3$

complex is 0.38 [17]. This parameter reflects the degree of trigonality. For a perfectly tetragonal geometry τ is equal to zero, while it becomes unity for a perfectly trigonal-bipyrimidal geometry. Then, the geometry of the five-coordinated copper(II) ions can be seen as possessing a strongly distorted tetragonal-pyramidal geometry, this geometry is not common for copper(II) surrounded by monodentate tetrazole ligands. All previously published [11, 12, 18–23] structures of copper(II)

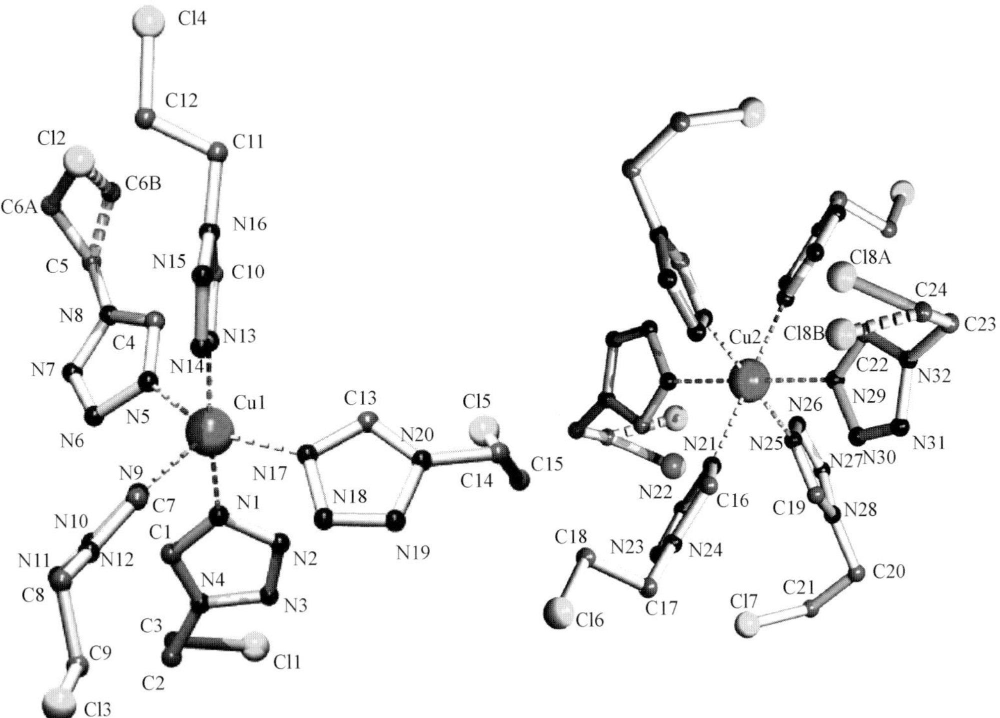

Fig. 3. Molecular structure of the copper centers; left: $[Cu(teec)_5]^{2+}$, right: $[Cu(teec)_6]^{2+}$ [28, 29]

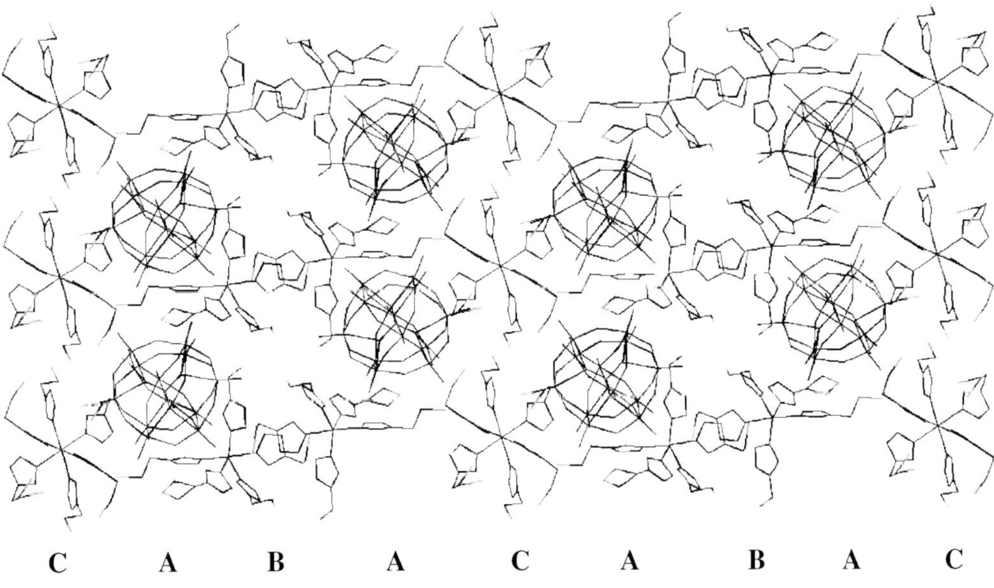

Fig. 4. Packing of the layers along the *b* axis [29]

with tetrazole ligands possess hexakis surrounded copper(II) centers of the form CuL_6 or CuL_2X_4. A closer look to the environment of Cu(1) results in the observation of a terminal oxygen atom of the polyoxometalate anion, O(20), at a distance of 3.152(10) Å of Cu(1) with an angle N(9)–Cu(1)–O(20) of 178.2(4)°, *i.e.* at a

Table 2. Selected angles (°) of $[Cu(teec)_5]_2[Cu(teec)_6][PMo_{12}O_{40}]_2 \cdot 2H_2O$

Atoms	Angle (°)	Atoms	Angle (°)
N21–Cu2–N25	87.5(5)	N5–Cu1–N9	108.5(4)
N21–Cu2–N29	91.0(5)	N5–Cu1–N13	89.5(4)
N25–Cu2–N29	90.5(4)	N5–Cu1–N17	150.9(4)
N1–Cu1–N5	90.7(4)	N9–Cu1–N13	92.6(4)
N1–Cu1–N9	92.3(4)	N9–Cu1–N17	100.6(5)
N1–Cu1–N13	174.(4)	N13–Cu1–N17	87.7(5)
N1–Cu1–N17	89.6(5)		

semi-coordination position. So that the reason for the unusual geometry in this complex must be the 'lack' of space in this cationic layer, which avoids the presence of the six teec ligand for Cu(1) and making it possible for the polyoxometalate to act as a semi-coordinating ligand.

The other copper center, located in layer C, (Cu(2)) shows a distorted octahedral surrounding (see Fig. 3, right). The N–Cu–N angles vary between 87.5(5)° and 91.0(5)°. In these copper(II) centers, all three copper-ligand distances are quite different. This observation is in contrast to already published mononuclear six-coordinated copper(II) tetrazole compounds $[Cu(teec)_6](Anion)_2(teec)$ (with Anion is BF_4 or ClO_4) [18], where four of the teec ligands are at equal distance, and two ligands are positioned on the elongated *Jahn-Teller* axis. In the $[Cu(teec)_6]^{2+}$ center of $[Cu(teec)_5]_2[Cu(teec)_6][PMo_{12}O_{40}]_2 \cdot 2H_2O$, the three unique teec ligands are each at different distances (see Table 3). Because all ligands are the same, this difference in bond length must be caused by internal pressure, induced by the crystal packing.

Spectroscopic and magnetic properties

In the infrared spectrum of **1**, C–H stretching vibrations of the ligand are visible at 3136 cm^{-1} (CH tetrazole ring), 3024 and 2969 cm^{-1} (CH ethyl chain). Moreover, a characteristic signal of the polyoxometallate is visible at 1060 cm^{-1} (P–O).

Table 3. Selected bond lengths (Å) of $[Cu(teec)_5]_2[Cu(teec)_6][PMo_{12}O_{40}]_2 \cdot 2H_2O$

Atoms	Bond length (Å)
Cu2–N21	2.160(13)
Cu2–N25	2.040(11)
Cu2–N29	2.276(15)
Cu1–N1	2.001(11)
Cu1–N5	2.015(10)
Cu1–N9	2.224(11)
Cu1–N13	2.013(11)
Cu1–N17	2.035(11)

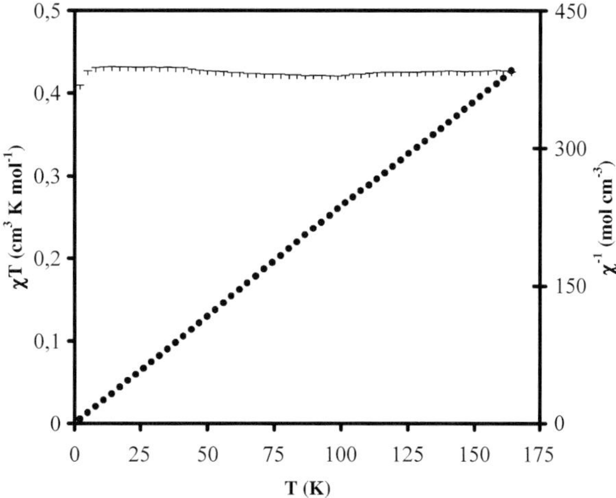

Fig. 5. Magnetic susceptibility of $[Cu(teec)_5]_2[Cu(teec)_6][PMo_{12}O_{40}]_2 \cdot 2H_2O$, with χT (+) and χ^{-1} (●)

In the EPR spectra of the solid at room temperature and at 77 K, only one anisotropic signal is seen with $g_{\perp} = 2.08$ and $g_{||} = 2.28$ ($g_{av} = 2.15$). The magnetic susceptibility has been determined in the temperature range 5–165 K. The plots of χ^{-1} versus the temperature and of χT versus the temperature (Fig. 5) show a *Curie* law, in full agreement with the structure which indicates that the Cu(II) ions are magnetically isolated. A *Curie* constant of $C = 0.425\,cm^3\,K\,mol^{-1}$ has been obtained. From this value one can calculate an average g parameter of 2.13 which is very close to that obtained from the EPR spectra.

Conclusions

The crystallization of teec with $Cu_3[PMo_{12}O_{40}]$ results in $[Cu(teec)_5]_2$ $[Cu(teec)_6][PMo_{12}O_{40}]_2 \cdot 2H_2O$, a layered structure containing both five- and six-surrounded copper(II) ions. This unusual set of Cu complexes in one compound is most likely caused by steric hindrance and crystal packing efficiency, induced by the large $[PMo_{12}O_{40}]^{3-}$ anions. To obtain a neutral complex, the ratio anion to cation must be 2 to 3. The lattice of the $[PMo_{12}O_{40}]^{3-}$ anions consists of layers at relatively equal distance, yielding alternating layers of distorted octahedrally surrounded $[Cu(teec)_6]^{2+}$ and distorted square-pyramidal surrounded $[Cu(teec)_5]^{2+}$ which also has an oxygen atom of the polyoxometalate semi-coordinated to this copper center.

Although the angles in $[Cu(teec)_6]^{2+}$ are close to octahedral, all three metal–ligand distances very considerably. In normal *Jahn–Teller* distorted octahedral CuL_6 systems, the four ligands positioned on the equatorial plane have all about the same metal–ligand distance, while the two axial ligands have a slightly longer bond length [19]. In $[Cu(teec)_5]_2[Cu(teec)_6][PMo_{12}O_{40}]_2 \cdot 2H_2O$, the three axes of the $[Cu(teec)_6]^{2+}$ center are of different length (see Table 3).

In EPR, IR and ligand field spectroscopy no difference is observed between the five- and six-coordinated copper(II) anions, because the signals are superimposed.

The compound shows a *Curie* behavior, in agreement with the lack of interactions between the Cu(II) ions.

The complex can be considered as a good model system for the design of new compounds combining polyoxometalates and iron(II) spin-crossover cations, as the bulky polyoxoanions can impose an arrangement of the cations in isolated layers, so the elastic cooperativity may be studied in only two dimensions [30].

Experimental

Physical methods

Vis-NIR spectra were obtained on a Perkin-Elmer Lambda 900 spectrophotometer using the diffuse-reflectance technique, with MgO as a reference. X-band powder EPR spectra were obtained on a Jeol RE2x electron spin resonance spectrometer using DPPH ($g = 2.0036$) as a standard. FTIR spectra were obtained on a Perkin Elmer Paragon 1000 FTIR spectrophotometer equipped with a Golden Gate ATR device (4000–300 cm^{-1}, res. 4 cm^{-1}). Magnetic susceptibility measurements (2–300 K) were carried out using a Quantum Design MPMS-5 5T SQUID magnetometer (measurements carried out at 1000 Gauss). Data were corrected for magnetization of the sample holder and for diamagnetic contributions, which were estimated from the *Pascal* constants [24]. C, H, N determinations were performed on a Perkin Elmer 2400 Series II analyzer. Microanalysis has been performed using an Environmental Scanning Electron Microscope (Philips, XL30 ESEM).

Synthesis

$Ag_3[PMo_{12}O_{40}]$

5.4 g of the *Keggin* acid $H_3[PMo_{12}O_{40}]$ (3.0 mmol) was dissolved in 40 ml of cold water. As soon as the solution was clear, a second solution of 1.7 g $AgNO_3$ (10.0 mmol) in approximately 5 ml of water has been added. A yellow precipitate formed immediately. The solution was stirred for another hour and then left at 4°C for 12 hours. The compound is obtained by filtration, washed with water and ether and dried on air. The yield consists of approximately 5.15 g (80%) of yellow powder.

From micro analysis, it has been concluded that all acidic protons are replaced by silver ions, and the molybdenum to silver ratio is 12:3. It has been concluded that the *Keggin* anion is still intact, as the phosphorous to molybdenum ratio is close to 1:12. In the infra-red spectrum, a characteristic signal is visible at 1059 cm^{-1}. Micro analysis (relative weight %): P 2.27, Mo 55.85, Ag 12.47%.

$Cu_3[PMo_{12}O_{40}]_2$

0.43 g of $Ag_3[PMo_{12}O_{40}]$ (0.2 mmol) were added in excess to a solution of 0.068 g of $CuCl_2 \cdot 2H_2O$ (0.4 mmol) in 7 ml of water. The poorly soluble *Keggin* salt reacts with the chloride ions, forming insoluble AgCl and very soluble $Cu_3[PMo_{12}O_{40}]_2$. This solution is, after filtration, added to a solution of 0.32 g teec (2.4 mmol) in 5 ml of alcohol. After one day, green, elongated plate-like crystals are formed.

Microanalysis confirms that both the polyoxometalate and copper are present. The copper to *Keggin* ratio is 3:2, whereas the copper to ligand ratio (determined by the amount of chloride) is 3:15, which is close to the expected 3:16. The ratio between the nitrogen and chloride confirms that no chloride ions, originating from copper(II) chloride, are present. Elemental analysis for $C_{48}H_{84}N_{64}Cu_3Cl_{16}P_2Mo_{24}O_{82}$, $\{[Cu(teec)_5]_2[Cu(teec)_6][PMo_{12}O_{40}]_2 \cdot 2H_2O\}$; found (calc.): C 9.9 (9.6), H 1.5 (1.4), N 14.8 (15.0) %. Microanalysis: relative ratio found: N 17.23, Cu 1.67, P 1.74, Mo 12.83, Cl 8.33%.

Crystal structure determination and refinement

The crystal structure analysis was carried out on a green plate-like single crystal of $[Cu(teec)_5]_2[Cu(teec)_6][PMo_{12}O_{40}]_2 \cdot 2H_2O$ with approximate dimensions $0.48 \times 0.2 \times 0.05\,mm^3$. Relevant crystallographic data and structure determination parameters are given in Table 1. Selected angles and bond lengths are given in Tables 2 and 3, respectively. Cell parameters were obtained by the least-squares refinement method of 25 reflections. Intensity data were measured at room temperature on an Enraf-Nonius CAD4 diffractometer with graphite-monochromated Mo $K\alpha$ radiation with the $\omega - 2\theta$ method. All calculations were carried out using the *WinGX* package [25]. The structure was solved by direct methods using the SIR 97 program [26], followed by *Fourier* synthesis, and refined of F^2 using the SHELXL-97 program [27]. *Lorentz*, polarization, and semiempirical absorption corrections (ψ-scan method) were applied to the intensity data.

One carbon atom of a teec ligand coordinated to Cu(1) (C(6A)) and a chloride atom of another teec ligand coordinated to Cu(2) (Cl(8A)) were found to be disordered over two positions with refined occupancies of 0.508/0.492 for C(6A)/C(6B) and 0.643/0.356 for Cl(8A)/Cl(8B). The occupancy factors of the two water molecules were refined, adopting values close to 0.5 and, in the next refinement they were fixed to these values and the O atoms were refined isotropically. All other non-H atoms were refined anisotropically. The positions of the hydrogen atoms were added in calculated positions and refined riding on the corresponding C atoms.

Crystallographic data (excluding structure factors) for the structures reported in this paper have been deposited with the Cambridge Crystallographic Data Centre as supplementary publication no. CCDC-196075 Copies of the data can be obtained free of charge on application to CCDC, 12 Union Road, Cambridge CB2 1EZ, UK [Fax: int. code + 44(1223)336-033; E-mail: deposit@ccdc.cam.ac.uk].

Acknowledgements

The work described in the present paper has been supported by the Leiden University Study Group WFMO. Financial support by the European Union, allowing regular exchange of preliminary results with several European colleagues in the TOSS network, under contract ERB-FMNRX-CT98-0199. Support by the ESF Programme Molecular Magnets (1998–2003) is kindly acknowledged.

References

[1] Palleux R, Schmalle HW, Huber R, Fischer P, Hauss T, Ouladdiaf B, Decurtins S (1997) Inorg Chem **36**: 2301
[2] Coronado E, Galan-Mascaros JR, Gomez-Garcia CJ, Laukhin V (2000) Nature **408**: 447
[3] Decurtins S, Schmalle HW, Schneuwly P, Pellaux R, Ensling J (1995) Mol Cryst Liq Cryst **273**: 167
[4] Hauser A, von Arx ME, Pellaux R, Decurtins S (1996) Mol Cryst Liq Cryst **286**: 225
[5] von Arx ME, Hauser A, Riesen H, Pellaux R, Decurtins S (1996) Phys Rev B: 15800
[6] Pope MT (1983) Heteropoly and Isopoly Oxometalates; Springer: Berlin
[7] Gouzern P, Proust A (1998) Chem Rev **98**: 77
[8] Rhule JT, Hill CL, Judd DA, Schinazi RF (1998) Chem Rev **98**: 327
[9] Müller A, Peters F, Pope MT, Gatteschi D (1998) Chem Rev **98**: 239
[10] Coronado E, Gümez-Garcia CJ (1998) Chem Rev **98**: 273
[11] Stassen AF, Driessen WL, Haasnoot JG, Reedijk J, Inorg Chim Acta (in press)
[12] Stassen AF, Kooijman H, Spek AL, de Jongh LJ, Haasnoot JG, Reedijk J, Inorg Chem (in press)
[13] Stassen AF, Roubeau O, Ferrero Gramage I, Linarès J, Varret F, Mutikainen I, Turpeinen U, Haasnoot JG, Reedijk J (2001) Polyhedron **20**: 1699
[14] Stassen AF, Dova E, Ensling J, Schenk H, Gütlich P, Haasnoot JG, Reedijk J (2002) Inorg Chim Acta **335**: 61

[15] Dova E, Stassen AF, Driessen AJR, Sonneveld E, Goubitz K, Peschar R, Haasnoot JG, Reedijk J, Schenk H (2001) Acta Cryst B **57**: 531
[16] Franke PL, Haasnoot JG, Zuur AP (1982) Inorg Chim Acta **59**: 5
[17] Addison AW, Rao TN, Reedijk J, vanRijn J, Verschoor GC (1984) J Chem Soc Dalton Trans 1349
[18] Stassen AF, Kooijman H, Spek AL, Haasnoot JG, Reedijk J (2001) J Chem Cryst 185
[19] Wijnands PEM, Wood JS, Reedijk J, Maaskant WJA (1996) Inorg Chem **35**: 1214
[20] Virovets AV, Podberezskaya NV, Lavrenova LG (1994) Polyhedron **13**: 2929
[21] Virovets AV, Podberezskaya NV, Lavrenova LG, Bikzhanova GA (1995) Acta Cryst C **6**: 1084
[22] Virovets AV, Baidina IA, Alekseev VI, Podberezskaya NV, Lavrenova LG (1996) J Struc Chem **37**: 288
[23] Virovets AV, Bikzhanova GA, Podberezshaya NV, Lavreneva LG (1997) Zh Strukt Khim **38**: 128
[24] Kolthoff IM, Elving PJ (1963) Treatise on Analytical Chemistry New York, Vol. 4
[25] Farrugia LJ (1999) J Appl Crystallorg **32**: 837
[26] Altomare A, Burla MC, Camalli M, Cascarano G, Giacovazzo C, Guagliardi A, Moliterni AGG, Polidori G, Spagna R (1999) J Appl Crystallogr 115
[27] Sheldrick GM SHELXL-97 (1997) Program for crystal structure refinement. Univ of Göttingen, Germany
[28] Carson C (ed), POVRAY (1996–1999) Rendering engine for Windows
[29] Spek AL (2000) PLATON, A multi-purpose crystallographic tool. Utrecht Univ, The Netherlands. Internet: http://www.cryst.chem.uu.nl/platon/
[30] Spiering H, Meissner E, Köppen H, Müller EW, Gütlich P (1982) Chem Phys **68**: 65

Synthesis and Characterization of a $[Mn_{12}O_{12}(O_2CR)_{16}(H_2O)_4]$ Complex Bearing Paramagnetic Carboxylate Ligands. Use of a Modified Acid Replacement Synthetic Approach

Philippe Gerbier[1], **Daniel Ruiz-Molina**[1], **Neus Domingo**[2], **David B. Amabilino**[1], **José Vidal-Gancedo**[1], **Javier Tejada**[2], **David N. Hendrickson**[3], and **Jaume Veciana**[1,*]

[1] Institut de Ciència de Materials de Barcelona (CSIC), Campus UAB, E-08193, Cerdanyola, Spain
[2] Facultat de Física, Universitat de Barcelona, E-08028 Barcelona, Spain
[3] Department of Chemistry and Biochemistry-0358, University of California at San Diego, La Jolla, California 92093-0358, USA

Received March 27, 2002; accepted May 2, 2002
Published online September 2, 2002

Summary. A new modified approach for the synthesis of Mn_{12} clusters, based on the use of complex $[Mn_{12}O_{12}(O_2C^tBu)_{16}(H_2O)_4]$ (**2**) as starting material to promote the acidic ligand replacement, is presented here. This new synthetic approach allowed us to obtain complex $[Mn_{12}O_{12}(O_2CC_6H_4N(O^\bullet)^tBu)_{16}(H_2O)_4]$ (**3**), whose preparation remained elusive by direct replacement of the acetate groups of $Mn_{12}Ac$ (**1**). Complex **3** bearing open-shell radical units, was prepared to increase the total spin number of its ground state, and consequently, to increase T_B, with the expectation that the radical ligands may couple ferromagnetically with the Mn_{12} core. Unfortunately, magnetic measurements of complex **3** revealed that the sixteen radical carboxylate ligands interact antiferromagnetically with the Mn_{12} core to yield a $S = 2$ magnetic ground state.

Keywords. Single-Molecule Magnet; Synthesis; Paramagnetic ligand; Pivalic acid.

Introduction

The rapid growth of high-speed computers and the miniaturization of magnetic technology have led to much interest in the field of nanoscale magnetic materials [1–3]. In the past decade, the data density for magnetic hard disk drives has

* Corresponding author. E-mail: vecianaj@icmab.es

increased at a phenomenal pace: doubling every 18 months and, since 1997, doubling every year, which is much faster than the *Moore*'s Law for integrated circuits. To maintain such miniaturization rates constantly, the development of new technologies based on lithographic and scanning probe microscopies, the so-called *top-down* approach, have been greatly enhanced. However, the ever-increasing demand of higher areal density magnetic storage media may find technological and economical limitations in a near future. Moreover, the continuous miniaturization of magnetic materials may lead to the observation of new phenomenologies such as superparamagnetic effects or quantum behavior.

The use of synthetic methodologies, the so-called *bottom-up* approach, offers a potential alternative to obtain monodispersed nanoscale magnetic materials of a sharply defined size. The discovery of large metal cluster complexes with interesting magnetic properties characteristic of nanoscale magnetic particles, such as out-of-phase ac magnetic susceptibility signals and stepwise magnetization hysteresis loops, represented an exciting breakthrough to access ultimate high-density information storage devices and quantum computing applications.

In 1993 it was discovered for the first time that $[Mn_{12}O_{12}(O_2CCH_3)_{16}(H_2O)_4] \cdot 4H_2O \cdot 2CH_3CO_2H$ (**1**), functions as a nanoscale molecular magnet and for this reason the term of Single-Molecule Magnet (SMM) was coined [4, 5]. Since then, a few more families of complexes that function as SMM's have been obtained including several other structurally related neutral or negatively charged dodecanuclear manganese complexes, $[Mn_{12}O_{12}(O_2CR)_{(16\text{-}X)}L_X(H_2O)_4]$, commonly known as Mn_{12}, where R can be a saturated or an unsaturated organic group and L a diphenylphosphinate ligand or a nitrate anion [6–11], Mn_4 mixed-valence cubane molecules [12, 13], tetranuclear vanadium(III) complexes with a butterfly structure [14, 15] and iron(III) complexes such as $[Fe_8O_2(OH)_{12}(tacn)_6]^{8+}$ [16, 17] and $[Fe_4(OMe)_6(dpm)_6]$ [18] where *tacn* and *dpm* stand for 1,4,7-triazacyclononane and dipivaloylmethane respectively.

Even though different families of SMMs have been synthesized (*vide supra*), all of them show low blocking temperatures (T_B) above which they behave as superparamagnets. The highest T_B (*ca.* 6 K) so far reported corresponds to the Mn_{12} family. Mn_{12} complexes can be described as a $[Mn_{12}(\mu_3\text{-}O)_{12}]$ core comprising a central $[Mn^{IV}{}_4O_4]^{8+}$ cubane unit held within a nonplanar ring of eight Mn^{III} ions by eight $\mu_3\text{-}O^{2-}$ ions. Peripheral ligation is provided by sixteen carboxylate groups and three or four H_2O ligands. As a consequence, Mn_{12} complexes have a high-spin ground state $S = 10$, which can be understood assuming that the Mn^{IV} $(S = 3/2)$ of the central $[Mn^{IV}{}_4O_4]^{8+}$ cubane are aligned with all the spins down that interact antiferromagnetically with all the Mn^{III} $(S = 2)$ of the external ring with all spin aligned up. Moreover, the strong uniaxial magnetic anisotropy of the molecule originated by the single-ion zero field splitting experienced by the Mn^{III} ions splits the $S = 10$ ground state into the different $m_s = \pm 10, \pm 9, \pm 8, \pm 7, \ldots 0$ levels. In zero field, the $m_s = 10$ levels are the lowest in energy followed by 9, 8, 7,… at higher energies and $m_s = 0$ is the highest energy. Then, an energy barrier for the interconversion from the *spin up* to the *spin down* state of the complex appears and slow magnetization relaxation processes are observed. Such barrier is the responsible of the SMM behavior of the Mn_{12} family.

There are basically two different synthetic procedures available for making new $[Mn_{12}O_{12}(O_2CR)_{16}(H_2O)_4]$ complexes [19]. The first involves the comproportionation between a Mn^{II} source and Mn^{VII} from MnO_4^- in the presence of the desired carboxylic acid (RCOOH). This was the original method used by *Lis* [20] to synthesize complex **1**.

$$44\,Mn^{2+} + 16\,Mn^{7+} \rightarrow 5[Mn_{12}]^{40+} \quad (1)$$

In the second synthetic approach, a variety of derivatives have been prepared by ligand substitution reactions, which are driven by the greater acidity of the added carboxylic acids RCO_2H and/or the removal by distillation of an azeotrope of acetic acid and toluene (see Eq. 2).

$$[Mn_{12}Ac] + 16\,RCO_2H \rightarrow [Mn_{12}O_{12}(O_2CR)_{16}(H_2O)_4] + 16\,MeCO_2H \quad (2)$$

It has to be emphasized that several treatments with the new carboxylic acid are sometimes needed to replace all acetate groups. The advantage is that reaction yields are generally larger than those obtained in the first approach.

More recently, a new functionalization of Mn_{12} SMM with ligands other than carboxylate or site-specific modifications to yield mixed-carboxylate $[Mn_{12}O_{12}(O_2CR)_8(O_2CR')_8H_2O)_4]$ complexes, have been achieved [8, 9]. The interest for the development of these new synthetic methodologies lies in the variety of reactivity studies and applications that can be achieved with an intrinsic SMM behavior.

In this work, we present a modification of the synthetic route shown in Eq. 2 for the convenient synthesis of a large variety of already known and new manganese complexes. Such modification, which is based on the use of complex $[Mn_{12}O_{12}(O_2C^tBu)_{16}(H_2O)_4]$ (**2**), as starting material for the substitution reaction, may possess several advantages: *i*) the presence of *tert*-butyl groups instead of methyl groups at the periphery of the Mn_{12} core should increase significantly its solubility in organic solvents, *ii*) the steric compression afforded by the presence of the bulky *tert*-butyl groups should help the substitution by less bulky acids, and *iii*) when compared with other carboxylic acids, the dissociation constant of the pivalic acid also should favor the displacement of the substitution equilibrium to completion [21]. In short, this approximation is expected to favor the formation of new and exotic Mn_{12} SMMs otherwise unrealizable by direct replacement of the acetate groups of $Mn_{12}Ac$ (**1**). As an experimental example for the convenience of this new synthetic route we report here the synthesis and characterization of the new Mn_{12} complex $[Mn_{12}O_{12}(O_2CC_6H_4N(O^\bullet)^tBu)_{16}(H_2O)_4]$ (**3**), which has the 1-[*N-tert*-butyl-*N*-(oxyl)amino]-4-benzoic acid radical (**4**) in the peripheral ligation. All the attempts to synthesize complex **3** by direct replacement of the acetate groups of $Mn_{12}Ac$ (**1**) invariably afforded ill-defined products whose preliminary analyses showed large amounts of Mn^{2+} ions. The interest to obtain such complex is clear; the open-shell character of the radical carboxylate ligands was expected to increase the high-spin ground state value in the most favorable case of a ferromagnetic interaction between the ligands and the Mn_{12} core, and consequently, to increase the blocking temperature T_B of the SMM.

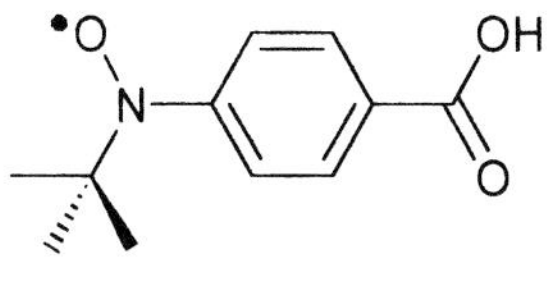

•O OH N O

4

Results and Discussion

Synthesis

Complex **2** was synthesized according to the conventional synthetic approach shown in Eq. 2 and fully characterized by elemental analysis, LDI/MALDI-TOF mass spectroscopy, FT-IR and UV-Vis spectroscopy, and SQUID measurements. To a slurry of complex **1** was added an excess of pivalic acid (HO_2C^tBu) and the resulting solution was allowed to stir overnight. Then, the mixture was concentrated under vacuum to remove the acetic acid. To fully substitute the acetate ligands, this procedure was repeated once more, yielding a microcrystalline material that was satisfactorily characterized as complex **2**. In contrast to **1**, which is poorly soluble, complex **2** is quite soluble in non polar organic solvents such as hexane, although it may be recrystallized from polar solvents such as acetonitrile. However, in spite of repeated efforts, no crystals suitable for X-ray structural determination were obtained.

^1H NMR spectroscopy was used to follow the formation of complex **2**. The ^1H NMR spectrum of a solution of complex **1** in deuterated acetonitrile, displays resonances at $\delta = 48.2$, 41.8 and 13.9 ppm with a 1:2:1 relative intensity. These signals have been ascribed to axial methyl groups linked to two Mn^{III}, equatorial methyl groups linked to one Mn^{III} and axial methyl groups linked to one Mn^{IV}, respectively. In the same solution, an excess of pivalic acid (1:40) was added and its evolution with time was monitored. According to the spectrum recorded after 20 min of reaction, an extensive ligand exchange reaction took place. The ligand exchange reaction was continued for 10 additional hours although no further changes were noticed in the corresponding ^1H NMR spectra, indicating that the exchange equilibrium has been reached at an early stage of the reaction.

The spectrum of a solution of complex **2** is shown in Fig. 1. The *tert*-butyl groups are split into three sets of resonances centered at $\delta = 11.6$, 5.0, and -2.0 ppm in a relative ratio of 1:1:2, the latter being further split into a triplet. The additional resonance at $\delta = 15.3$ ppm was ascribed to the coordinated water molecules.

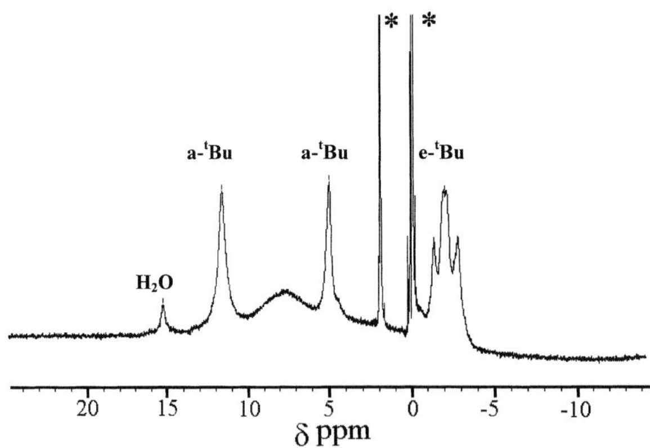

Fig. 1. ^1H NMR spectrum of a CD_3CN solution of complex $[Mn_{12}O_{12}(O_2C^tBu)_{16}(H_2O)_4]$ **2**. Signals associated to solvent and TMS, when present, are marked with an asterisk; a = axial, e = equatorial

As described recently this spectrum is consistent with an effective D_{2d} molecular symmetry in solution [9]. This makes the eight equatorial $^{t}BuCO_2$ groups virtually equivalent, but the axial groups are of two types. Therefore, the singlets at $\delta = 11.6$ and 5.0 ppm are assigned to axial *tert*-butyl groups whereas the triplet at $\delta = -2.0$ ppm is ascribed to equatorial ones. The multiplicity of the latter peak should arise from the fact that the equatorial groups of Mn_{12} complexes are diastereoscopic [9] giving a 1:2:1 distribution for a *tert*-butyl group. From these experiments, one can firstly assess that the ligand exchange reaction should start with the replacement of the more labile axial carboxylates. Moreover, if we consider both the relative intensity of the peaks at $\delta = 9.6$ and 5.1 ppm (axial ^{t}Bu) observed after 20 min of reaction, and the structures of the mixed ligand $[Mn_{12}O_{12}(NO_3)_4(O_2CCH_2{}^{t}Bu)_{12}(H_2O)_4]$ and $[Mn_{12}O_{12}(O_2CCH_3)_4(O_2CCH_2CH_3)_{12}(H_2O)_4]$ complexes [9], one can secondly assess that the resonance at ca. 10 ppm, which is the more prominent, should be ascribed to the axial pivalate ligands linked to two Mn^{III}. Finally, if we consider that these four axial positions are fully occupied by pivalate ligands, this gives for the pivalate ligands an occupancy of 0.25 for the four remaining axial positions (linked to one Mn^{III} and one Mn^{IV}) and an occupancy of 0.50 for the eight equatorial positions at the exchange equilibrium.

As previously mentioned, direct reaction of the acidic radical **4** with the $Mn_{12}Ac$ complex yielded different ill-defined Mn(II)-based products, most probably due to a thermal/acidic-promoted side reaction. So, once complex **2** was obtained and fully characterized, the next step was the reaction of complex **2** with radical **4**. However, prior to the reaction and to fully assess the stability of radical **4** in front complex **2**, a methylene chloride solution of radical **4** and complex **2** (1:1) was prepared and its evolution wit time was followed by cyclic voltammetry. After 1 min, the cyclic voltammogram displays three electrochemical processes: two quasi-reversible processes at 350 mV and 860 mV and one strongly irreversible process at 1140 mV. As previously described [5], the first two electrochemical processes are attributed to the redox couples $[Mn_{12}O_{12}]/[Mn_{12}O_{12}]^{-}$ and $[Mn_{12}O_{12}]^{+}/[Mn_{12}O_{12}]$, respectively. The last electrochemical process has been assigned to the irreversible oxidation of the nitroxide radical to an unstable oxoammonium ion [22] appearing at similar potential to that found for a free solution of radical **4**. After 9 min there is a displacement of *ca.* 30 mV of the anodic peaks corresponding to the oxidation process of **4**. No further evolution or changes on the voltammograms were observed. Therefore, this result shows the chemical stability of the $Mn_{12}O_{12}$ core under the reaction conditions.

To give more insight into the origin of the anodic peak displacement, a methylene chloride solution of radical **4** and complex **2** (1:1) was prepared and its evolution wit time was followed by X-band EPR spectroscopy. Initially, the EPR spectrum shows the same pattern and hyperfine coupling constants than those observed for free radical **4**. After four days, the EPR spectrum of the mixture remains very similar with the only variation of a decrease of the hyperfine coupling constants associated to the nitrogen nuclei (Table 1). This fact indicates that there is a delocalization effect of the spin density onto the aromatic ring [23, 24] probably due to the enhancement of the electron-withdrawing capacity of the carboxylic group when passing from the protonated form of the free acid to the anionic carboxylate once linked to the Mn_{12} complex. The same arguments can

Table 1. EPR hyperfine coupling constants

Compd.	a_N	a_{Hortho}	a_{Hmeta}
4	11.57	2.13	0.92
3	11.70	2.09	0.84

be used to explain the displacement towards higher potentials of the redox process associated to the oxidation process of **4**.

From these results, it can be inferred that radical **4** quickly exchanges the pivalate ligand for the radical carboxylate and, secondly, its stability towards any side redox reaction. Therefore, Mn_{12} complex **3** was prepared by layering a dichloromethane solution of an excess (200%) of the acidic radical **4** and complex **2** with hexane. Complex **3** was collected as a microcrystalline brown-orange powder and fully characterized by elemental analysis, LDI/MALDI-TOF mass spectroscopy, FT-IR and UV-Vis spectroscopy, and SQUID measurements. It has to be emphasized that despite the use of recurrent crystallization experiments the obtaining of single crystals suitable for X-ray studies remained elusive. Elemental analysis, IR spectroscopy and the total absence of any signal in the ^1H NMR spectrum, which is due to the fully paramagnetic nature of the complex, are consistent with a total replacement of the pivalate ligands by the carboxylate radicals.

Magnetochemical characterization

$[Mn_{12}O_{12}(O_2C^tBu)_{16}(H_2O)_4]$ (**2**). Complex **2** exhibits the characteristic single-molecule magnetism behavior of Mn_{12} complexes. Ac magnetic susceptibility data were obtained for a polycrystalline sample of complex **2** in the 1.8–10 K range with a 1 Oe ac field oscillating in the frequency range of 1–1000 Hz (see Fig. 2) and with an external magnetic field held at zero. Frequency-dependent signals in

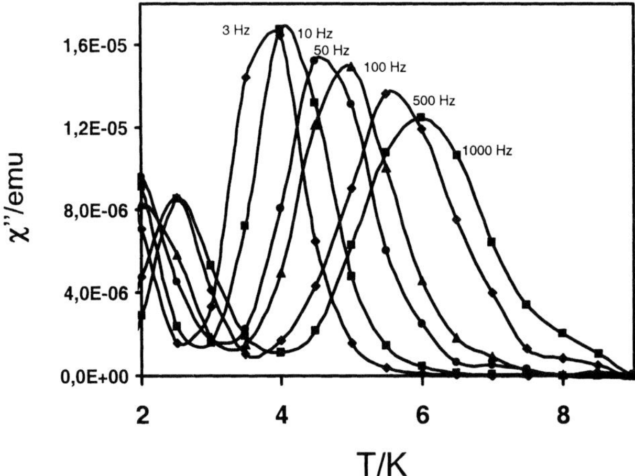

Fig. 2. Ac out-of-phase signals (x″) of complex $[Mn_{12}O_{12}(O_2C^tBu)_{16}(H_2O)_4]$ **2**. The lines are visual guides

the out-of-phase ac magnetic susceptibility are seen, which indicates that complex **2** retains the single-molecule magnetic behavior. Remarkable is the observation of two frequency dependent peaks in the temperature range of 2–4 K and 4–6 K, as previously observed for other Mn_{12} complexes, which may be attributed to the presence of at least two different magnetization relaxation processes.

Magnetization relaxation times (τ) are obtained from the relationship $\varpi\tau = 1$ at the maxima of the χ''_M *vs.* temperature curves [25], which can be determined by fitting the χ''_M *vs.* temperature data to a *Lorentzian* function. Indeed, the ac susceptibility data for complex **2** were least-squares fit to the *Arrhenius* law (Eq. 3):

$$\frac{1}{\tau} = \frac{1}{\tau_0} \exp(-U_{eff}/kT) \tag{3}$$

where U_{eff} is the effective anisotropy energy barrier, k is the *Boltzmann* constant and T is the temperature at which the maximum occurs. The least-squares fit of the ac susceptibility data for the low-temperature and high-temperature out-of-phase signals gave an energy barrier of 23.8 K and 58.1 K, with an attempt frequency of $1.0 \cdot 10^{-7}$ s and $8 \cdot 10^{-8}$ s, respectively.

Magnetization hysteresis data were obtained for a polycrystalline sample of complex **2** at three different temperatures between 1.8 and 2.5 K employing a SQUID magnetometer (see Fig. 3).

The sample is first magnetically saturated in a $+2.0$ T field, and then the field is swept down to -2.0 T, and cycled back to $+2.0$ T. As the field is decreased from $+2.0$ T, the first pronounced step appears at zero field consistently with the observation of two out-of-phase frequency dependent peaks in the ac magnetic susceptibility data. The lower effective barrier, corresponding to the low temperature peak, is still not active at the temperature of measure, and contributes with a superparamagnetic behavior at this temperature, that dominates the magnetic

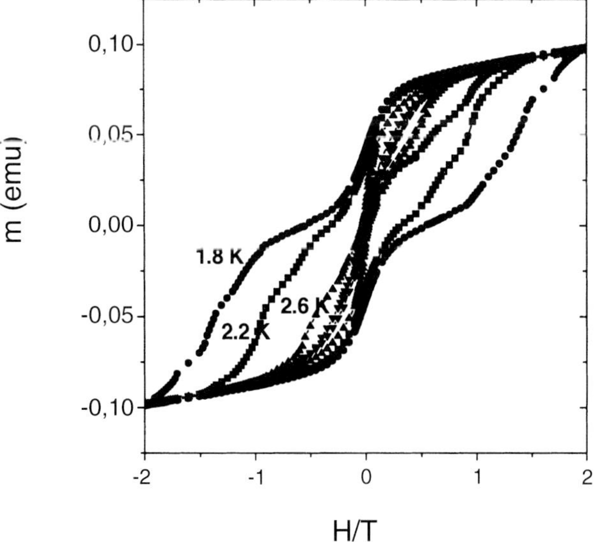

Fig. 3. Magnetization hysteresis loops measured at 1.8 K (•), 2.2 K (■), 2.6 K (▲) and 3.0 K (▼). The sample was aligned by external magnetic field and fixed with eicosane

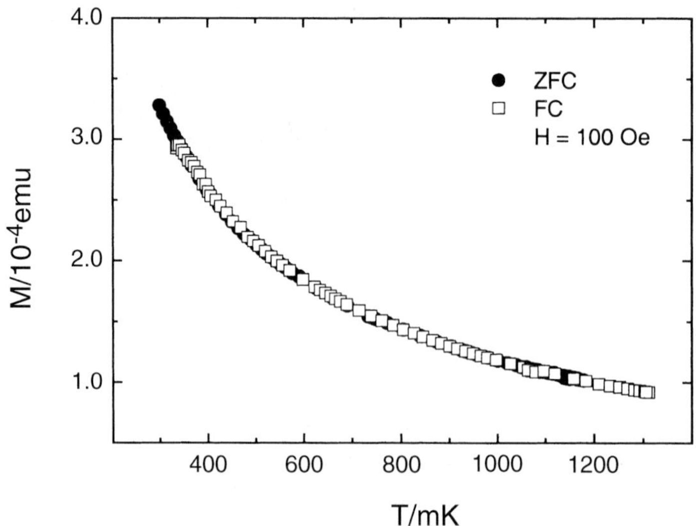

Fig. 4. ZFC-field cooled FC magnetization experiments at 100 Oe down to 300 mK for $[Mn_{12}O_{12}(O_2CC_6H_4N(O^\bullet)^tBu)_{16}(H_2O)_4]$ complex **1c**

relaxation at zero field. In addition, there are successive steps observed at a field interval of approximately 4.5 kOe, which can be explained in terms of resonant spin tunneling relaxation.

$[Mn_{12}O_{12}(O_2CC_6H_4N(O^\bullet)^tBu)_{16}(H_2O)_4]$ (**3**). Magnetic measurements were performed in the temperature range of 0.3 K to 20 K. The zero field cooled (ZFC)-field cooled (FC) magnetization experiments at 100 Oe down to 300 mK are shown in Fig. 4. As it can be observed, there is good matching between the experimental data of both, ZFC and FC magnetization, indicating that complex **3** exhibits a superparamagnetic behavior in all the temperature range studied.

Figure 5 shows the field dependence of the magnetization at five different temperatures ranging from 1.8 to 20 K where no hysteresis loop is observed even at the lowest temperature of 1.8 K. Fitting of the experimental data to the *Brillouin* function indicates that the ground state of the complex is $S = 2$.

To explain the resulting low effective magnetic moment, first it is convenient to revise the magnetic core of Mn_{12} clusters. Mn_{12} complex possesses a $[Mn_{12}(\mu_3\text{-}O)_{12}]$ core comprising a central $[Mn^{IV}{}_4O_4]^{8+}$ cubane held within a non-planar ring of eight Mn^{III} ions. Assuming the presence of diamagnetic carboxylate ligands, Mn_{12} complexes must have a $S = 10$ state, which can be loosely described setting all the Mn^{III} spins up ($S = 8 \cdot 2 = 16$) and all the Mn^{IV} spins down ($S = 4 \cdot -3/2 = -6$). If we now include additional 16 free radicals ($S = 1/2$) from the peripheral ligation interacting antiferromagnetically with the Mn_{12} core ($S = 10$), a $S = 2$ magnetic ground state should result, as in fact it was experimentally observed. Assuming that complex **3** maintains constant the magnetic anisotropy arising from the single-ion zero-field splitting of Mn^{III}, the low $S = 2$ value may explain why no blocking temperature is observed. However, we cannot preclude that an hypothetical SMM behavior may remain hidden by the enhancement of the magnetic relaxation afforded by the paramagnetic ligands, as previously

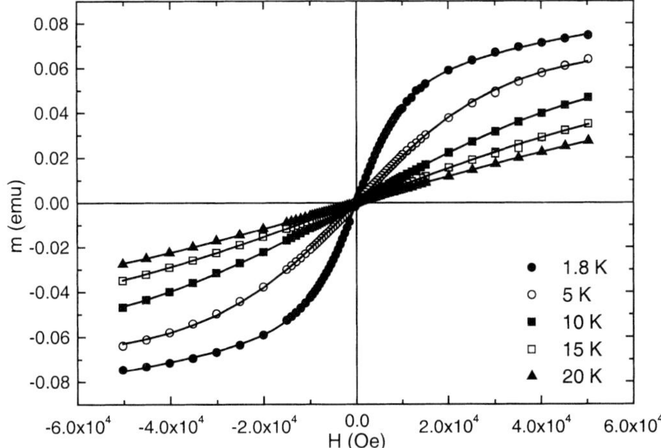

Fig. 5. Field strength dependence of the magnetization in the temperature range of 1.8–20 K showing the superparamagnetic behavior of a microcrystalline sample of complex **1c**. Solid lines represent the fit of experimental data to a *Brillouin* function assuming an $S = 2$ magnetic ground state

observed with the organic radical cation of a Mn_{12}^{-} complex [11]. Indeed, the presence of additional paramagnetic species may promote a fast magnetic relaxation process in spite the presence of an energy barrier for the interconversion from the *spin up* to the *spin down* state, which is expected to promote slow magnetization relaxation processes. Further high-field ESR experiments are currently underway to fully discard such possibility.

Conclusion

We have presented a new modified synthetic approach, based on the use of complex $[Mn_{12}O_{12}(O_2C^tBu)_{16}(H_2O)_4]$ (**2**) as starting material, for the convenient synthesis of the new manganese complex $[Mn_{12}O_{12}(O_2CC_6H_4N(O^{\bullet})^tBu)_{16}(H_2O)_4]$ (**3**).

Table 2. Series of Mn_{12} complexes prepared following the modified synthetic approach based on the use of complex **2** as starting material

Ligand formula	Yield (%)	Complex formula
OH, O	96	$[Mn_{12}O_{12}(O_2CCH_3)_{16}(H_2O)_4] \cdot 4H_2O \cdot 2CH_3CO_2H$
OH, O	90	$[Mn_{12}O_{12}(O_2CCH{=}CHCH_3)_{16}(H_2O)_4] \cdot H_2O$
OH, O	86	$[Mn_{12}O_{12}(O_2CC{\equiv}CH)_{16}(H_2O)_4] \cdot 4H_2O$

(*continued*)

Table 2 (*continued*)

Ligand formula	Yield (%)	Complex formula
O OH	97	$[Mn_{12}O_{12}(O_2CC_6H_5)_{16}(H_2O)_4]$
O OH S	91	$[Mn_{12}O_{12}(O_2CC_4H_3S)_{16}(H_2O)_4]$
OH O	98	$[Mn_{12}O_{12}(O_2CC{\equiv}CC_6H_5)_{16}(H_2O)_4] \cdot 4H_2O$
O HO tBu N O•	78	$[Mn_{12}O_{12}(O_2CC_6H_4N(O^{\bullet})^tBu)_{16}(H_2O)_4]$

The preparation of complex **3** remained elusive with the conventional synthetic procedure shown in Eq. 2. Complex **3** bearing open-shell radical units was prepared to increase the total spin number of its ground state, and consequently to increase T_B, with the expectation that radical ligands may be coupled ferromagnetically with the Mn_{12} core. Unfortunately, magnetic measurements of complex **3** revealed that the sixteen radical carboxylate ligands interact antiferromagnetically with the Mn_{12} core to yield a $S=2$ magnetic ground state, which proved to be negative to achieve a SMM behavior. Finally, it is important to emphasize that this new synthetic approach not only allowed the preparation of complex **3** but other Mn_{12} complexes (shown in Table 2), otherwise unrealizable by direct replacement of the acetate groups of $Mn_{12}Ac$ (**1**). Further work to fully characterize all the Mn_{12} complexes shown in Table 2 is currently in progress.

Experimental

Solvents were distilled prior to use. THF was distilled over sodium/benzophenone under Argon atmosphere whereas CH_2Cl_2 was distilled over P_2O_5 under nitrogen atmosphere. All the reagents were used as received. Microanalyses were performed by the Servei d'Analisi of the Universitat de Barcelona. Manipulations involving organometallic reagents were done using the standard *Schlenck* techniques. $[Mn_{12}O_{12}(O_2CCH_3)_{16}(H_2O)_4] \cdot 4H_2O \cdot 2CH_3CO_2H$ (**1**) was prepared using the method originally described by *Lis* [20]. Radical 1-[*N*-*tert*-butyl-*N*-(oxy)amino]-4-benzoic radical (**4**) was prepared as previously described [26].

Physical measurements

DC magnetic measurements were collected on oriented powder samples restrained in eicosane to prevent torquing on a Quantum Design MPMS2 SQUID (rf) magnetometer equipped with a 5 T (50 kOe) magnet and capable of achieving temperatures from 1.8 to 350 K. Sample alignment in eicosane was performed while keeping the samples in a 5 T field at a temperature above the melting point (312 K) of eicosane for 15 min, and then decreasing the temperature gradually to constrain the sample. Measurements below 1.8 K were performed in a SQUID (dc) magnetometer placed in a $^3He + ^4He$ dilution cryostat, which can achieve temperatures from 100 to 1500 mK. Cyclic voltammetry was carried out on a EG&G Instrument potentiostat/galvanostat, model 263A. Commercial tetrabutylammonium hexafluorophosphate was used as supporting electrolyte (0.1 M). A platinum spiral was used as the working electrode a platinum thread as the counter electrode and Ag/AgCl electrode as the reference electrode. EPR spectra were recorded on degassed solutions using a Bruker ESP-300E spectrometer operating in the X-band (9.3 GHz). Liquid state ^1H NMR spectra were recorded at room temperature on a Bruker Advance DPX 200 spectrometer operating at 200.13 MHz. IR spectra were taken on a Perkin Elmer 1600 FT by using the standard KBr dispersion method. Matrix Assisted LASER Desorption Ionization-Time of Flight (MALDI-TOF) mass spectra were recorded using a KRATOS ANALYTICAL KOMPACT MALDI-2 K-PROBE instrument, equipped with a nitrogen laser ($\lambda = 337$ nm) for the charactrization of Mn_{12} complexes [27].

$[Mn_{12}O_{12}(O_2C^tBu)_{16}(H_2O)_4]$ (**2**; $C_{80}H_{152}O_{48}Mn_{12}$)

To a slurry of complex **1** (1.0 g, 0.49 mmol) in 50 ml of toluene was added HO_2C^tBu (2.0 g, 19.6 mmol). The solution was allowed to stir overnight. Then, the mixture was concentrated under vacuum to remove the acetic acid. The resulting mixture were dissolved in toluene (50 ml), and then concentrated under vacuum. To fully substitute the acetate ligands, this procedure was repeated once more. The resulting brown semi-solid was recrystallized in acetonitrile. The resulting black crystals of **2** (1.0 g, 80%) were collected on a frit and washed with cold acetonitrile. ^1H NMR δ (CD_3CN, ppm): 15.3 (8H, H_2O), 11.6 (36H, axial ^tBu), 5.0 (36H, axial ^tBu), -2.0 (72H, equatorial ^tBu). FTIR (KBr, cm^{-1}): 3436 (broad, OH str); 2963 (medium, C–H str); 1587, 1558, 1529, 1426 (strong, CO_2^- str); 1484 (strong, ^tBu bend); 720 (medium, $Mn_{12}O_{12}$ str). LDI-TOF MS (negative-ion mode): $m/z = 2266$ $[Mn_{12}O_{12}(O_2C^tBu)_{14}]^-$ (20%). Elemental analysis calcd for $C_{80}H_{152}O_{48}Mn_{12}$: C 37.80, H 5.98. Found: C 37.87, H 5.79.

$[Mn_{12}O_{12}(O_2CC_6H_4N(O^\bullet)^tBu)_{16}(H_2O)_4]$ (**3**; $C_{176}H_{216}N_{16}O_{64}Mn_{12}$)

To a solution of **2** (0.100 g, 0.04 mmol) in dichloromethane (5 ml) was added the desired carboxylic acid (1.6 mmol, 40 eq) and the resulting solution was stirred for few minutes. Recrystallization was achieved by slow diffusion of hexane (5 ml) into this solution. The resulting crystals or solids were collected, washed with hexane and dried on the frit orange-brown microcrystals (78%). FTIR (KBr, cm^{-1}): 3422 (broad, OH str); 2976 (medium, C–H str); 1594, 1545, 1404 (strong, CO_2^- str); 1430 (weak, ^tBu bend); 603 (medium, $Mn_{12}O_{12}$ bend). LDI-TOF MS (negative-ion mode): $m/z = 3558$ $[Mn_{12}O_{12}(O_2CC_6H_4N(O^\bullet)^tBu)_{13}]$ (20%). Elemental analysis calcd for $C_{176}H_{216}N_{16}O_{64}Mn_{12}$: C 49.86, H 5.10, N 5.29. Found: C 49.91, H 5.22, N 4.74.

Acknowledgments

This work was supported by the *Information Society Technologies* Programme of the European Commission, under project NANOMAGIQC, from DGI (MAT 2000-1388-C03-01), CIRIT (2001SGR 00362) and the 3MD Network of the TMR program of the E.U. (contract ERBFMRXCT 980181). Ph. G. is grateful to the CSIC and to the Région Languedoc-Roussillon for their financial support.

References

[1] Leuenberger MN, Loss D (2001) Nature **40**: 789

[2] Tejada J, Chudnovsky EM, Del Barco E, Hernández JM, Spiller TP (2000) Nanotechnology **12**: 181

[3] Richter HJ (1999) J Phys D: Appl Phys **32**: R147
[4] Sessoli R, Gatteschi D, Caneschi A, Novak M (1993) Nature **365**: 149
[5] Sessoli R, Tsai H-K, Schake AR, Wang S, Vincent JB, Folting K, Gatteschi D, Christou G, Hendrickson DN (1993) J Am Chem Soc **115**: 1804
[6] Aubin SMJ, Sun Z, Eppley HJ, Rumberger EM, Guzei IA, Folting K, Gantzel PK, Rheingold AL, Christou G, Hendrickson DN (2001) Inorg Chem **40**: 2127
[7] Soler M, Artus P, Folting K, Huffman JC, Hendrickson DN, Christou G (2001) Inorg Chem **40**: 4902
[8] Boskovic C, Pink M, Huffman JC, Hendrickson DN, Christou G (2001) J Am Chem Soc **123**: 9914
[9] Artus P, Boskovic C, Yoo J, Streib WE, Brunel L-C, Hendrickson DN, Christou G (2001) Inorg Chem **40**: 4199
[10] Eppley HJ, Tsai H-L, De Vries N, Folting K, Christou G, Hendrickson DN (1995) J Am Chem Soc **117**: 301
[11] Takeda K, Awaga K (1997) Phys Rev B **56**: 14560
[12] Aubin SMJ, Dilley NR, Wemple MW, Maple MB, Christou G, Hendrickson DN (1998) J Am Chem Soc **120**: 839
[13] Aubin SMJ, Dilley NR, Pardi L, Kryzstek J, Wemple MW, Brunel L-C, Maple MB, Christou G, Hendrickson DN (1998) J Am Chem Soc **120**: 4991
[14] Sun Z, Grant CM, Castro SL, Hendrickson DN, Christou G (1998) Chem Commun 721
[15] Castro SL, Sun Z, Grant CM, Bollinger JC, Hendrickson DN, Christou G (1998) J Am Chem Soc **120**: 2365
[16] Barra A-L, Debrunner P, Gatteschi D, Schulz CE, Sessoli R (1996) Europhys Lett **35**: 133
[17] Sangregorio C, Ohm T, Paulsen C, Sessoli R, Gatteschi D (1997) Phys Rev Lett **78**: 4645
[18] Barra A-L, Caneschi A, Cornia A, Fabrizi de Biani F, Gatteschi D, Sangregorio C, Sessoli R, Sorace L (1999) J Am Chem Soc **121**: 5302
[19] Ruiz-Molina D, Christou G, Hendrickson DN (2002) Single-Molecule Magnets. In: Sasabe H (ed) Hyperstructured Materials, Gordon-Breach, in press
[20] Lis T (1980) Acta Cryst **B36**: 2042
[21] Chrystiuk E, Jusoh A, Santafianos D, Williams A (1986) J Chem Soc, Perkin Trans 2, 163
[22] Baur JE, Wang S, Brandt MC (1996) Anal Chem **68**: 3815
[23] Shultz DA, Gwaltney KP, Lee H (1998) J Org Chem **63**: 769
[24] Barbarella G, Rassat A (1969) Bull Soc Chim Fr 2378
[25] Paulsen C, Park J-G. In: Quantum Tunneling of Magnetization-QTM'94; Gunther L, Barbara B, Kluwer Academic Publishers, Dordrecht, 1995; pp 171–188
[26] Maspoch D, Catala L, Gerbier Ph, Ruiz-Molina D, Vidal-Gancedo J, Wurst K, Rovira C, Veciana J (2002) Chem Eur J, in press
[27] Ruiz-Molina D, Gerbier Ph, Rumberger E, Amabilino DB, Guzei IA, Folting K, Huffman JC, Rheingold A, Christou G, Veciana J, Hendrickson DN (2002) J Mater Chem **12**: 1152

Size Effect on Local Magnetic Moments in Ferrimagnetic Molecular Complexes: An XMCD Investigation

Guillaume Champion[1,2], **Marie-Anne Arrio**[3], **Philippe Sainctavit**[2,3], **Michele Zacchigna**[4], **Marco Zangrando**[4], **Marco Finazzi**[4], **Fulvio Parmigiani**[5], **Françoise Villain**[1,2], **Corine Mathonière**[6], and **Christophe Cartier dit Moulin**[1,2,*]

[1] Laboratoire de Chimie Inorganique et Matériaux Moléculaires, Université Pierre et Marie Curie, F-75252 Paris cedex 05, France

[2] Laboratoire pour l'Utilisation du Rayonnement Electromagnétique, BP34, Université Paris-Sud, F-91898 Orsay cedex, France

[3] Laboratoire de Minéralogie Cristallographie de Paris, Universités Paris 6 et 7, F-75252 Paris cedex 05, France

[4] Laboratorio TASC, INFM c/o Sincrotrone Trieste, S.S. 14 Km. 163.5, Trieste, Italy

[5] INFM-TASC, Trieste and Catholic University, Dept. of Mathematic and Physics, I-25121 Brescia, Italy

[6] Institut de Chimie de la Matière Condensée, 87, F-33608 Pessac cedex, France

Received September 4, 2002; accepted September 6, 2002
Published online November 21, 2002

Summary. Molecular chemistry allows to synthesize new magnetic systems with controlled properties such as size, magnetization or anisotropy. The theoretical study of the magnetic properties of small molecules (from 2 to 10 metallic cations per molecule) predicts that the magnetization at saturation of each ion does not reach the expected value for uncoupled ions when the magnetic interaction is antiferromagnetic. The quantum origin of this effect is due to the linear combination of several spin states building the wave function of the ground state and clusters of finite size and of finite spin value exhibit this property. When single crystals are available, spin densities on each atom can be experimentally given by Polarized Neutron Diffraction (PND) experiments. In the case of bimetallic MnCu powdered samples, we will show that X-ray Magnetic Circular Dichroism (XMCD) spectroscopy can be used to follow the evolution of the spin distribution on the Mn^{II} and Cu^{II} sites when passing from a dinuclear MnCu unit to a one dimensional $(MnCu)_n$ compound.

Keywords. X-ray absorption spectroscopy (XAS); XMCD; Quantum size effect.

* Corresponding author. E-mail: Christophe.Cartier@lure.u-psud.fr

Introduction

Molecular magnetism appears as one of the most fruitful area of research in the study of new effects such as the now famous quantum tunneling effect of Mn_{12} compounds [1]. This improvement is mainly due to the flexibility of molecular chemistry that is able to confer original properties to various new objects. In this research area, for many molecular prospects, having the most precise description of the ground state of the molecular species is crucial for the understanding of the magnetic macroscopic properties. This description is provided by the spin density map generally obtained from Polarized Neutron Diffraction (PND) experiments. But this spectroscopy is limited to samples for which large single crystals can be obtained. The goal of our work is to show that X-ray Magnetic Circular Dichroism (XMCD) spectroscopy, for which no single crystals are necessary, can also get unique information on the spin distribution in such molecular systems.

We choose to report here the theoretical and experimental study of a quantum size effect of local magnetization loss that affects finite magnetic clusters presenting open shell transition metal ions in antiferromagnetic exchange interaction. In such systems, whereas the macroscopic magnetization reaches the expected value at saturation, spin Hamiltonian calculations predict that the magnetization of each cation is lower at saturation than the one of the corresponding free ion [2]. This effect has been recently evidenced by *Kahn* [3] by performing PND experiments on two systems:

(i) the one-dimensional MnCu(1,3-propylenebis(oxamato))$(H_2O)_3 \cdot 2H_2O$ [4] $[(MnCu)_n]$ ferrimagnetic chain where no local magnetization loss is expected;
(ii) the $[Mn^{II}((\pm)$-5,7,7,12,14,14-hexamethyl-1,4,8,11-tetraazacyclotetradecane) Cu^{II}(*N*,*N'*-bis(3-aminopropyl)oxamido)] $(CF_3SO_3)_2$ [5] [MnCu] dinuclear compound for which spin Hamiltonian calculations predict such an effect.

We will first focus on the origin of this quantum size effect. Then, we will briefly introduce the principle of XMCD spectroscopy. The X-ray absorption and XMCD spectra obtained at the Cu and Mn $L_{2,3}$-edges for the two previously described compounds will be presented and analyzed. Finally, we will show that XMCD can be used for powdered samples, as PND for single crystals, to follow the evolution of the spin distribution on the Mn^{II} and Cu^{II} sites when passing from a dinuclear MnCu unit to a one dimensional $(MnCu)_n$ compound.

Results and Discussion

Quantum Size Effect

The quantum size effect of local magnetization loss is evidently not restricted to the bimetallic systems composed of Mn^{II} and Cu^{II} ions. But in order to simplify the explanation of this phenomenon, we will consider an hypothetical system composed of two spins, S_A and S_B, antiferromagnetically coupled with $S_A = 5/2$ and $S_B = 1/2$ (in $\hbar$ units). The Hamiltonian operator $\hat{\boldsymbol{H}}_{ham}$ describing this system is composed of two terms: the first one is the phenomenological *Heisenberg* Hamiltonian and the second one corresponds to the *Zeeman* effect. Therefore, if we assume that

$g = g_{Mn} = g_{Cu} = 2$, it can be written as follows:

$$\hat{H}_{ham} = -J\hat{\mathbf{S}}_{\mathbf{A}} \cdot \hat{\mathbf{S}}_{\mathbf{B}} + g\mu_B \hat{\mathbf{H}} \cdot \hat{\mathbf{S}}$$

Where $\hat{\mathbf{S}}_{\mathbf{A}}$ and $\hat{\mathbf{S}}_{\mathbf{B}}$ are the spin operators related to S_A and S_B, $\hat{\mathbf{S}} = \hat{\mathbf{S}}_{\mathbf{A}} + \hat{\mathbf{S}}_{\mathbf{B}}$, J is the exchange constant, $\hat{\mathbf{H}}$ the applied magnetic field and μ_B stands for *Bohr* magneton.

If we suppose an antiferromagnetic interaction between S_A and S_B, the ground state in the presence of a magnetic field can be written as $|(S_A = 5/2,\ S_B = 1/2), S = 2, Ms = -2>$. When the magnetic saturation is reached, this ground state is the only one populated and one can neglect the contribution of the excited spin states. Since we also neglect in this model any orbital contribution to magnetic moments, the value of the magnetization at saturation, M_{Sat}, of such a system is $2\,g\,\mu_B = 4\,\mu_B$. This value is the one that is expected and experimentally observed for such complexes [6].

This ground state is an eigenfunction of $\hat{\mathbf{S}}_{\mathbf{z}}$, the projection of $\hat{\mathbf{S}}$ on the quantification axis. But if we are interested in the local magnetic moments carried by the Mn^{II} ($S_{Mn} = 5/2$) and Cu^{II} ($S_{Cu} = 1/2$) cations, the adequate quantum numbers are M_{SMn} and M_{SCu}, respectively eigenvalues of $\hat{\mathbf{S}}_{\mathbf{zMn}}$ and $\hat{\mathbf{S}}_{\mathbf{zCu}}$, projection of $\hat{\mathbf{S}}_{\mathbf{Cu}}$ and $\hat{\mathbf{S}}_{\mathbf{Mn}}$ on the quantification axis. The description of the ground state of each cation cannot be directly obtained because it has to be projected on the basis formed by the eigenfunctions of $\hat{\mathbf{S}}_{\mathbf{zMn}}$ and $\hat{\mathbf{S}}_{\mathbf{zCu}}$. Using the *Clebsch-Gordan* coefficients, the ground state of the system may be written as a linear combination of two wave functions:

$$\begin{aligned}
&|(S_{Mn} = 5/2, S_{Cu} = 1/2), S = 2, M_S = -2> = \\
&\qquad \sqrt{5/6}|S_{Mn} = 5/2, M_{SMn} = -5/2>|S_{Cu} = 1/2, M_{SCu} = 1/2> \\
&\qquad - \sqrt{1/6}|S_{Mn} = 5/2, M_{SMn} = -3/2>|S_{Cu} = 1/2, M_{SCu} = -1/2>
\end{aligned}$$

As these two wave functions involved in the ground state are orthogonal, the local magnetic moments carried by the Mn^{II} and the Cu^{II} cations at saturation, respectively M_{SatMn} and M_{SatCu}, can be easily obtained [2]:

$$\begin{aligned}
M_{SatMn} &= (5/6 * 5/2 + 1/6 * 3/2)\, g\mu_B = 4.66\,\mu_B \\
M_{SatCu} &= (5/6 * (-1/2) + 1/6 * 1/2)\, g\mu_B = -0.66\,\mu_B
\end{aligned}$$

It is important to notice that Mn^{II} and Cu^{II} cations do not present strong magnetic anisotropy. This is why the simple model introduced earlier where $g = 2$ and where the magnetic moment has only a spin contribution is convenient for this study.

We can note here that in the case of ferromagnetic exchange interactions, this effect would not occur. The ground state arises indeed from only one wave function centred on the cations and therefore the combination of local states responsible for this magnetization loss does not take place. This effect can only be observed on systems exhibiting an antiferromagnetic exchange interaction.

The magnitude of the local magnetization loss, compared to the expected values for the free ions (5 μ_B for Mn^{II} and 1 μ_B for Cu^{II}) is not negligible since it corresponds to a loss of 7% for the Mn^{II} and 34% for the Cu^{II}. Moreover, the distribution of the *Clebsch-Gordan* coefficients depends on the value of the total spin of the system. One can show that the bigger the total spin is, the smaller the

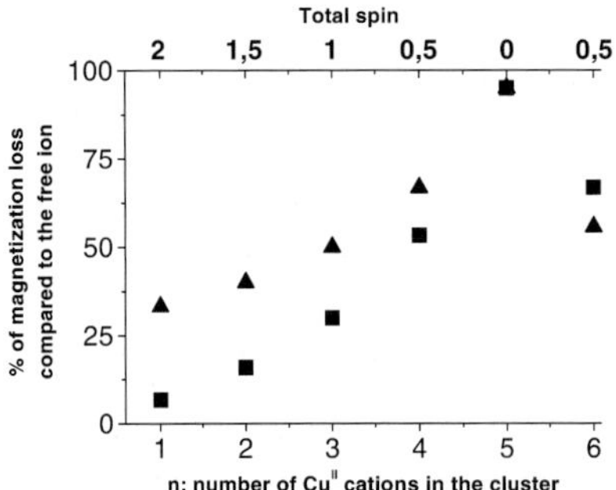

Fig. 1. Change of the magnetization loss with the number of Cu^{II} cations in the MnCu$_n$ clusters

effect is. Considering for example a molecule built with a Mn^{II} central cation, in antiferromagnetic interaction with n Cu^{II} cations. For each n value, we calculated the magnetization loss on each cation. The results are reported on Fig. 1.

The local magnetization loss is important: always above 30% for the Cu^{II} ions, it increases until $n = 5$ and decreases after. The case corresponding to $n = 5$ is particular with a magnetization loss of 100% on the Mn^{II} and on the Cu^{II} cations. This situation corresponds to a system where the 5 *Bohr* magnetons of the manganese ion are antiferromagnetically coupled with $5 * 1$ *Bohr* magnetons of the copper ions. Therefore, the resulting spin of the cluster is equal to zero and there is no more any privileged direction and any resulting local moments on the cations.

Moreover, these calculations show that the size of the cluster and more precisely the size of its spin is predominant in the magnitude of the phenomenon and this is the reason why it can be called a quantum size effect. For the infinite chain composed of numerous alternating Mn^{II} and Cu^{II} cations in an antiferromagnetic interaction, no such effect have been evidenced by PND [3].

We will show now that XMCD as PND experiments are able to evidence such an effect by choosing two compounds previously studied by PND experiments [3]. The first one is a dinuclear molecule, where the Mn^{II} cation is bridged by an oxamide ligand to one Cu^{II} cation, in an antiferromagnetic interaction ($J = -31.1\,\text{cm}^{-1}$) [5]. The second compound consists of an infinite bimetallic (Mn^{II}Cu^{II})$_n$ chain where the copper and manganese ions alternate. The bridge is an oxamate bridge and the exchange constant is $-23.4\,\text{cm}^{-1}$ [4]. The local magnetic moments carried by the Mn^{II} and Cu^{II} cations obtained by PND measurements are reported in Table 1 and compared to the calculated values obtained earlier in this work.

The values given by PND are lower than those given by the previously described calculations. That traduces the partial delocalization of the magnetic moments carriers on the ligands, expected in such compounds and not taken into account in the calculations. We will present now the XAS and XMCD results.

Determination of the Local Magnetic Moments by XMCD

X-ray Absorption Spectroscopy (XAS) is an atomic selective technique that measures the absorption cross section. During the absorption process, the atom

Table 1. Comparison of the calculated magnetic moments carried by the Mn^{II} and the Cu^{II} cations in the MnCu and $(MnCu)_n$ compounds with the experimental ones obtained by PND and by XMCD

		Calculated magnetic moments (μ_B)	Magnetic moments from PND measurements (μ_B)	Integrated area of the normalized L_3-edge signal XMCD
MnCu	Mn	4.67	4.32	$-1.07 \cdot 10^{-1}$
	Cu	−0.67	−0.47	$1.27 \cdot 10^{-2}$
	/Ratio/	6.97	*9.19*	*8.40*
$(MnCu)_n$	Mn	5	4.93	$-5.25 \cdot 10^{-2}$
	Cu	−1	−0.75	$9.31 \cdot 10^{-3}$
	/Ratio/	5	*6.57*	*5.64*

undergoes a transition from an initial state to a final state, and the cross section is expressed in the electric dipole approximation [7]. XMCD is performed when the cross section of a magnetic sample is registered for circularly polarized light [8]. When applying an external magnetic field on ferro or ferrimagnetic materials, the sample does not absorb in the same way right and left circularly polarized light, given the selection rules in the electric dipole approximation. The difference between the two absorption spectra is the XMCD signal. It is, in theory, possible to extract the quantitative value of the local magnetic moment from the XMCD signal, and to separate spin and orbital contributions, using the sum rules [9]. But it has been pointed out, for transition metal $L_{2,3}$ edges, a serious discrepancy between the magnetic moments extracted from XMCD sum rules and the expected ones [10–11]. This is often attributed to surface effect. Nevertheless, the integrated area of the XMCD signal is directly proportional to the local magnetic moment carried by the probed atom. So, this spectroscopy is well-adapted to characterize spin densities in the two materials we want to study.

We recorded the absorption and dichroic spectra at the Mn and Cu $L_{2,3}$-edges for the two compounds. At these edges, we probed selectively the 3d levels since the transitions that occur are $2p^6\,3d^n \rightarrow 2p^5\,3d^{n+1}$. The experimental conditions (detailed in the Experimental section) have been chosen in order to populate only the ground state and to reach the saturation of the macroscopic magnetization. XAS and XMCD spectra are reported on Fig. 2.

The XAS spectra are characteristic for Mn^{II} and Cu^{II} ions in high spin states, with numerous and well resolved structures for the Mn $L_{2,3}$-edges predicted by multiplet calculations [2, 12] and only one single peak for the copper spectrum due to the unique $3d^{10}$ electronic configuration of the final state. To be compared, the spectra presented on Fig. 2 have been normalized using the calculated spectra in the multiplet approach (not shown here). Renormalized to fully circularly polarized light for the Cu $L_{2,3}$-edges (See Experimental section), the cross sections presented here are the absolute values given by the calculations and correspond to one atom of manganese and copper.

For both compounds, whereas the Mn^{II} dichroic signal is mainly negative at low energy (L_3 edge) then positive at high energy (L_2 edge), the Cu^{II} dichroic

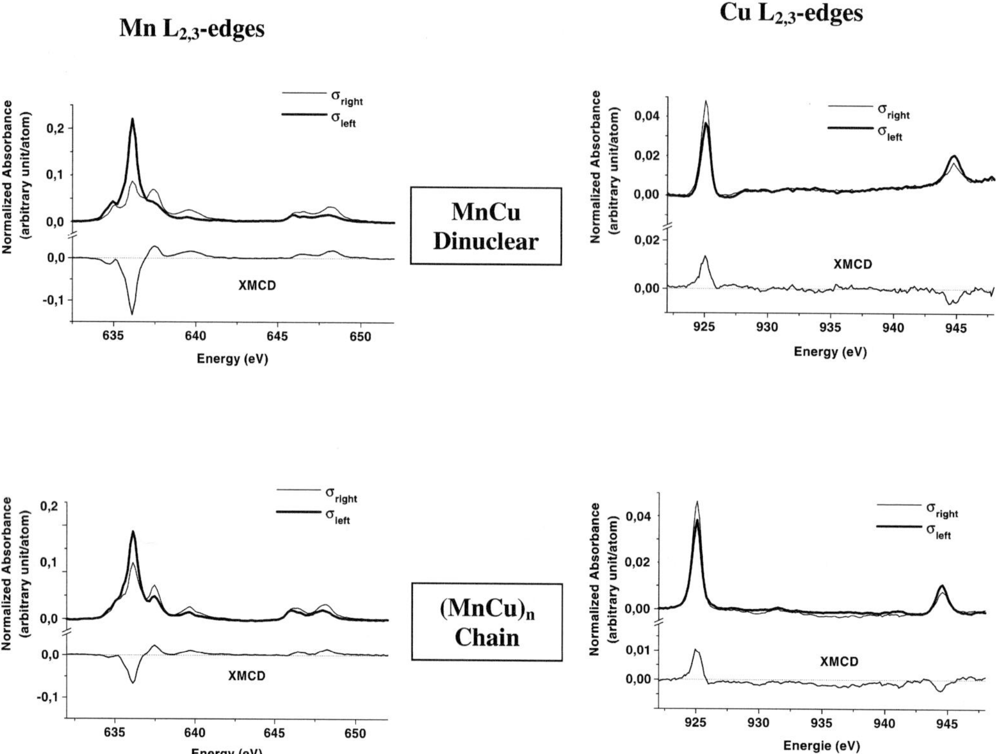

Fig. 2. Normalized XAS and XMCD spectra at the Mn and Cu $L_{2,3}$-edges for the MnCu dinuclear and the $(MnCu)_n$ chain ($T = 2$ K, $H = \pm 5$ T)

signal is inverse: positive at the L_3-edge and then negative at the L_2-edge. This inversion evidences directly the antiferromagnetic exchange interaction between the manganese and the copper ions, expected for these systems. Table 1 compiles the values of the integrated areas of the dichroic signals at the Mn and Cu L_3-edges for both compounds.

By considering the values reported in Table 1, one can remark that the dichroic signals at the two edges for the chain are less intense than the one of the dinuclear compound, which is surprising because we expect higher local magnetic moments. We checked experimentally that the saturation magnetization of the local magnetic moments was reached for both compounds (See Experimental section). We propose two possible origins for the low intensity of the XMCD signals obtained for the chain. Since the total electron yield detection mode is surface sensitive, one cannot discard the hypothesis that the ions at the surface of the powder grain have reduced number of magnetic neighbours or present strong magnetic anisotropy that would make the magnetization perpendicular to the surface negligible [10].

For these reasons and the difficulty to use the sum rules to extract quantitative local magnetic moments from XMCD signals, the direct comparison of PND and XMCD results is not feasible. Anyway, the comparison of the ratios of the XMCD signals integrated areas obtained at the two cations L_3-edges for each compound with the ratios of spin densities given by PND has a physical meaning. The results are reported in Table 1. The Mn/Cu ratio given by PND experiments is very different for the two compounds, due to the magnetization loss effect observed for the dinuclear compound. For the chain as for the dinuclear compound, the results found by XMCD are in agreement with those obtained by PND: 5.64 (XMCD) compared to 6.57 (PND) for the $(MnCu)_n$ chain, 8.40 (XMCD) compared to 9.19 for the MnCu dinuclear compound.

Conclusion

This work is the first experimental evidence of magnetization loss by XMCD. These results show that XMCD experiments coupled with multiplet calculations allows to evaluate precisely magnetic moments localization in a molecule, for powdered samples for which PND cannot be used. Moreover, XMCD allows to separate spin and orbital contributions to the local magnetic moments, important for the characterization of the anisotropy, which are not given by PND. The calculations of these quantities are in progress, performing multiplet calculations with ligand field to simulate the experimental data.

After this work on reference compounds characterized by PND and XMCD, we will start XMCD experiments on new compounds of the series for which no single crystals are available. The goal is to follow by XMCD the magnetization loss of each cation in function of the molecule size, and to evaluate the spin delocalization on the ligands using this approach.

Experimental

Sample Preparation

The synthesis of the two compounds [Mn(($\pm$)-5,7,7,12,14,14-hexamethyl-1,4,8,11-tetra-azacyclotetradecane)Cu(*N,N'*-bis(3-aminopropyl)oxamido)] $(CF_3SO_3)_2$ $(Mn^{II}Cu^{II})$ and MnCu(1,3-propylenebis(oxamato))$(H_2O)_3 \cdot 2H_2O$ $(Mn^{II}Cu^{II})_n$ were made as described in Refs. [4] and [5].

XMCD Data Collection

The data were collected at the Beamline for Advanced diCHroism (BACH) [13] at the ELETTRA Synchrotron Radiation Source in Trieste, Italy. The radiation source was an Apple II helical undulator [14]. The first (Mn-edge) and third (Cu-edge) undulator harmonic were used, giving a circular light polarization rate equal to 100% at the Mn $L_{2,3}$ edges and 85 at the Cu $L_{2,3}$ edges. Highly monochromatic light was obtained through a variable-integrated-angle monochromator employing spherical gratings. The entrance and exit slits of the monochromator were chosen in order to set the photon energy resolution power equal to 10000 at the two edges. The sample was cooled down to 1.5 K with a $H = 5$ Tesla and 6 Tesla applied magnetic field, to check the saturation of the magnetization of the samples. Four spectra were recorded to obtain the dichroic signals, changing the circular polarization (right or left) and reversing the magnetic field applied.

Acknowledgements

We would like to thank Pr. *Michel Verdaguer* for all the fruitful discussions, and for his constant support for our work.

References

[1] Sessoli R, Gatteschi D, Caneschi A, Novak MA (1993) Nature **365**: 141

[2] Arrio MA, Scuillier A, Sainctavit P, Cartier dit Moulin C, Mallah T, Verdaguer M (1999) J Am Chem Soc **121**: 6414

[3] Kahn O, Mathonière C, Srinivasan B, Gillon B, Baron V, Grand A, Ohrstrom L, Ramasesha S (1997) New J Chem **21**: 1037

[4] Pei Y, Verdaguer M, Kahn O, Sletten J, Renard JP (1987) Inorg Chem **26**: 138

[5] Mathonière C, Kahn O, Daran J, Hilbig H, Kohler FH (1993) Inorg Chem **32**: 4057

[6] Baron V, Gillon B, Plantevin O, Cousson A, Mathonière C, Kahn O, Grand A, Ohrstrom L, Delley B (1996) J Am Chem Soc **118**: 11822

[7] Koningsberger DC, Prins R (eds) (1987) X-ray absorption: principles, applications, techniques of EXAFS, SEXAFS and XANES vol 92, Wiley, New York

[8] Beaurepaire E, Scheurer F, Krill G, Kappler JP (eds) (2001) Magnetism and synchrotron radiation. Springer, Berlin

[9] Thole BT, Carra P, Sette F, van der Laan G (1992) Phys Rev Lett **68**: 1943

[10] Arrio MA, Sainctavit P, Cartier dit Moulin C, Brouder C, de Groot FMF, Mallah T, Verdaguer M (1996) J Phys Chem **100**: 4679

[11] Sainctavit P, Cartier dit Moulin C, Arrio M-A (2001) Magnetic Measurements at the Atomic Scale in Molecular Magnetic and Paramagnetic Compounds. In: Magnetism: Molecules to Materials, Miller JS, Drillon M (eds), Wiley VCH Verlag, Berlin, p 131

[12] Arrio MA, Sainctavit P, Cartier dit Moulin C, Mallah T, Verdaguer M, Pellegrin E, Chen CT (1996) J Am Chem Soc **118**: 6422

[13] Zangrando M, Finazzi M, Paolucci G, Comelli G, Diviacco B, Walker RP, Cocco D, Parmigiani F (2001) Rev Sci Instrum **72**: 1313

[14] Sasaki S (1994) Nucl Instrum Methods Phys Res A **347**: 83

Polymorphism and Pressure Driven Thermal Spin Crossover Phenomenon in $[Fe(abpt)_2(NCX)_2]$ ($X = S$, and Se): Synthesis, Structure and Magnetic Properties

Ana B. Gaspar[1], **M. Carmen Muñoz**[2], **Nicolás Moliner**[1], **Vadim Ksenofontov**[3], **Georgii Levchenko**[4], **Philipp Gütlich**[3], and **José Antonio Real**[1,*]

[1] Departament de Química Inorgánica/Institut de Ciència Molecular, Universitat de València, E-46100 Burjassot, València, Spain
[2] Departament de Física Aplicada, Universitat Politècnica de València, Camino de Vera s/n, E-46071 València, Spain
[3] Institut für Anorganische und Analytische Chemie, Johannes Gutenberg Universität, D-55099 Mainz, Germany
[4] Donetsk Physico-Thecnical Institute, NAS of Ukraine, UA-83114 Donetsk, Ukraine

Received June 12, 2002; accepted July 1, 2002
Published online November 7, 2002

Summary. The monomeric compounds $[Fe(abpt)_2(NCX)_2]$ ($X = S$ (**1**), Se (**2**) and *abpt* = 4-amino-3,5-bis(pyridin-2-yl)-1,2,4-triazole) have been synthesized and characterized. They crystallize in the monoclinic $P2_1/n$ space group with $a = 11.637(2)$ Å, $b = 9.8021(14)$ Å, $c = 12.9838(12)$ Å, $\beta = 101.126(14)^\circ$, and $Z = 2$ for **1**, and $a = 11.601(2)$ Å, $b = 9.6666(14)$ Å, $c = 12.883(2)$ Å, $\beta = 101.449(10)^\circ$, and $Z = 2$ for **2**. The unit cell contains a pair mononuclear $[Fe(abpt)_2(NCX)_2]$ units related by a center of symmetry. Each iron atom, located at a molecular inversion center, is in a distorted octahedral environment. Four of the six nitrogen atoms coordinated to the Fe(II) ion belong to the pyridine-N(1) and triazole-N(2) rings of two abpt ligands. The remaining *trans* positions are occupied by two nitrogen atoms, N(3), belonging to the two pseudo-halide ligands. The magnetic susceptibility measurements at ambient pressure have revealed that they are in the high-spin range in the 2 K–300 K temperature range. The pressure study has revealed that compound **1** remains in high-spin as pressure is increased up to 4.4 kbar, where an incomplete thermal spin crossover appears at around $T_{1/2} = 65$ K. Quenching experiments at 4.4 kbar have shown that the incomplete character of the conversion is a consequence of slow kinetics. Relatively sharp spin transition takes place at $T_{1/2} = 106$, 152 and 179 K, as pressure attains 5.6, 8.6 and 10.5 kbar, respectively.

* Corresponding author. E-mail: jose.a.real@uv.es

Keywords. Iron(II) complexes; Spin crossover; Polymorphism; X-ray structure determination; Pressure-induced spin transition.

Introduction

The spin crossover phenomenon deals with molecular materials that undergo phase transitions induced by temperature and/or pressure, as well as by light irradiation. The phase transition provokes drastic changes in their magnetic and optical properties conferring them a bistable character, which could be useful for the design of molecular devices [1]. Most spin crossover compounds are constituted by metallic ions with $3d^4$–$3d^7$ electronic configurations in pseudo-octahedral surroundings. Generally, two different arrangements for the d electrons, in the e_g and t_{2g} orbital subsets, may be envisaged according to whether the magnitude of the ligand field strength is smaller or greater than the mean inter-electronic repulsion energy. In the former case the 3d electrons adopt the high-spin (HS) configuration in order to minimize the inter-electronic repulsion. In the latter case the ligand field counter-balance the inter-electronic repulsion and the fundamental state becomes the low-spin state (LS) violating the *Hund*'s rule of maximum spin multiplicity. For intermediate ligand field strengths, a reversible thermal spin conversion between the LS ground state and the HS excited state may be observed when the energy difference between the HS state and the LS state is close to thermal energy. Then, an intra-ionic electron transfer takes place between the e_g and t_{2g} orbitals. Because the antibonding nature of the e_g orbitals, population and depopulation of these orbitals provokes a structural reorganization both at the molecular and the crystalline level. At the molecular level the main structural feature involves the metal-to-ligand bond distances change, being 0.1–0.2 Å shorter in the LS state.

The thermal spin crossover phenomenon is an entropy-driven process [2, 3]. Entropy stems from the difference in spin multiplicity and in vibrational density of the LS and HS states. Hence, at high temperatures molecules have sufficient energy as to populate nearly completely the HS excited state. In contrast, in a piezo-induced spin conversion the increase of pressure makes the metal-to-ligand bond distances shorter and, consequently, destabilizes the HS state in favour of the LS state.

Recently, we have reported the structural, magnetic, calorimetric and photomagnetic characterization of the spin crossover system $[Fe(abpt)_2(NCX)_2]$ ($X = S$ (**3**), Se (**4**), and NCN (**5**) prepared in methanol–chloroform solutions; *abpt* = 4-amino-3,5-bis(pyridin-2-yl)-1,2,4-triazole system [4, 5]. Compounds **3** and **4** are isostructural as they crystallize in the $P2_1/n$ space group and present very similar crystal and molecular parameters. The iron atom is surrounded by two abpt ligands, which occupy the equatorial positions, while the pseudohalide ligands complete the remaining positions of the $[FeN_6]$ core. Compound **5** crystallizes in the triclinic P-1 space group. Despite this difference the crystal packing is not very different to that of **3** and **4**. In fact, the $[Fe(abpt)_2(NCX)_2]$ mononuclear units interact via π-stacking defining infinite chains, which self-organize to define parallel sheets, in the three compounds. Compounds **3** and **4** undergo thermal induced spin conversion at $T_{1/2} = 180$ (1) and 224 (2) K. $T_{1/2}$ is the temperature at 50% of spin conversion where the number of HS and LS molecules is the same. Compound

5 undergoes a two-step spin transition at very low temperature ($T_{1/2} = 86\,K$). Light-induced excited spin-state trapping (LIESST) is observed when irradiating samples of **3–5** with green light at 10 K.

During the synthesis of the red single crystals of **3** and **4** the simultaneous crystallization of a second type of orange single crystals was observed depending on the nature of the solvent we used. For instance, methanol–chloroform solutions of abpt and Fe/NCX^- (1:2) usually afforded **3** and **4**, while in methanol–water mixtures the main compound was an orange form, compounds **1** (S) and **2** (Se). Both kinds of compounds can be easily separated by hand with the help of binocular lens. Their chemical analyses, crystal diffraction patterns, and magnetic properties indicate that we were faced with two different polymorphs of [Fe(*abpt*)$_2$(NC*X*)$_2$] with $X = $ S, and Se. In the following text compounds **3** and **4** represent the polymorphs A and compounds **1** and **2** correspond to the polymorphs B. In this report we present and discuss the synthesis, the crystal structure and the thermal dependence of the magnetic properties of the polymorphs B at different pressures.

Results

Crystal Structure of [Fe(abpt)$_2$(NCX)$_2$] (X = S (**1**), *Se* (**2**))

Compounds **1** and **2** crystallize in the monoclinic $P2_1/n$ space group (see Table 1). The unit cell contains a pair of mononuclear [Fe(*abpt*)$_2$(NC*X*)$_2$] units related by a center of symmetry. Each iron atom, located at a molecular inversion center, is in a

Table 1. Crystallographic data for [Fe(*abpt*)$_2$(NC*X*)$_2$] (*X*: S, Se)

Compound **1**			
empirical formula	$C_{26}H_{20}FeN_{14}S_2$	V, Å^3	1417.6(4)
fw	648.53	Z	2
space group	$P2_1/n$	T, K	293(2)
a, Å	11.601(2)	λ, Å	0.71073
b, Å	9.6666(14)	μ, mm^{-1}	0.725
c, Å	12.883(2)	ρ_{calc}, g/cm^3	1.519
α, deg	90	$R1^a$	0.0405
β, deg	101.126(14)	$wR2^a$	0.0984
γ, deg	90		
Compound **2**			
empirical formula	$C_{26}H_{20}FeN_{14}Se_2$	V, Å^3	1451.5(3)
fw	742.33	Z	2
space group	$P2_1/n$	T, K	293(2)
a, Å	11.637(2)	λ, Å	0.71073
b, Å	9.8021(14)	μ, mm^{-1}	3.072
c, Å	12.9838(12)	ρ_{calc}, g/cm^3	1.698
α, deg	90	$R1^a$	0.0391
β, deg	101.449(10)	$wR2^a$	0.0731
γ, deg	90		

[a] $R1 = \Sigma||F_o|-|F_c||/\Sigma|F_o|$; $wR2 = \Sigma[w(F_o^2 - F_c^2)^2/\Sigma[w(F_o^2)^2]]^{1/2}$. $w = 1/[\sigma^2(F_o^2) + (0.0522P)^2 + 0.1930P]$ where $P = (F_o^2 + 2F_c^2)/3$

distorted octahedral environment. Four of the six nitrogen atoms coordinated to the Fe(II) ion belong to the pyridine-N(1) and triazole-N(2) rings of two abpt ligands. The remaining trans positions are occupied by two nitrogen atoms, N(3), belonging to the two pseudo-halide ligands (see Fig. 1). Interatomic bond distances and angles are collected in Table 2. The Fe–N bond distances involving the pyridine

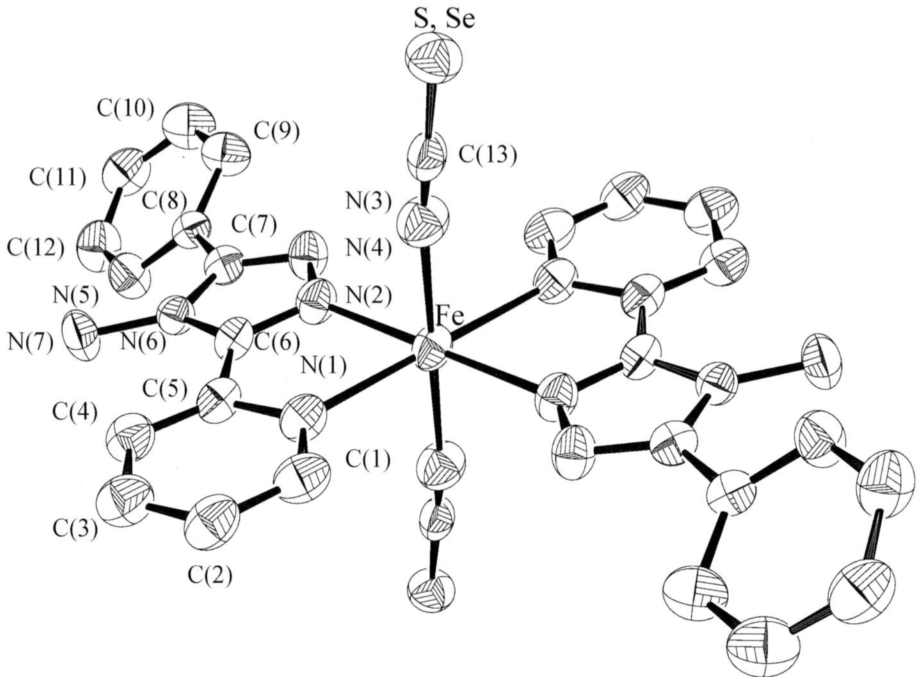

Fig. 1. Molecular structure of polymorphs B [Fe(*abpt*)$_2$(NC*X*)$_2$] (X = S, Se) at 293 K. Hydrogen atoms are omitted for clarity

Table 2. Selected bond distances (Å) and angles (deg)[a] for [Fe(*abpt*)$_2$(NC*X*)$_2$] (*X*: S, Se) polymorphs B

Compound **1**			
Fe–N(1)	2.226(4)	S(1)–C(13)	1.639(6)
Fe–N(2)	2.162(4)	N(3)–C(13)	1.165(6)
Fe–N(3)	2.125(5)	N(7)–H(7B)	0.90(6)
N(3)–C(13)	1.165(6)	N(7)–H(7A)	0.87(8)
N(1)–Fe–N(2)	75.18(14)	N(2)–Fe–N(3)	86.8(2)
N(1)–Fe–N(3)	89.2(2)	N(3)–C(13)–S(1)	179.7(5)
Compound **2**			
Fe–N(1)	2.217(5)	Se(1)–C(13)	1.798(7)
Fe–N(2)	2.171(5)	N(3)–C(13)	1.159(7)
Fe–N(3)	2.120(6)	N(7)–H(7B)	0.93(7)
N(3)–C(13)	1.159(7)	N(7)–H(7A)	0.79(6)
N(1)–Fe–N(2)	74.8(2)	N(2)–Fe–N(3)	86.9(2)
N(1)–Fe–N(3)	88.9(2)	N(3)–C(13)–Se(1)	179.3(6)

[a] Numbers in parentheses are estimated standard deviations in the least significant digit

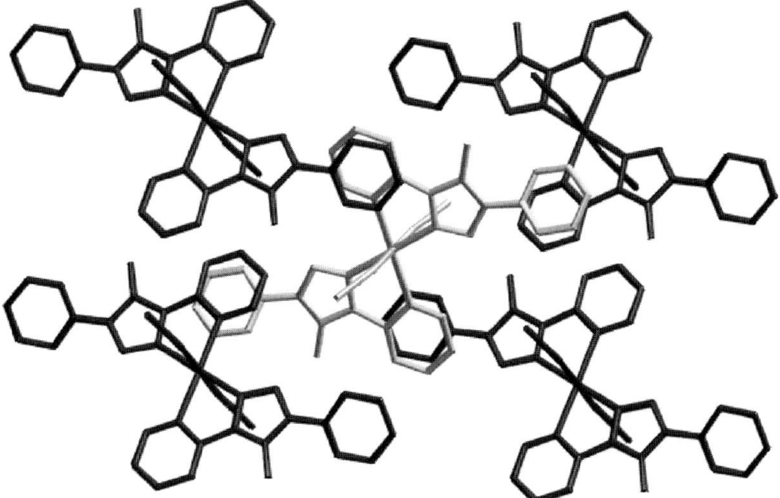

Fig. 2. Perspective view of the crystal packing of compounds **1** and **2**

ring (Fe–N(1) = 2.226(4) and 2.217(5) Å for **1** and **2**, respectively) are larger than those involving the triazole ring (Fe–N(2) = 2.162(4) and 2.171(5) Å for **1** and **2**, respectively) and the pseudo-halide groups (Fe–N(3) = 2.125(5) and 2.10(6) Å for **1** and **2**, respectively).

The NCX^- groups are almost linear (N(3)–C(13)–X = 179.7(5)° and 179.3(6)° for **1** and **2**, respectively) whereas the linkages Fe–NCX are bent (Fe–N(3)–C(13) = 169.6(4)° and 169.7(5)° for **1** and **2**, respectively). The coordinated pyridine and triazole rings almost remain in the same plane (dihedral angle 125.4(4)°). In contrast, the uncoordinated pyridine group and the chelate pyridine-triazole system define a dihedral angle of 34.6° (**1**) and 34.4° (**2**).

As can be seen in Fig. 2 the crystal packing of **1** and **2** is defined by the π-stacking of mononuclear [Fe($abpt$)$_2$(NCX)$_2$] units. Each unit interacts with four surrounding neighbor units in such a way that the uncoordinated pyridine ring of one abpt ligand in [Fe($abpt$)$_2$(NCX)$_2$] interacts with the coordinated pyridine ring of the adjacent [Fe($abpt$)$_2$(NCX)$_2$] unit. The average distance between overlapping ligands is ca. 3.6 Å.

Magnetic Properties under Pressure

Figure 3 shows the temperature dependence of the $\chi_M T$ product for **1** (open rhombuses) and **2** (full rhombuses), χ_M being the molar magnetic susceptibility and T the temperature. At room temperature and $P = 1$ bar, $\chi_M T$ is equal to 3.68 and 3.79 $cm^3\,K\,mol^{-1}$ for **1** and **2** respectively, which is in the range of the values expected for an iron(II) ion in the HS state. As the temperature is lowered, $\chi_M T$ practically remains constant for both compounds, the dropping of $\chi_M T$ at temperatures below 25 K corresponds most probably to the occurrence of zero-field splitting of the HS iron(II) ions.

The magnetic properties of compound **1** were measured at different pressures in the range of temperatures 300–4.2 K. The $\chi_M T$ product is displayed in the Fig. 4 at

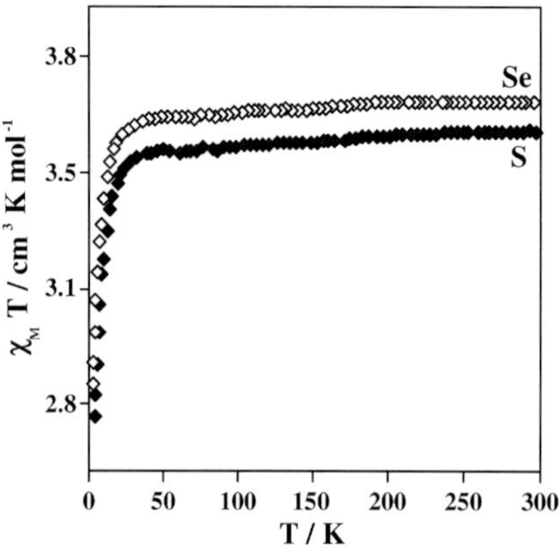

Fig. 3. $\chi_M T$ versus T plots for compounds **1** and **2**. Samples were cooled from 300 to 2 K at 2 K min^{-1}

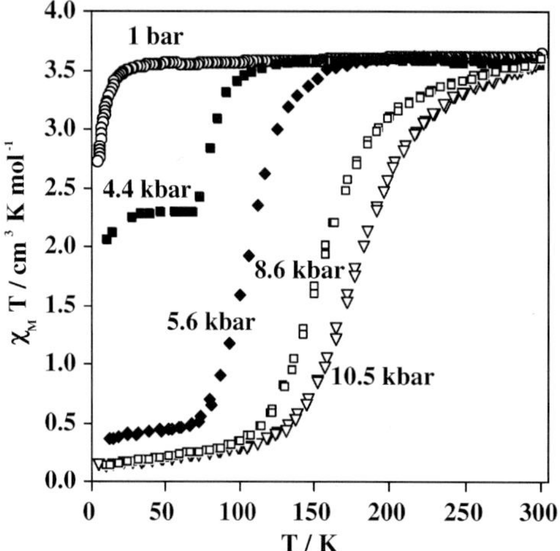

Fig. 4. Temperature dependence of $\chi_M T$ at different pressures for [Fe(*abpt*)$_2$(NCS)$_2$] polymorph B: 1 bar (open circles), 4.4 kbar (black squares), 5.6 kbar (black rhombuses), 8.6 kbar (open squares), 10.5 kbar (open triangles)

various pressures. Compound **1** does not reveal the spin crossover behaviour at ambient pressure and the iron(II) ion is in the HS state in the whole range of temperatures. This behaviour remains as pressure is increased up to 4.4 kbar, where an incomplete thermal spin crossover appears around $T_{1/2} = 65$ K. This $T_{1/2}$ value is one of the lowest temperatures observed for an iron(II) spin-crossover compound. Observation of considerable amounts of trapped HS molecules at low temperature is usually attributed to the occurrence of two different sites in the crystal [6]. One of the sites feels weaker ligand field strength and, consequently, stabilizes HS molecules whereas the other with stronger ligand field stabilizes spin-crossover centres. Texture effects or occurrence of different polymorphs, one of them being paramagnetic in the whole range of temperatures, has also been claimed [1b–d]. However, it is more reasonable to consider here that slow kinetics could block the HS–LS equilibrium, due to the low temperatures involved in the spin transition of

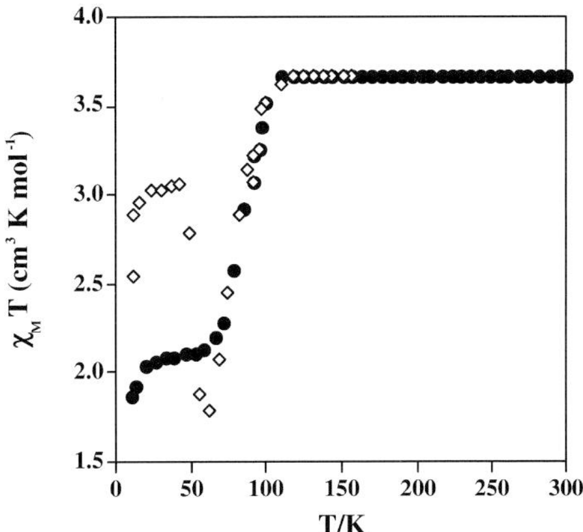

Fig. 5. $\chi_M T$ versus T plots for [Fe(*abpt*)$_2$(NCS)$_2$] polymorph B. Sample was cooled from 300 to 2 K at 2 K min^{-1} (black circles). Thermal quenching at 4.4 kbar: sample was cooled from 300 to 5 K at 100 K min^{-1} and then warmed slowly (open rhombuses, see text)

1 at 4.4 kbar. A relatively sharp spin transition takes place at $T_{1/2} = 106$, 152 and 179 K, as pressure attains 5.6, 8.6 and 10.5 kbar, respectively. In the slow cooling and heating modes with the rate 0.1 K min^{-1} providing the thermodynamic equilibrium conditions, the transitions are accompanied by a 2 K wide thermal hysteresis at all pressures studied.

Dynamics of the Spin Conversion. Thermal Quenching at 4.4 kbar

We have performed thermal quenching experiments at 4.4 kbar by cooling the sample rapidly (100 K min^{-1}) from room temperature down to 5 K to demonstrate that slow kinetics effects are present in **1**. Figure 5 displays the magnetic behaviour of the quenched sample at increasing temperatures. $\chi_M T$ is equal to 2.5 cm^3 K mol^{-1}, at 5 K. This value represents apparently 68% of trapped HS molecules. A progressive increase of $\chi_M T$ was observed as the sample was slowly warmed (2 K min^{-1}). The thermal dependence of $\chi_M T$ attains a maximum value of 3.0 cm^3 K mol^{-1} in the range of temperatures 20–40 K. The increase of $\chi_M T$ corresponds, most probably, to the occurrence of zero-field splitting in the S = 2 ground state of the trapped HS molecules. Hence, the effective thermal quenching at low temperatures actually involves around 84% of molecules in the HS state. In the range of temperatures 50–65 K, $\chi_M T$ diminishes as a consequence of HS–LS relaxation. In this range of temperatures, $\chi_M T$ was registered every 2 K after waiting sufficiently long at each temperature in order to reach thermodynamic equilibrium. The sequence with delays given in parentheses, was the following: 48 K (60 min), 50 K (120 min), 52 K (120 min), 54 K (60 min), 56 K (60 min), 58 K (30 min), 60 K (30 min) and 62 K (15 min). As can be seen in Fig. 5, $\chi_M T$ falls down to ca. 1.7 cm^3 K mol^{-1} at 62 K. Obviously, greater delays would produce smaller $\chi_M T$ values according to a greater extent of the HS-LS relaxation. For the temperatures higher than 62 K, the molecules overcome the energy barrier between the two LS and HS potential wells. Consequently, both, normal regime and thermal quenching experiments coincide above $T = 72$ K.

Discussion

Polymorphs B (compounds **1** and **2**) have a molecular structure very similar to that previously reported for $[Fe(abpt)_2(NCX)_2]$ ($X=$S, Se), polymorphs A. The average Fe–N bond distances, $R[FeN_6]$, are very similar in both polymorphs being larger for polymorph B (see Table 2). The main structural difference of **1** and **2** with respect to polymorphs A, is the intermolecular hydrogen bonding between the amine group of the triazole ring and the nitrogen atom of the uncoordinated pyridine ring that does not take place in the polymorphs B [4, 5, 7]. This is the reason why the crystal packing of **1** and **2** differs from that of polymorphs A. The absence of intramolecular hydrogen bonding in **1** and **2** favours the formation of two-dimensional nets stacking in the [101] direction. The cohesion force inside the plane is determined by the π–π intermolecular interactions between the coordinate and uncoordinated pyridine rings.

The different crystal packing in polymorphs B with respect to polymorphs A, provokes that the Fe–N bond distances in **1** and **2** are longer than in **3** and **4**. This is probably due to steric hindrance induced by the proximity of the uncoordinated pyridine ring of the adjacent abpt ligand to the Fe–N bond of the coordinated pyridine group.

As previously reported $[Fe(abpt)_2(NCX)_2]$ ($X=$S, Se), polymorphs A show thermal spin transition with $T_{1/2}=180(1)$ and 224(2) K [4]. In contrast, the compounds **1** and **2** do not undergo spin transition, they remain in the HS state in the whole range of temperature. The thermal dependence of the magnetic properties of compound **1** has been studied at different pressures. The compound undergoes an incomplete thermal spin transition at 4.4 kbar. The trapping experiments have shown that, in the warming mode of the sample, at rates as slow as 0.028 K min^{-1} the metastable HS state relaxes back to the LS ground state. The relaxation takes place with greater efficiency than in the standard warming procedure where 62% of molecules remain in the HS state at low temperatures. For temperatures greater than 65 K the population of the HS state becomes more favourable and $\chi_M T$ increases. As pressure is increased the spin conversion becomes complete. It should be mentioned that the magnetic properties of **1** at 10.5 kbar are very similar to those of **3** at ambient pressure. Also, it should be noted the linear behaviour of the pressure dependence of $T_{1/2}$ for the spin conversion of **1** (see Fig. 6). The slope of the $T_{1/2}$ *vs* P straight line,

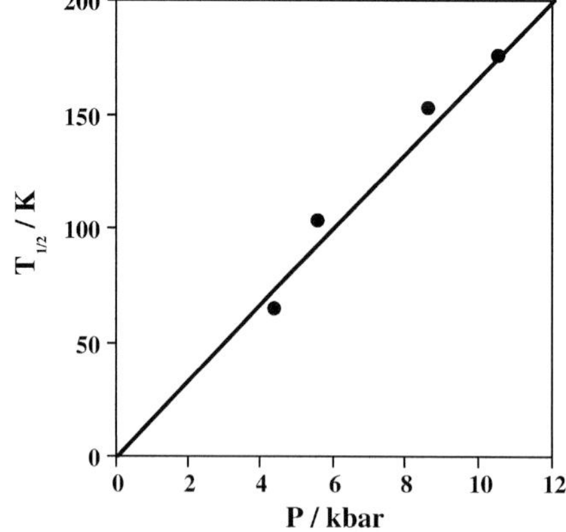

Fig. 6. $T_{1/2}$ versus P plot for **1**

$dT_{1/2}/dP = 17.6\,K\,kbar^{-1}$, is very close to that observed for the mononuclear compound [Fe(2-*pic*)$_3$]$Cl_2 \cdot EtOH$ (2-*pic* = 2-picolylamine) [8] where $dT_{1/2}/dP = 15\,K\,kbar^{-1}$. However, these values of $dT_{1/2}/dP$ are significantly smaller than those observed for the one-dimensional coordination spin crossover polymer [Fe(*hyetrz*)$_3$]$X_2 \cdot 3H_2O$ (*hyetrz* = 4-(2′-hydroxyethyl)-1,2,4-triazole, X = 3-nitrophenylsulphonate), which is around $24\,K\,kbar^{-1}$ [9].

Conclusion

We have reported here a new example of the effect of polymorphism on the spin transition. Two crystal forms, so-called polymorphs A and B, of the same complex exhibit different magnetic properties: polymorphs A undergo thermal spin transition at ambient pressure while polymorphs B are paramagnetic in the whole range of temperature at 1 bar, however, they undergo thermal spin transition at higher pressures. Comparison of the crystal structures of the two polymorphs showed that the differences are mostly due to crystal packing effects rather than to changes in the complex geometry, even though small differences in the Fe–N intramolecular distances and in the conformations of the chelate rings have been evidenced.

Experimental

4-amino-3,5-bis(pyridin-2-yl)-1,2,4-triazole (Aldrich), $FeSO_4 \cdot 7H_2O$ (Panreac) and KNC*X* (X = S, Se) (Aldrich) where purchased from commercial sources and used as received.

[Fe(abpt)$_2$(NCX)$_2$] (X = S (**1**), *Se* (**2**))

To a solution of abpt (0.5 mmol, 119 mg) in methanol (20 ml) was added a water/methanol (1:1) solution (40 ml) of $FeSO_4 \cdot 7H_2O$ (0.25 mmol). The resulting solution was mixed with a solution of KNC*X* (X = S, Se) (0.5 mmol of KNC*X*) in water (20 ml). The final orange solution was filtered and allowed to evaporate for a week, giving orange crystals of **1** and **2** suitable for X-rays studies. All manipulations were performed in an argon atmosphere. Yield = 60%. Anal. calc. for $C_{26}H_{20}N_{14}S_2Fe$ (**1**): C 48.16, H 3.09, N 30.25; found: C 47.95, H 3.07, N 30.08. Anal. calc. for $C_{26}H_{20}N_{14}Se_2Fe$ (**2**): C 42.06, H 2.69, N 26.42; found: C 42.58, H 2.62, N 25.98.

Crystallographic Data

Diffraction data for compounds **1** and **2** were collected at room temperature with an Enraf-Nonius CAD-4 diffractometer using graphite monochromated Mo Kα radiation with the $\omega - 2\theta$ scan method (see Table 1). The unit cell parameters were determined from least squares refinement on the setting angles from 25 centered reflections in the range $12 < \theta < 20°$. No significant fluctuations were observed in the intensities of three standard reflections monitored periodically throughout data collections. The structures were solved by standard *Patterson* methods and refined by the full-matrix least-squares method on F^2. The computations were performed by using SHELXS97 and SHELXL97 [10]. All non-hydrogen atoms were refined anisotropically. The final full-matrix least-squares refinement, minimizing $\Sigma[w(F_o^2 - F_c^2)^2]$ converged at the values of $R1$ and $wR2$ are listed in Table 1. The molecular plots were drawn with the ORTEP program [11].

Magnetic Susceptibility Measurements

The variable-temperature magnetic susceptibility measurements at atmospheric pressure were carried out on microcrystalline samples using a Quantum Design MPMS2 SQUID susceptometer equipped with a 55 kG magnet and operating in the range 0.1–1T and 1.8–300 K. The susceptometer was calibrated with $(NH_4)Mn(SO_4)_2 \cdot 12H_2O$. The set of magnetic measurements performed at variable pressure was carried out in the temperature range 5–320 K on a Foner magnetometer (Pricenton Applied Research PAR 151) equipped with a Bruker electromagnet operating at 1T and a CryoVac liquid helium cryostat. A prototype of the actual construction of a hydrostatic high-pressure cell (HPC) made of hardened beryllium bronze with silicon oil as a pressure-transmitting media was described elsewhere [12, 13]. The cell has the following characteristics: weight 8 g, range of pressure up to 13 kbar, accuracy $\pm$ 0.25 kbar, nonhydrostaticity of pressure is less than 0.5 kbar, sample space is 1 mm in diameter and 5–7 mm in length. The pressure was calibrated using the transition temperature of superconducting tin of purity 99.99%. Experimental susceptibilities were corrected for diamagnetism of the constituent atoms by the use of *Pascal*'s constants.

Acknowledgements

This work was funded by the European Commission trough TMR-Network "Thermal and Optical Switching of Spin States (TOSS)", Contract No. ERB-FMRX-CT98-0199EEC/TMR, the Deutsche Forschungsgemeinschaft, Schwerpunktprogramm "Molekularer Magnetismus", the Fonds der Chemischen Industrie and the Ministerio Español de Ciencia y Tecnología (project BQU 2001-2928).

References

[1] a) Goodwin HA (1976) Coord Chem Rev **18**: 293; b) Gütlich P (1981) Struct Bonding (Berlin) **44**: 83; c) Gütlich P, Hauser A, Spiering H (1994) Angew Chem Int Ed Engl **33**: 2024; d) König E, Ritter G, Kulshreshtha SK (1985) Chem Rev **85**: 219; e) König E (1991) Struct Bonding (Berlin) **76**: 51; f) Lawthers I, McGarvey JJ (1984) J Am Chem Soc **106**: 4280; g) Decurtins S, Gütlich P, Hasselbach KM, Spiering H, Hauser A (1985) Inorg Chem **24**: 2174; h) Decurtins S, Gütlich P, Köhler CP, Spiering H, Hauser A (1984) Chem Phys Lett **105**: 1

[2] Sorai M, Seki S (1974) J Phys Chem Solids **35**: 555

[3] Slichter CP, Drickamer HG (1972) J Chem Phys **56**: 2142

[4] Moliner N, Muñoz MC, Létard S, Létard J-F, Solans X, Burriel R, Castro M, Kahn O, Real JA (1999) Inorg Chim Acta **291**: 279

[5] Moliner N, Gaspar AB, Muñoz MC, Niel V, Cano J, Real JA (2001) Inorg Chem **40**: 3986

[6] a) Poganiuch P, Decurtins S, Gütlich P (1990) J Am Chem Soc **112**: 3270; b) Poganiuch P, Gütlich P (1988) Hyperfine Interact **40**: 331; c) Buchen T, Poganiuch P, Gütlich P (1994) J Chem Soc, Dalton Trans 2285; d) Hinek R, Spiering H, Schollmeyer D, Gütlich P, Hauser A (1996) Chem Eur J **2**: 1127

[7] a) Cornelissen JP, Diemen JH, Groeneveld LR, Haasnoot JG, Spek AL (1992) Inorg Chem **31**: 198; b) Faulmann C, Koningsbruggen PJ, Graaff RAG, Haasnoot JG, Reedijk J (1990) Acta Crystallogr **C46**: 2357; c) García MP, Manero JA, Oro LA, Apreda MC, Cano FH, Foces-Foces C, Haasnot JG, Prins R, Reedijk J (1986) Inorg Chim Acta **122**: 235

[8] Köhler CP, Jakobi R, Meissner E, Wiehl H, Spiering H, Gütlich P (1990) J Phys Chem Solids **51**: 239

[9] Garcia Y, Van Koningsbruggen P, Lapouyade R, Fournés L, Rabardel L, Kahn O, Ksenofontov V, Levchenko G, Gütlich P (1998) Chem Mater **10**: 2426

[10] Sheldrick GM, SHELXS97 (1990) Acta Crystallogr **A46**: 467; SHELXL97 (1997) Program for the Refinement of Crystal Structures, Univ of Götingen, Germany

[11] Johnson CK, ORTEP (1971) Report ORNL-3794, Oak Ridge National Laboratory, Oak Ridge, TN

[12] Dyakonov VP, Levchenko G (1983) Sov J Pribori Texnika Experimenta **5**: 236

[13] Baran M, Levchenko G, Dyakonov VP, Shymchack G (1995) Physica **C241**: 383

Substituent Effects on the Spin-Transition Temperature in Complexes with Tris(pyrazolyl) Ligands

Hauke Paulsen[1,*], **Lars Duelund**[2], **Axel Zimmermann**[3], **Frédéric Averseng**[1], **Michael Gerdan**[1], **Heiner Winkler**[1], **Hans Toftlund**[3], and **Alfred X. Trautwein**[1]

[1] Institut für Physik, Universität zu Lübeck, D-23538 Lübeck, Germany

[2] MEMPHYS – Center for Biomembrane Physics, Department of Chemistry, University of Southern Denmark, DK-5230 Odense, Denmark

[3] Department of Chemistry, University of Southern Denmark, 5230 Odense, Denmark

Received June 26, 2002; accepted July 22, 2002
Published online January 7, 2003

Summary. Iron (II) complexes with substituted tris(pyrazolyl) ligands, which exhibit a thermally driven transition from a low-spin state at low temperatures to a high-spin state at elevated temperatures, have been studied by *Mössbauer* spectroscopy and magnetic susceptibility measurements. From the observed spectra the molar high-spin fraction and the transition temperature have been extracted. All substituents, except for bromine, lead to a decrease of the transition temperature. Density functional calculations have been carried out to compare the experimentally observed shifts of the transition temperature with those derived from theory.

Keywords. Density functional calculations; Magnetic properties; *Mössbauer* spectroscopy; Spin crossover.

Introduction

Spin-crossover complexes exhibit a transition from a low-spin (LS) to a high-spin (HS) state that can be reversibly induced by changing temperature or pressure or by irradiation with light [1]. These complexes are therefore very promising materials for display and memory devices. Gradual transitions, where the molar HS fraction $\gamma_{HS}(p, T)$ at given pressure p changes smoothly over a large interval of temperature T, can be explained with a simple model [2], that explicitly neglects cooperativity:

$$\gamma_{HS}(p,T) = 1/[1 + \exp(\Delta G/k_B T)]. \tag{1}$$

Here ΔG denotes the difference (HS-LS) of the *Gibbs* free energy,

$$G = E_{el} + E_{vib} + pV - TS, \tag{2}$$

* Corresponding author. E-mail: paulsen@physik.uni-luebeck.de

depending on the total electronic energy E_{el}, the vibrational energy E_{vib}, pressure p, volume V, temperature T, entropy S, and, implicitly, on γ_{HS}. This means that Eq. (1) has to be solved iteratively. The molar HS fraction γ_{HS} is used for the determination of many properties of spin-crossover materials, for instance the transition temperature $T_{1/2}$, which is defined by

$$\gamma_{HS}(p, T_{1/2}) = 1/2. \tag{3}$$

The purpose of the present study is the investigation of the influence of substituents on the spin-transition temperature of iron(II) complexes with tris(pyrazolyl) ligands. The starting point for this investigation is a complex formed by an iron(II) center and two tris(pyrazol-1-yl)methane ligands (compound **1**[a], Fig. 1) [3]. A variety of ligands can be obtained by substituting the hydrogens of the pyrazol rings (Scheme 1). In the present study four different ligands have been

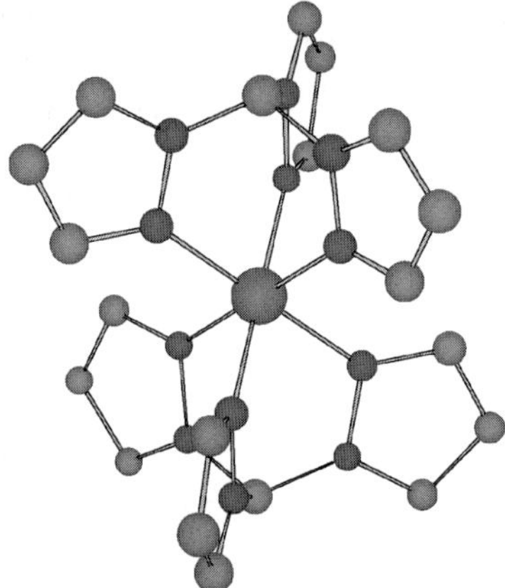

Fig. 1. Molecular structure of the HS isomer of complex **1** derived from geometry optimization; hydrogen atoms have been omitted for clarity

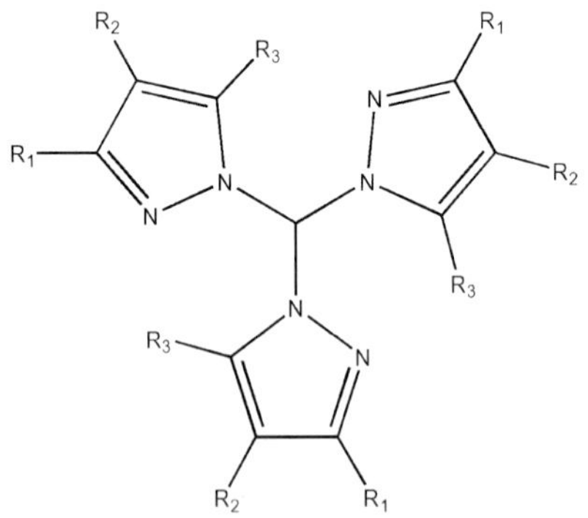

Scheme 1. Drawing of the ligands tris(pyrazol-1-yl)methane (R_1,R_2,R_3 = H), tris(3-methylpyrazol-1-yl)methane ($R_1 = CH_3$, R_2,R_3 = H), tris(4-methyl-pyrazol-1-yl)methane (R_1,R_3 = H, $R_2 = CH_3$), tris(4-bromo-pyrazol-1-yl)methane (R_1,R_3 = H, R_2 = Br), and tris(3,5-dimethylpyrazol-1-yl)methane (R_1,$R_3 = CH_3$, R_2 = H) used for complexes **1**, **2**, **3**, **4**, and **5**, respectively

used: tris(3-methylpyrazol-1-yl)methane (compounds $\mathbf{2}^a$, $\mathbf{2}^b$, and $\mathbf{2}^c$), tris(4-methyl-pyrazol-1-yl)methane (compound $\mathbf{3}^a$), tris(4-bromo-pyrazol-1-yl)methane (compound $\mathbf{4}^b$), and tris(3,5-dimethylpyrazol-1-yl)methane (compound $\mathbf{5}^b$). The superscripts denote the counteranions PF_6^- (a), ClO_4^- (b), and BF_4^- (c).

One purpose of such substitutions is to shift $T_{1/2}$. A general understanding of such substituent effects could help to rationalize the design of spin-crossover complexes with certain properties which are useful for technical applications (e.g. $T_{1/2}$ at ambient temperature and wide hysteresis of the temperature-dependent molar HS fraction).

Mössbauer transmission experiments and magnetic susceptibility measurements have been performed at various temperatures in order to find out whether these complexes exhibit spin-crossover behaviour and to determine $T_{1/2}$. At the same time electronic structure calculations have been performed to estimate $T_{1/2}$ from theory. These calculations have been carried out for isolated complexes *in vacuo*. This approximation neglects interactions between neighbouring complexes and interactions between the complexes and their counterions. Both interactions are known to considerably influence $T_{1/2}$. Calculations which do not regard these interactions can therefore at best be in qualitative agreement with the experiment. Nevertheless, such calculations for isolated complexes *in vacuo* may reveal information about the molecular contribution to substituent-induced shifts of $T_{1/2}$. This information can hardly be gained experimentally since any experiment with a solid sample will only reflect the combined influence of intra- and intermolecular interactions. Calculation of the normal modes of vibrations applying density functional theory (DFT) [4, 5] have demonstrated that the molecular approximation can be useful also for spin-crossover complexes in the solid state. In future attempts we also plan to include intramolecular interactions in our calculations.

Results and Discussion

Mössbauer spectra

The measured *Mössbauer* spectra for compounds $\mathbf{2}^a$, $\mathbf{2}^b$, $\mathbf{2}^c$, $\mathbf{3}^a$, $\mathbf{4}^b$, and $\mathbf{5}^b$ have been fitted using subspectra for the HS and the LS isomers (Fig. 2), *Mössbauer* spectra for compound $\mathbf{1}^a$ have been published earlier [3]. For the low-spin isomers at low temperature the measured quadrupole splittings are in a narrow range between 0.3 and 0.4 mm/s, the observed isomer shifts are in the range between 0.47 and 0.53 mm/s (Table 1). The influence of the substituents on the *Mössbauer* parameters is weak, and no significant influence of the counterions could be observed for compounds $\mathbf{2}^a$, $\mathbf{2}^b$, and $\mathbf{2}^c$. The calculated quadrupole splittings for the LS isomers range from 0.01 to 0.12 mm/s. This is most probably due the fact that the calculations have been performed for molecules *in vacuo* with D_{3d} symmetry. In the solid sample small distortions of the D_{3d} symmetry are expected.

The *Mössbauer* parameters for the HS isomer are less straightforward to compare. At low temperatures usually no or only a small fraction of HS isomers are present leading to large error margins for the measured *Mössbauer* parameters. At higher temperatures, where a large HS fraction is present, a temperature-dependent

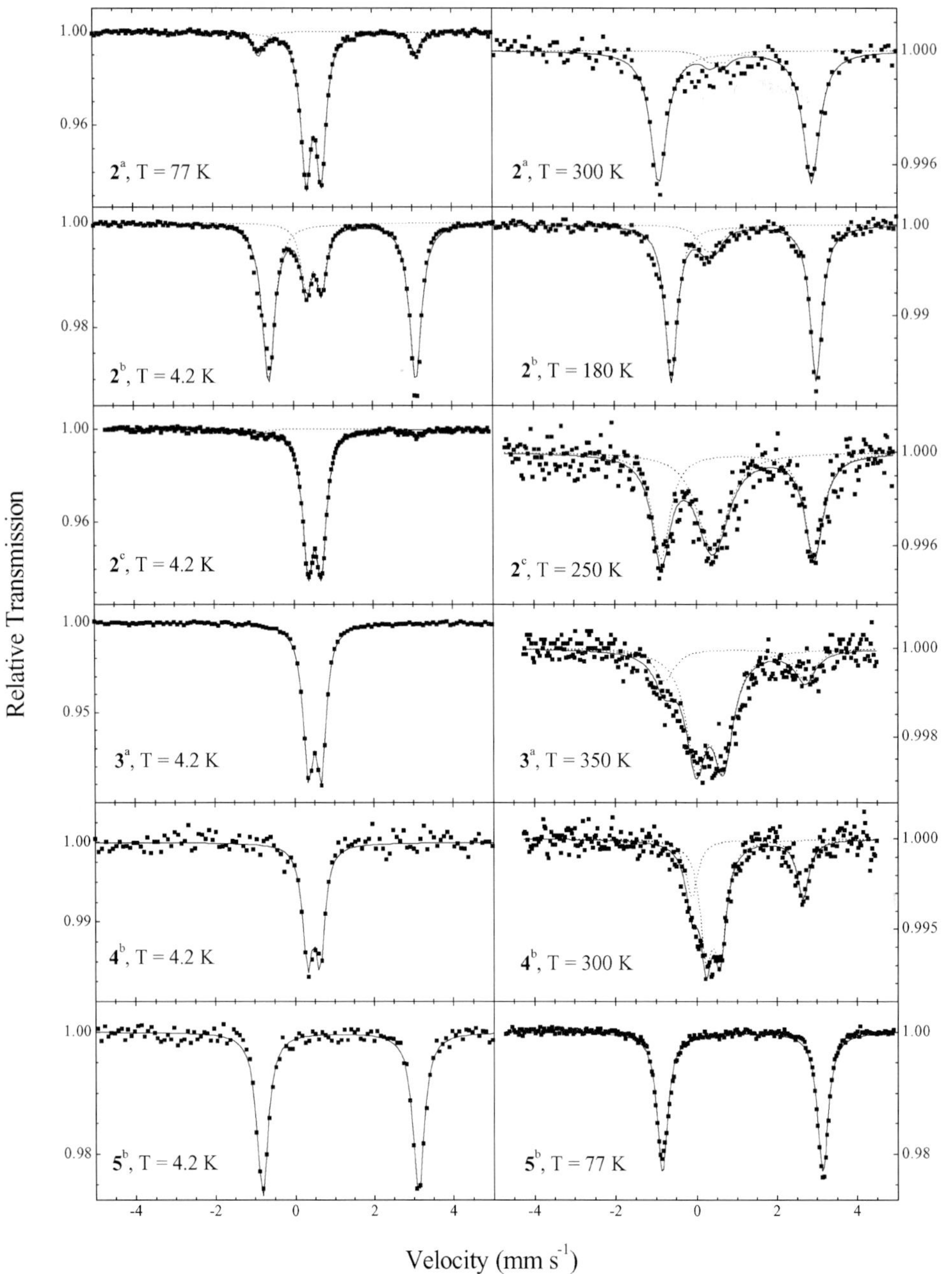

Fig. 2. Mössbauer spectra of different complexes at temperatures as indicated

reduction of ΔE_Q may be observed. This is due to thermal population of low-lying excited electronic states. The observed quadrupole splittings of HS isomers range between 3 and 4 mm/s, except for compound **1**[a], where a quadrupole splitting of 2.2 mm/s is reported [3]. However, due to the high transition temperature of about 355 K the low-temperature value is experimentally very difficult to access. The calculated quadrupole splitting of 3.81 mm/s suggests that the low-temperature quadrupole splitting of compound **1**[a] is quite similar to the

Table 1. Measured quadrupole splitting ΔE_Q and isomer shift δ in mm/s at low temperatures (calculated values for ΔE_Q using the BLYP//6-311G method are given in brackets)

Complex	ΔE_Q		δ	
	LS	HS	LS	HS
1	0.30 (0.10)	2.20[d] (3.81)	0.47	0.85[d]
2[a]	0.39 (0.01)	3.96 (3.75)	0.53	1.13
2[b]	0.43 (0.01)	3.71 (3.75)	0.53	1.24
2[c]	0.36 (0.01)	3.99 (3.75)	0.53	1.13
3[a]	0.32 (0.09)	3.55 (–)	0.51	0.97
4[b]	0.35 (0.12)	3.12 (3.34)	0.48	1.41
5[b]	– (–)	3.99 (–)	–	1.15

[a] With counterion PF_6^-; [b] with counterion ClO_4^-; [c] with counterion BF_4^-; [d] measured at room temperature

corresponding values of the other compounds. For the isomer shift values between 0.9 and 1.4 mm/s have been observed. The calculated quadrupole splittings for the HS isomers are in quantitative agreement with the measured values, considering the uncertainty of the influence of spin-orbit coupling and thermal population of excited states.

The molar HS fraction γ_{HS} can be retrieved in principle from the relation of the area of the HS subspectrum to the area of the LS subspectrum (Fig. 2). For an accurate determination of γ_{HS} the measured areas have to be corrected for the *Lamb-Mössbauer* factors which can be quite different for the HS and the LS isomers. Even if a small fraction of spin-crossover complexes is randomly distributed in a host lattice of other complexes, the molecular contribution to the *Lamb-Mössbauer* factor of the LS isomer can be different from the respective HS value [4, 6].

Magnetic susceptibility

The second experimental technique that has been applied here to monitor the temperature-dependent molar HS fraction are magnetic susceptibility measurements. At room temperature an effective magnetic moment $\mu_{eff} = 4.9\ \mu_B$ is expected for an iron(II) HS isomer, whereas μ_{eff} should vanish for a pure LS isomer. Due to spin-orbit coupling the measured μ_{eff} for HS isomers is usually larger than 5 μ_B, and in case of near degeneracy of the electronic ground state even effective magnetic moments larger than 6 μ_B have been measured [7, 8]. From the temperature-dependent effective magnetic moment the molar HS fraction has been derived as a function of temperature (Fig. 3). The resulting transition temperatures are in the range between 175 and 355 K (Table 2). For the compounds **2**[a] and **5**[b] the transition temperature determined by magnetic susceptibility measurements is close to zero and it might be that these complexes undergo only a partial or no spin-transition. Except for compound **4**[b] which has roughly the same transition temperature as compound **1**[a], all other compounds exhibit a lower transition temperature. Substituting a hydrogen atom of the pyrazol ring by a methyl group, like

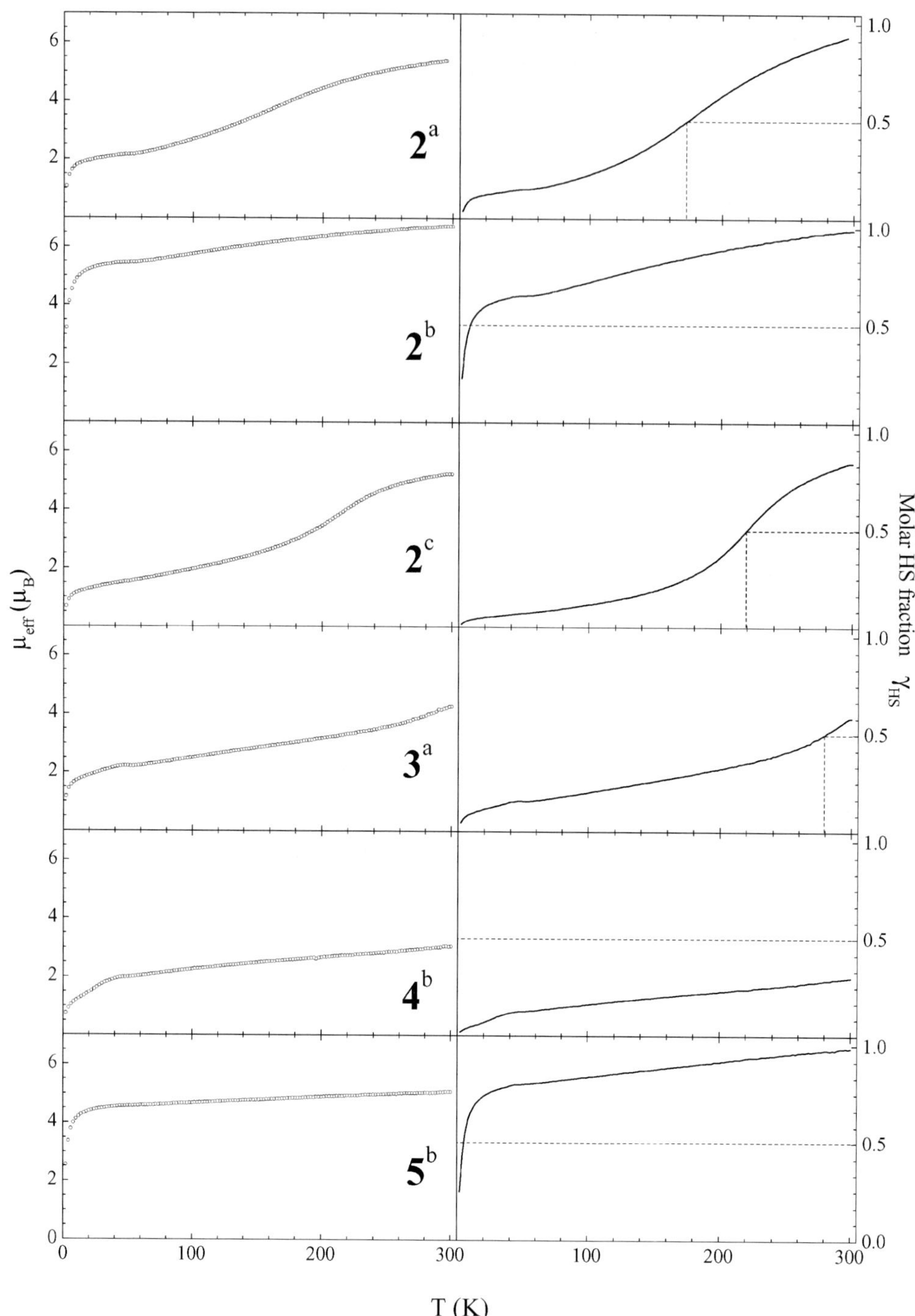

Fig. 3. Temperature dependent effective magnetic moment and derived molar HS fraction of compounds as indicated

in compounds $\mathbf{2}^a$, $\mathbf{2}^b$, $\mathbf{2}^c$, and $\mathbf{3}^a$, decreases $T_{1/2}$. The decrease is even stronger if two hydrogens are substituted as is the case for compound $\mathbf{5}^b$. Comparison of compounds $\mathbf{2}^a$, $\mathbf{2}^b$, and $\mathbf{2}^c$ reveals a significant influence of the counteranion on $T_{1/2}$.

Table 2. $T_{1/2}$ in Kelvin as derived from magnetic susceptibility measurements (Fig. 3). The theoretical values were calculated with BLYP//LANL2DZ and a constant was added afterwards in order to match the experimental value for compound **1**[a]

Complex	Experiment	Theory
1[a]	≈355[a]	355[d]
2[a], **2**[b], **2**[c]	≈0[a], 175[b], 220[c]	<0[e]
3[a]	≈430[a]	
4[b]	≈355[b]	335[e]
5[b]	≈0[b]	

[a] With counterion PF_6^-; [b] with counterion ClO_4^-; [c] with counterion BF_4^-; [d] by definition; [e] see also Eq. (4) and text below

Density functional calculations

From the DFT calculations for complexes **1** to **5** the difference of electronic energy ΔE_{el}, of vibrational energy ΔE_{vib}, and of entropy ΔS can be retrieved. It has been assumed here, that the lowest electronic excitation energy is large compared to $k_B T$, and hence ΔE_{el} is regarded to be constant in the temperature range of interest. The term $p\Delta V$ can not be calculated in the molecular approximation applied here. A typical value of $\Delta V \approx 7\,\text{cm}^3\,\text{mol}^{-1}$ [1] leads to a contribution to the free energy of $p\Delta V \approx 0.7\,\text{J}\,\text{mol}^{-1}$ at ambient pressure and temperature. This is far less than the error margin of the other contributions to ΔG, and the neglect of $p\Delta V$ is therefore justified.

All calculated terms of the free energy depend significantly on the chosen method and basis set. For the temperature-dependent entropy difference of complex **1** the largest observed deviations are between the pure DFT methods like BLYP or PW91 on one side and the *Hartree-Fock* method on the other side (Fig. 4). The calculated curves $\Delta S(T)$ for the other compounds are affected in a similar way by the choice of the computational method. The same is valid for the temperature-dependent vibrational energy difference $\Delta E_{vib}(T)$. The curves calculated for

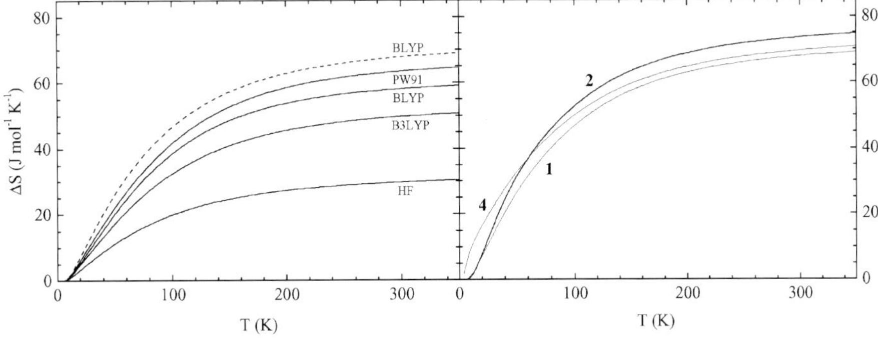

Fig. 4. ΔS for complex **1** as a function of temperature calculated with different methods as indicated (left panel, solid lines refer to the 6-311G basis, the dashed line refers to the LANL2DZ basis) and for complexes **1**, **2**, and **4** calculated with BLYP//LANL2DZ (right panel)

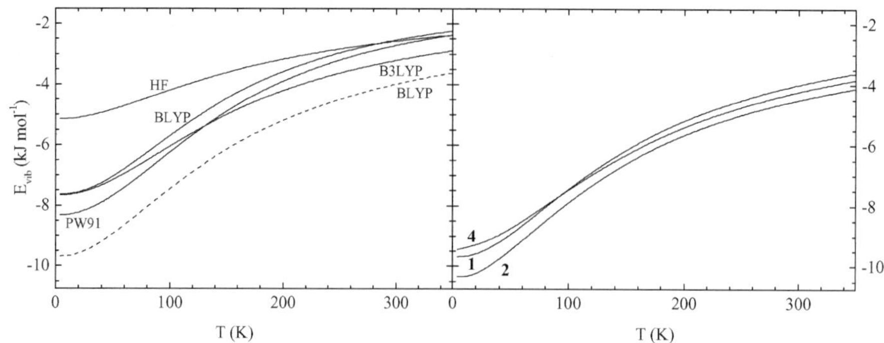

Fig. 5. ΔE_{vib} for complex **1** calculated with different methods as indicated (left panel, solid lines refer to 6-311G basis, dashed line refers to LANL2DZ basis) and for complexes **1**, **2**, and **4** calculated with BLYP//LANL2DZ (right panel)

complexes **1**, **2**, and **4** using the BLYP//LANL2DZ method are almost identical (Fig. 5).

The largest contributions to the free energy arise from the electronic energy difference ΔE_{el} (Table 3). The *Hartree-Fock* method gives negative energy differences and thus fails to predict the correct LS ground state at low temperature. To a lesser extent this is also true for the hybrid method B3LYP which includes about 20% HF contribution to the exchange energy. The reason for this behaviour is that the HF methods, which per definition does not include any electron correlation, favours higher spin-states where the *Pauli* principle forces the electrons to avoid each other [9, 10]. It is interesting to note that the calculated shifts of ΔE_{el}, when going from the unsubstituted complex **1** to the substituted complexes, seem to be of similar order of magnitude for the HF methods as for the DFT methods (Table 3). From the computational methods used in this study the pure DFT methods BLYP and PW91 seem to be the most reliable ones for the calculation of the free energy difference, since these methods always give the correct LS ground state.

Unfortunately the term ΔE_{el}, which dominates the free energy G, is the term with the largest error margin. The reason is that the difference ΔE_{el} is more than five orders of magnitude smaller than the absolute values of E_{el}. Qualitatively correct values of ΔE_{el} can, therefore, only be obtained if the systematic errors

Table 3. ΔE_{el} in kJ mol^{-1} calculated with 6-311G and LANL2DZ basis sets and with the *Hartree-Fock* and different DFT methods (values in brackets give the difference to complex **1**)

Complex	HF		B3LYP		BLYP		PW91	
	6-311G	LANL2DZ	6-311G	LANL2DZ	6-311G	LANL2DZ	6-311G	LANL2DZ
1	−300.158	−280.561	−7.302	9.607	81.721	84.953	110.768	103.552
2			−39.511	−23.853	39.117	43.253		60.423
			(32.210)	(33.459)	(42.604)	(41.701)		(43.129)
4		−280.385				83.252		
		(−0.176)				(−1.701)		

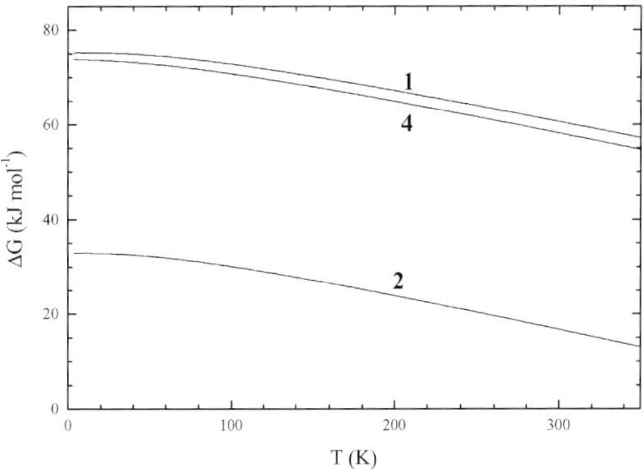

Fig. 6. ΔG for complexes **1**, **2**, and **4** calculated with BLYP//LANL2DZ

inherent in the individually calculated values E_{el} for HS and LS states cancel when the difference ΔE_{el} is formed. For ΔE_{vib} and ΔS the relation between the differences and the absolute values is by more than three orders of magnitude more favourable than for ΔE_{el} and the absolute values E_{el} for HS and LS states, and it has been demonstrated in the past that the normal modes of molecular vibrations, from which ΔE_{vib} and ΔS are derived, can be calculated for such complexes with reasonable accuracy [4–6].

The temperature-dependent free energy difference $\Delta G(T)$ (Fig. 6) is calculated by summing up the terms given in Eq. (2) except for the volume change which is neglected. Ideally, the points of intersection of the curves $\Delta G(T)$ with the abscissa should yield the transition temperatures $T_{1/2}$, according to its definition in Eq. (1). However, it is obvious that the calculated curves are shifted due to errors of ΔE_{el}, which prevents the determination of absolute values for $T_{1/2}$. A rough estimate for the difference $\Delta T_{1/2}$ of transition temperatures when comparing two complexes **a** and **b** can be obtained by the expression

$$\Delta T_{1/2} \approx (\Delta E_{el}{}^{\mathbf{b}} - \Delta E_{el}{}^{\mathbf{a}})/\Delta S(T_{1/2}{}^{\mathbf{a}}), \tag{4}$$

where $T_{1/2}{}^{\mathbf{a}}$ must be known from experiment and $\Delta S^{\mathbf{a}}(T) \approx \Delta S^{\mathbf{b}}(T)$ is assumed. Comparison of the shift of the transition temperature estimated in this way with experimental values (Table 2) yields agreement for the direction of the shift and for the order of magnitude of the shift.

Conclusions

It has been shown for the example of bis-tripodal chelates of iron(II) with tris(pyrazolyl)methane ligands that the temperature of spin transition can be influenced significantly if hydrogen atoms of the pyrazol rings are substituted by methyl groups. These ligands are promising materials for the finetuning of technically interesting properties of spin-crossover complexes. Because of the large variety of possible substitutions it would be very helpful if the effect of these substitutions could be tested by computer simulations with at least qualitative accuracy. The calculations presented here suggest that it should be possible with currently available density functional methods to predict the direction and the order of magnitude of a shift of the transition temperature. The dominating source of error is the calculation of the electronic energy difference. Future calculations with periodic boundary conditions will be performed to study the influence of intermolecular interactions and of counterions.

Materials and Methods

Synthesis

The ligands were synthesized by the method described in Refs. [11–13], which consists of reacting the substituted pyrazol with chloroform under solid liquid phase transfer conditions. The ligands where synthesised by refluxing the substituted pyrazol, finely powdered potassium carbonate and tetrakis(n-butyl)ammonium bromide in chloroform for approximately 30 h. After filtering of the solid potassium carbonate the solution was purified by different methods. Tris(3,5-dimethylpyrazol-1-yl)methane was purified by stirring the solution with charcoal, washing with hexane and after removing the solvent sublimating (420 K, 0.1 Torr). Tris(3-methylpyrazol-1-yl)methane, tris(4-methyl-pyrazol-1-yl)methane, and tris(4-bromo-pyrazol-1-yl)methane were purified by removing the chloroform under reduced pressure and crystallization from diethyl ether at 5°C.

^{1}H-NMR ($CDCl_3$, ppm relative to TMS): tris(3,5-dimethylpyrazol-1-yl)methane: 8.07, 5.87, 2.18, 2.01, tris(3-methylpyrazol-1-yl)methane: 8.21, 7.55, 7.32, 6.11, 2.40, 2.28, tris(4-methyl-pyrazol-1-yl)methane: 8.17, 7.48, 7.32, 2,08, tris(4-bromo-pyrazol-1-yl)methane: 8.22, 7.65, 7.27, 2.17.

^{13}C-NMR ($CDCl_3$, ppm) tris(3,5-dimethylpyrazol-1-yl)methane: 148.77, 140.77, 107.62, 80.77, 13.82, 10.76, tris(3-methylpyrazol-1-yl)methane: 141.11, 120.73, 107.05, 80.75, 14.15, 10.75, tris(4-methyl-pyrazol-1-yl)methane: 142.42, 127.71, 117.64, 83.21, 8.85, tris(4-bromo-pyrazol-1-yl) methane: 142.9, 129.71, 96.24, 83.84.

The iron(II) complexes were synthesized in ethanol by mixing stochiometric amounts of the ligand and either $Fe(ClO_4)_2$ for the perchlorate salt, or $Fe(SO_4)_2 \cdot 7H_2O$ for the other salts. The appropriate anion was then added and the complex was obtained as a powder, which was washed with ethanol and water. For all complexes satisfactory elementary analysis where obtained.

Spectroscopy

Conventional *Mössbauer* spectra were obtained in transmission geometry. The ^{57}Co[Rh] source was driven with constant acceleration. The energy calibration was performed with α-iron at room temperature and the isomer shift is relative to this standard.

For the magnetic measurements a SQUID magnetometer of the type MPMS Quantum Design was used. The applied field was 1 T for all temperatures. The magnetic susceptibility measurements were performed at various temperatures between 2 K and 300 K by decreasing and increasing the temperature in steps of 2 K. The effective magnetic moment μ_{eff} expressed in units of the *Bohr* magneton μ_{B} was derived from the experimental data by

$$\mu_{\text{eff}} = \mu_{\text{B}} \sqrt{8T(w_{\text{mol}}\chi_{\text{mass}} - \chi_{\text{D}})}. \qquad (5)$$

In this expression χ_{D} denotes the molar diamagnetic susceptibility which was estimated from *Pascals* constants [14], w_{mol} is the molecular weight of the compounds and χ_{mass} the experimental mass susceptibility at temperature T.

NMR spectra where recorded on a 300 MHz Bruker AC 300.

Density functional calculations

Several different density funtional methods were used: (i) *Perdew* and *Wang*'s exchange functional and their gradient corrected correlation functional [15] (PW91), (ii) *Becke*'s exchange functional [16] together with *Perdew*'s gradient corrected correlation functional [17] (BP86), (iii) *Becke*'s exchange functional [16] using the correlation functional of *Lee*, *Yang*, and *Parr* [18, 19] (BLYP), and (iv) *Becke*'s three parameter hybrid functional [20] using *Lee*, *Yang*, and *Parr*'s correlation functional [18, 19] (B3LYP). We used the following basis sets: (i) the 6-311 + G(2d,p) basis for H, C, and N and the *Wachters-Hay* double-ζ basis for Fe [21, 22] (6-311 for short), (ii) the *Dunning-Huzinaga* all-electron double-ζ basis for H, C, and N and the Los Alamos effective core potential plus double-ζ basis set on Fe [23, 24] (LANL2DZ). The calculations for the anion were performed with the program packages Gaussian 98 [25] and Turbomole [26].

All calculations were performed for molecules *in vacuo*. The total energy E_{el} for HS and LS states was calculated after full geometry optimization for the respective spin states. For part of the complexes the vibrational frequencies were calculated in order to determine $E_{\mathrm{vib}}(T)$ and $S(T)$ according to the relations

$$E_{\mathrm{vib}}(T) = k_{\mathrm{B}}T \sum_i x_i \coth(x_i) \tag{6}$$

and

$$S(T) = k_{\mathrm{B}} \sum_i \{x_i \coth(x_i) - \ln[2\sinh(x_i)]\} \tag{7}$$

using the definition $x_i = \hbar\omega_i/2k_{\mathrm{B}}T$ and denoting the angular frequency of the ith vibrational normal mode by ω_i.

Acknowledgements

Financial support by the Deutsche Forschungsgemeinschaft (DFG) within the priority programme *Molecular Magnetism* is gratefully acknowledged. MEMPHYS – Center for Biomembrane Physics is supported by the Danish National Research Foundation.

References

[1] Gütlich P, Hauser H, Spiering H (1994) Angew Chem, Int Ed Engl **33**: 2024
[2] Gütlich P, Köppen H, Link R, Steinhäuser HG (1979) J Chem Phys **70**: 3977
[3] Winkler H, Trautwein AX, Toftlund H (1992) Hyperfine Interact **70**: 1083
[4] Paulsen H, Winkler H, Trautwein AX, Grünsteudel H, Rusanov V, Toftlund H (1999) Phys Rev B **59**: 975
[5] Paulsen H, Benda R, Herta C, Schünemann V, Chumakov AI, Duelund L, Winkler H, Toftlund H, Trautwein AX (2001) Phys Rev Lett **86**: 1351
[6] Paulsen H, Grünsteudel H, Meyer-Klaucke W, Gerdan M, Winkler H, Toftlund H, Trautwein AX (2001) Eur Phys J B **23**: 463
[7] Ding X-Q, Paulsen H, Grodzicki M, Butzlaff Ch, Trautwein AX, Hartung R, Wieghardt K (1994) Hyperfine Interact **90**: 485
[8] Paulsen H, Ding X-Q, Grodzicki M, Butzlaff Ch, Trautwein AX, Hartung R, Wieghardt K (1994) Chem Phys **184**: 1
[9] Harris D, Loew GH, Kormonicki A (1997) J Phys Chem A **101**: 3959
[10] Paulsen H, Duelund L, Winkler H, Toftlund H, Trautwein AX (2001) Inorg Chem **40**: 2201
[11] Juliá S, del Mazo JM, Avila L, Elguero J (1984) Organic preparations and procedures int **16**: 299
[12] Reger DL, Little CA, Rheingold AL, Lam K-C, Concolino T, Mohan A, Long GJ (2000) Inorg Chem **39**: 4674
[13] Anderson PA, Astley T, Hitchman MA, Keene FR, Moubaraki B, Murray KS, Skelton BW, Tiekink ERT, Toftlund H, White AH (2000) J Chem Soc Dalton Trans 3505
[14] Kahn O (1993) Molecular Magnetism, VCH Publishers, New York

[15] Perdew JP, Burke K, Wang Y (1996) Phys Rev B **54**: 16533
[16] Becke AD (1988) Phys Rev A **38**: 3098
[17] Perdew JP (1986) Phys Rev B **33**: 8822
[18] Lee C, Yang W, Parr RG (1988) Phys Rev B **37**: 785
[19] Miehlich B, Savin A, Stoll H, Preuss H (1989) Chem Phys Lett **157**: 200
[20] Becke AD (1993) J Chem Phys **98**: 5648
[21] Wachters AJH (1970) J Chem Phys **52**: 1033
[22] Hay PJ (1977) J Chem Phys **66**: 4377
[23] Dunning Jr TH, Hay PJ (1976) in Modern Theoretical Chemistry, Ed Schaefer III HF, Plenum, New York
[24] Hay PJ, Wadt WR (1985) J Chem Phys **82**: 270
[25] Gaussian 98, Revision A7, Frisch MJ, Trucks GW, Schlegel HB, Scuseria GE, Robb MA, Cheeseman JR, Zakrzewski VG, Montgomery Jr JA, Stratmann RE, Burant JC, Dapprich S, Millam JM, Daniels AD, Kudin KN, Strain MC, Farkas O, Tomasi J, Barone V, Cossi M, Cammi R, Mennucci B, Pomelli C, Adamo C, Clifford S, Ochterski J, Petersson GA, Ayala PY, Cui Q, Morokuma K, Malick DK, Rabuck AD, Raghavachari K, Foresman JB, Cioslowski J, Ortiz JV, Baboul AG, Stefanov BB, Liu G, Liashenko A, Piskorz P, Komaromi I, Gomperts R, Martin RL, Fox DJ, Keith T, Al-Laham MA, Peng CY, Nanayakkara A, Gonzalez C, Challacombe M, Gill PMW, Johnson B, Chen W, Wong MW, Andres JL, Gonzalez C, Head-Gordon M, Replogle ES, Pople JA (1998) Gaussian Inc, Pittsburgh PA
[26] Ahlrichs R, Bär M, Häser M, Horn H, Kölmel C (1989) Chem Phys Lett **162**: 165

Magneto-Structural Correlations in Trinuclear Cu(II) Complexes: A Density Functional Study

Antonio Rodríguez-Fortea[1,3], **Eliseo Ruiz**[1,3,*], **Pere Alemany**[2,3], and **Santiago Alvarez**[1,3]

[1] Departament de Química Inorgànica, Universitat de Barcelona, E-08028 Barcelona, Spain
[2] Departament de Química Física, Universitat de Barcelona, E-08028 Barcelona, Spain
[3] Centre de Recerca en Química Teòrica (CeRQT), Universitat de Barcelona, E-08028 Barcelona, Spain

Received June 28, 2002; accepted July 22, 2002
Published online January 8, 2003

Summary. Density functional theoretical methods have been used to study magneto-structural correlations for linear trinuclear hydroxo-bridged copper(II) complexes. The nearest-neighbor exchange coupling constant shows very similar trends to those found earlier for dinuclear compounds for which the Cu–O–Cu angle and the out of plane displacement of the hydrogen atoms at the bridge are the two key structural factors that determine the nature of their magnetic behavior. Changes in these two parameters can induce variations of over $1000\,cm^{-1}$ in the value of the nearest-neighbor coupling constant. On the contrary, coupling between next-nearest neighbors is found to be practically independent of structural changes with a value for the coupling constant of about $-60\,cm^{-1}$. The magnitude calculated for this coupling constant indicates that considering its value to be negligible, as usually done in experimental studies, can lead to considerable errors, especially for compounds in which the nearest-neighbor coupling constant is of the same order of magnitude.

Keywords. Broken symmetry; Density functional calculations; Exchange coupling; Magnetic properties; Trinuclear transition metal complexes.

Introduction

One of the most prominent families of compounds in modern coordination chemistry is that of the polynuclear Cu(II) complexes which have been extensively studied in recent decades, specially in regard with their magnetic properties and the dependence of those properties on the molecular structure [1–8]. Although

* Corresponding author. E-mail: eliseo.ruiz@qi.ub.es

most attention has been focused either on the simplest representatives of this family, *i.e.*, dinuclear compounds, or on polymeric compounds with infinite chains, the presence of trinuclear or tetranuclear arrays of copper atoms at the active sites of enzymes has lead to an increased interest in compounds with intermediate nuclearity [9–15]. From the point of view of molecular magnetism these complexes are also particularly interesting since their intermediate position between the simplest dinuclear species and bulk materials may result in completely new magnetic properties.

Within the large family of polynuclear Cu(II) compounds that have been magnetically characterized, hydroxo-bridged dinuclear complexes have featured prominently in these studies, with *Hodgson* and *Hatfield* noting a correlation between the exchange coupling constant and the Cu–O–Cu angle [16, 17]. Evidences that other structural parameters may influence both the magnitude and the sign of the coupling constant continues to stimulate the interest in these compounds, especially those which feature unusual $Cu_2(\mu\text{-}OH)_2$ core geometries. All these studies have led to now well-established magneto-structural correlations for dinuclear compounds that allow, at least in a semiquantitative manner, to predict the sign and the magnitude of exchange coupling.

However, the situation for compounds of higher nuclearity is still quite obscure. The usual procedure adopted in most of the published studies is to neglect next-nearest neighbor coupling constants, and to assume that the dominant nearest-neighbor constants behave with respect to structural changes in a very similar manner as in analogous dinuclear compounds [18–23]. In this contribution we adopt a computational approach to shed some light on magneto-structural correlations for trinuclear compounds. Extensive studies of this kind on hydroxo-bridged dinuclear Cu(II) compounds have been successfully employed in the evaluation of exchange coupling constants and their relation to molecular structure [24, 25]. We think that even if there are, up to the present, no reported experimental studies for their trinuclear analogues, these compounds form an excellent testing ground for magneto-structural correlations in compounds of higher nuclearity.

Results and Discussion

Evaluation of Coupling Constants in Trinuclear Systems

Exchange coupling in a polynuclear compound is usually described at a phenomenological level by a *Heisenberg* Hamiltonian:

$$\hat{H} = -\sum_{i>j} J_{ij}\hat{S}_i\hat{S}_j \tag{1}$$

where J_{ij} are the exchange coupling constants between each pair of coupled paramagnetic centers and $\hat{S}_i$, $\hat{S}_j$ are the spin moments of the different paramagnetic centers. The set of coupling constants involved in this expression can be obtained experimentally by fitting the measured magnetic susceptibility *versus* temperature curve to the theoretical expression derived for a system that obeys Eq. (1). In the following we will briefly describe a simple strategy to evaluate computationally the values for the set of n different coupling constants in a polynuclear compound using standard DFT calculations. For this purpose we will need first the energies corresponding to $n+1$ different spin distributions of our compound.

As an illustration of this procedure we will start with the simplest possible case, a dinuclear compound with two identical paramagnetic centers with $S_i = 1/2$. The *Heisenberg* Hamiltonian used to fit the experimental data in this case contains a single coupling constant J_{12}:

$$\hat{H} = -J_{12}\hat{S}_1\hat{S}_2 \tag{2}$$

For this system we need to consider the two spin distributions shown schematically in **1**, which correspond to a ferromagnetic high spin (Φ_{HS}) and an antiferromagnetic low spin state (Φ_{LS}), respectively. The diagonal elements of the *Heisenberg* Hamiltonian are:

$$E_{HS} = \langle\Phi_{HS}|\hat{H}|\Phi_{HS}\rangle = -\frac{J_{12}}{4} \tag{3a}$$

and

$$E_{LS} = \langle\Phi_{LS}|\hat{H}|\Phi_{LS}\rangle = +\frac{J_{12}}{4} \tag{3b}$$

By taking the difference of these two equations we can evaluate the exchange coupling constant for this case:

$$E_{HS} - E_{LS} = -\frac{J_{12}}{2} \tag{4}$$

The calculation of E_{HS} and E_{LS} using quantum chemical methods is, however, not straightforward. The main problem arises when one wants to evaluate the energy for the spin distribution corresponding to the antiferromagnetic state. The single determinant Φ_{LS} is not a proper wave function of the $\hat{S}^2$ operator and it corresponds to a broken symmetry solution. Within the *Hartree-Fock* methodology it has been demonstrated that using spin-projection techniques one arrives at an expression that is identical to [4]. However, in previous works we have shown that the use of DFT-based calculations in this equation leads to poor results. Excellent agreement with experimental values for a large variety of compounds is obtained if no spin projection is applied. The resulting equation in this case is:

$$E_{HS}^{DFT} - E_{LS}^{DFT} = -J_{12} \tag{5}$$

The reader interested in the origin of the differences between both approaches may find an extensive discussion on this topic in references [24, 26, 27].

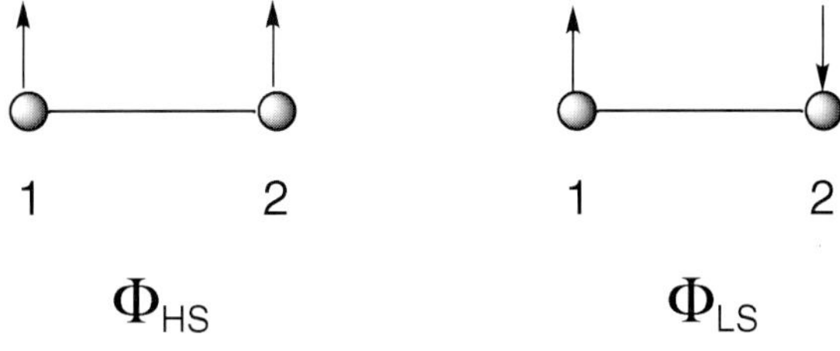

1

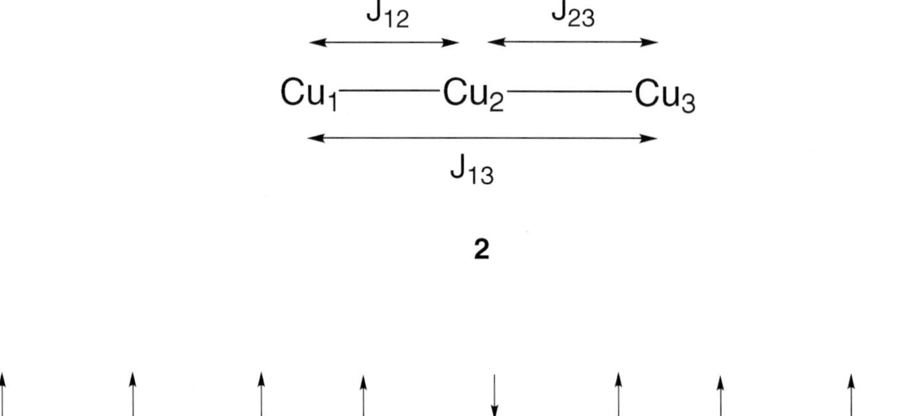

2

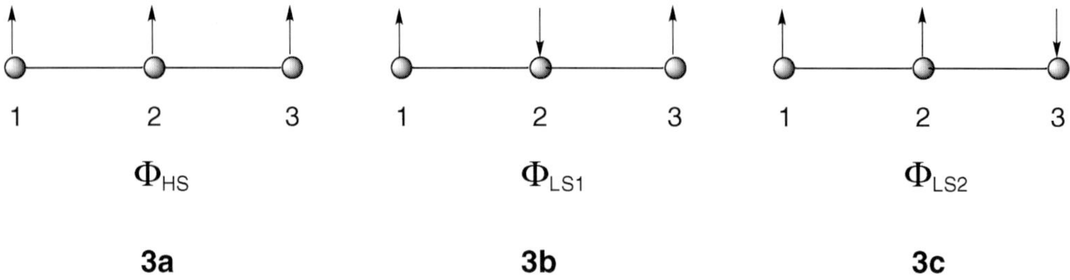

3a **3b** **3c**

Generalization of these equations to the case in which the two paramagnetic centers bear different spins, S_1 and S_2, respectively, can be obtained by introducing a new factor in Eqs. (4) and (5):

$$E_{\mathrm{HS}} - E_{\mathrm{LS}} = -(2S_1S_2)J_{12} \tag{6}$$

and

$$E_{\mathrm{HS}}^{\mathrm{DFT}} - E_{\mathrm{LS}}^{\mathrm{DFT}} = -(2S_1S_2 + S_2)J_{12} \tag{7}$$

respectively, where $S_1 > S_2$ [28].

Extension of this procedure to obtain the set of coupling constants for a polynuclear compound with more than two paramagnetic centers is quite straightforward and it will be illustrated here in the case of trinuclear Cu(II) compounds in which each paramagnetic center bears a single unpaired electron ($S_i = 1/2$). For a symmetric compound in which two of the three copper atoms are equivalent (**2**), the exchange coupling is defined by two independent coupling constants $J_{12} = J_{23}$ and J_{13}. In this case we must evaluate the energy for three different spin distributions in order to obtain two equations that relate the values of the coupling constants with the evaluated energies. The most convenient procedure to find these two equations is based on the fact that the *Heisenberg* Hamiltonian for the polynuclear system [1] is built as a simple sum of pairwaise spin interactions. If we consider the high-spin state shown schematically in **3a** and the low-spin states **3b** it is easy to see that the energy difference between these two states will have only two contributions. The first one will arise from the pair $S_1 \cdots S_2$ which is in a parallel alignment in **3a** and in an antiparallel one in **3b**. The same occurs for the spin pair $S_2 \cdots S_3$. Since the energy difference between the parallel and antiparallel alignments within each pair is given by Eq. (5) using a DFT-based method, the total energy difference between the two spin arrangements will be

$$E_{\mathrm{HS}}^{\mathrm{DFT}} - E_{\mathrm{LS1}}^{\mathrm{DFT}} = -2J_{12} \tag{8}$$

It is important to note that J_{13} does not appear in this equation because we have the same alignment for the $S_1 \cdots S_3$ pair in both spin distributions.

The equation needed to evaluate J_{13} can be obtained from the high-spin solution **3a** and the low-spin distribution **3c**. In this case the two spin pairs that contribute to the energy difference are $S_2 \cdots S_3$ and $S_1 \cdots S_3$. Using Eq. (5) for each pair, the total energy difference is:

$$E_{\mathrm{HS}}^{\mathrm{DFT}} - E_{\mathrm{LS2}}^{\mathrm{DFT}} = -J_{12} - J_{13} \tag{9}$$

Note that the coupling constant between nearest neighbors, J_{12}, is independent from the energy of the intermediate state Φ_{LS2}. In case we want to obtain only this value (J_{13} is usually taken to be strictly zero in many experimental studies) we would need only to compute the energies of the high spin and the low-spin state Φ_{LS1}.

Computational Details

All DFT calculations were performed using the hybrid B3LYP functional [29] as implemented in the Gaussian98 package [30]. Basis sets proposed by *Schaefer et al.* were employed throughout, of triple-ζ quality for the transition metal atoms [31] and double-ζ quality for all other atoms [32]. Due to the small magnitude of the exchange coupling constants all energy calculations must be performed including the SCF = *Tight* option of Gaussian to ensure sufficiently well converged values for the state energies.

Magneto-Structural Correlations for Trinuclear Hydroxo-Bridged Cu(II) Compounds

In previous works we have performed a detailed computational analysis of the influence of the molecular geometry on the exchange coupling constants for hydroxo- and alkoxo-bridged Cu(II) dinuclear compounds using the model compound shown in **4** (for the alkoxo-bridged compounds the hydrogen atoms on the bridges are replaced by methyl groups) [24, 25]. The reader must be aware that in those publications we employed a slightly different *Heisenberg* Hamiltonian using $2J$ instead of J as multiplicative factor. The most important conclusion obtained in those works is that two key structural parameters dictate the nature of the exchange coupling in these compounds: the Cu–O–Cu angle θ, and the out-of-plane displacement of the hydrogen atom (or alkoxo group) at the bridge, measured by the angle τ. The general conclusion is that for large values of the Cu–O–Cu angle ($\theta > 100°$) and the hydrogen atoms lying in the plane ($\tau = 0°$) strong antiferromagnetic coupling is expected ($J_{12} < -400\,\mathrm{cm}^{-1}$). If θ is decreased or if the hydrogen atoms are pulled out of the Cu_2O_2 plane, the antiferromagnetic character is decreased. Ferromagnetic coupling can be achieved, for instance, in hydroxo-bridged compounds with $\theta < 98°$ and $\tau = 60°$. A detailed analysis of the variation of the ground state energy with the molecular geometry at the Cu_2O_2 ring shows, however, that both parameters are indeed not independent, so that small values of θ imply large out of plane deviations

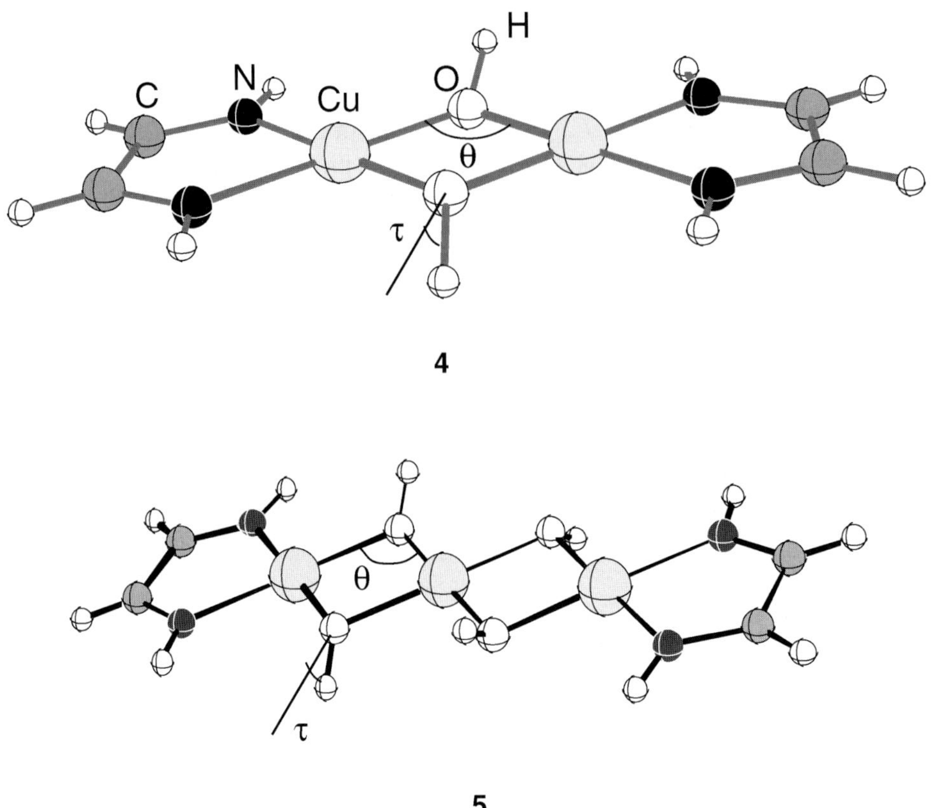

of the hydrogen atom or the alkyl groups at the bridge. Although the same qualitative behavior is predicted for hydroxo and alkoxo-bridged compounds, the latter show stronger antiferromagnetic coupling for geometries with the same pair of (θ, τ) values.

The main aim of this work is to find out if these magneto-structural correlations still hold for analogous linear trinuclear compounds. In this case the additional coupling constant between next-nearest neighbors and its variation with the molecular geometry must be also considered. The model employed for the calculations on trinuclear compounds **5** is just an extension of that used earlier for dinuclear compounds (see **4**), keeping the nature of the terminal ligands and all geometric parameters the same [24, 25].

The nearest-neighbor exchange coupling constant J_{12} for the hydroxo-bridged trinuclear complex shows a very similar dependence on θ (Fig. 1) to that found for the dinuclear model. Antiferromagnetic coupling is expected for all values of θ if the hydrogen atoms on the bridge are kept on the Cu_2O_2 plane. The absolute values of J_{12} are very similar to those calculated for the dinuclear model, with the largest deviation found for large values of θ. Pulling the hydrogen atoms out of the plane decreases the antiferromagnetic coupling between nearest neighbors, in the same way as in dinuclear compounds. Nearest-neighbor ferromagnetic coupling is expected for $\theta < 99°$ and $\tau > 60°$.

Exchange coupling between next-nearest neighbors is practically unaffected by the distortion of the molecular geometry (Fig. 2). An average value of

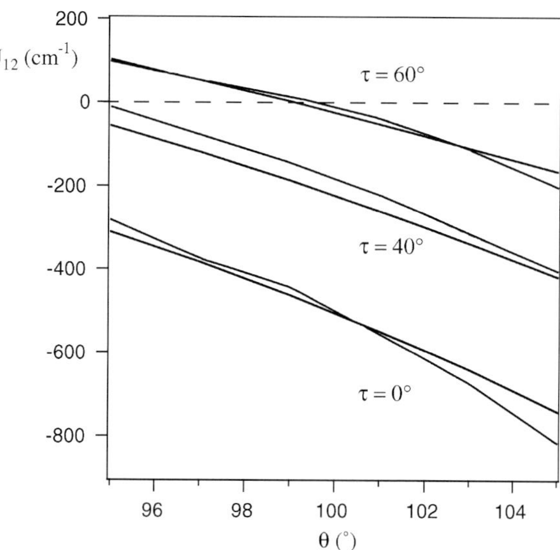

Fig. 1. Nearest-neighbor coupling constant J_{12} calculated for the trinuclear model compound with different values of τ; for comparison, the values obtained for the coupling constant in a dinuclear model with the same geometry are also indicated with dashed lines

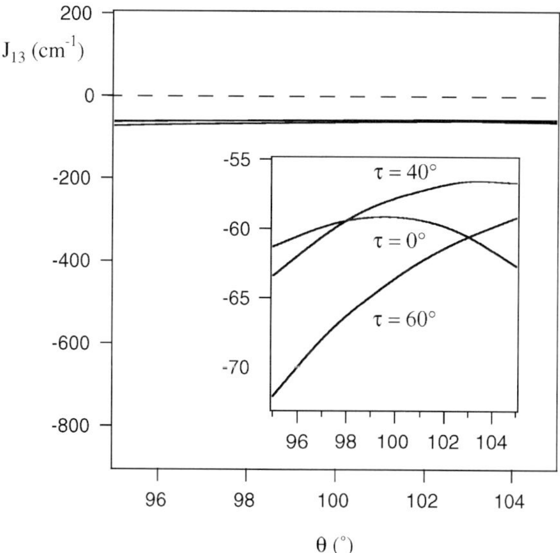

Fig. 2. Next-nearest-neighbor coupling constant J_{13} calculated for the trinuclear model compound with different values of τ; the same curves are shown in the inset using an expanded scale for the coupling constant

approximately $-60\,cm^{-1}$ is observed for J_{13} in all cases. The magnitude of this parameter indicates that although roughly constant, the usual approximation in experimental studies which consider this value as negligible is, at least, doubtful. A detailed analysis shows that J_{13} is more sensitive to geometry changes for cases with larger τ values. It is interesting to note that in trinuclear compounds with the

Table 1. Calculated exchange coupling constants (cm^{-1}) using trinuclear models in which one of the Cu(II) ions is replaced by a diamagnetic Zn(II) ion. Coupling constants obtained for the trinuclear Cu(II) model with the same molecular geometry are given in parenthesis

θ	τ	J_{12}	J_{13}
95°	60°	+ 82.7 (+ 97.8)	– 90.3 (– 72.1)
101°	60°	– 88.5 (– 53.4)	– 90.8 (– 62.4)
105°	0°	– 833.4 (– 740.0)	– 129.8 (– 62.7)

hydrogen atoms far from the Cu_2O_2 plane, the situation in which the values of J_{12} and J_{13} are similar is possible. For the compound with $\theta = 101°$ and $\tau = 60°$, for example, the computed values of J_{12} and J_{13} are -88.5 and $-90.8\,cm^{-1}$, respectively.

An alternative way to estimate the exchange coupling constants in a trinuclear compound is the replacement of one of the three Cu(II) paramagnetic centers by a diamagnetic ion such as Zn(II). These Cu_2Zn models are equivalent in their magnetic behavior to dinuclear Cu(II) complexes for which the coupling constant can be evaluated using Eq. (5). If one of the terminal Cu atoms is replaced by Zn in the model, the calculation yields an approximation to the nearest neighbor coupling constant J_{12}. On the other hand, if the central Cu atom is replaced, the computed exchange coupling constant is comparable to J_{13} in the trimer. Calculations of the coupling constants by this alternative procedure for three significative structures (Table 1) confirm the general trends found for the coupling constants evaluated for the copper trinuclear compounds, although slightly stronger antiferromagnetic coupling is found in all cases. The only case in which a significant discrepancy between the two methods is found is that of $\theta = 105°$ and $\tau = 0°$, for which the value for J_{13} in the model with Zn is twice that found for the trinuclear copper compound.

Concluding Remarks

Nearest-neighbor exchange coupling in trinuclear hydroxo-bridged copper(II) complexes shows very similar trends to those found earlier for dinuclear compounds for which the Cu–O–Cu angle and the out of plane displacement of the hydrogen atoms at the bridge are the two key structural factors that determine their magnetic behavior. A variation of more than $1000\,cm^{-1}$ can be obtained by a suitable change in these two parameters. Coupling between next-nearest neighbors is, on the contrary, found to be nearly independent from structural changes in the Cu_2O_2 core. The usual approach of considering a practically negligible coupling constant between the terminal atoms in linear trinuclear compounds is, however, questioned by the relatively important value of the coupling constant predicted by our calculations. The good agreement between calculated and experimental coupling constants obtained using the same computational scheme for other dinuclear and tetranuclear compounds gives us confidence on the general trends found for the case of linear trinuclear hydroxo-bridged copper(II) compounds for which no experimental data have been published.

Acknowledgments

Financial support of this work was provided by *Dirección General de Enseñanza Superior* (*DGES*) through project number PB98-1166-C02-01 and by *Comissió Interdepartamental de Ciència i Tecnologia* (*CIRIT*) through grant SGR99-0046. The computing resources at *CESCA/CEPBA* were generously made available through grants from *CIRIT* and *Universitat de Barcelona*.

References

[1] Coronado E, Delhaes P, Gatteschi D, Miller JS (1996) Magnetic Molecular Materials: From Molecular Assemblies to the Devices, Kluwer, Dordrecht
[2] Gatteschi D, Kahn O, Miller JS, Palacio F (1991) Magnetic Molecular Materials, Kluwer Academic, Dordrecht
[3] Gatteschi D, Sessoli R, Cornia A (2000) Chem Commun: 725
[4] Kahn O, Pei Y, Journaux Y (1996) Molecular Inorganic Magnetic Materials. In: Bruce DW, O'Hare D (eds) Inorganic Materials, 2nd ed, John Wiley & Sons, Chichester
[5] Kahn O (1993) Molecular Magnetism, VCH Publishers, New York
[6] Kahn O (1996) Magnetism: A Supramolecular Function, Kluwer, Dordrecht
[7] Miller JS, Drillon M (eds) (2001) Magnetism: Molecules to Materials, vols 1–3, Wiley-VCH, Weinheim
[8] Muller A, Peters F, Pope MT, Gatteschi D (1998) Chem Rev **98**: 239
[9] Solomon EI (1988) Metal Cluster in Proteins, ACS Symposium Series 372, Washington DC
[10] Huber R (1989) Angew Chem, Int Ed Engl **101**: 848
[11] Haase W (1984) J Mol Catal **23**: 331
[12] Itoh T, Hisada H, Sumiya T, Hosono M, Usui Y, Fujii Y (1997) Chem Commun: 677
[13] Tan XS, Fujii Y, Sato T, Nakano Y, Yashiro M (1999) Chem Commun: 881
[14] Kitajima N, Moro-oka Y (1994) Chem Rev **94**: 737
[15] Kaim W, Rall J (1996) Angew Chem, Int Ed Engl **35**: 43
[16] Hodgson DJ (1975) Prog Inorg Chem **19**: 173
[17] Crawford VH, Richardson HW, Wasson JR, Hodgson DJ, Hatfield WE (1976) Inorg Chem **15**: 2107
[18] Chaudhuri P, Winter M, Della Védova BPC, Bill E, Trautwein A, Gehring S, Fleischhauer P, Nuber B, Weiss J (1991) Inorg Chem **30**: 2149
[19] Figgis BN, Martin DJ (1972) J Chem Soc, Dalton Trans: 2174
[20] Haase W, Gehring S (1985) J Chem Soc, Dalton Trans: 2609
[21] Okawa H, Koikawa M, Kida S, Luneau D, Oshio H (1990) J Chem Soc, Dalton Trans: 469
[22] Otieno T, Rettig SJ, Thompson RC, Trotter J (1995) Inorg Chem **34**: 1719
[23] Ruiz R, Sanz J, Cervera B, Lloret F, Julve M, Bois C, Faus J, Muñoz MC (1993) J Chem Soc, Dalton Trans: 1623
[24] Ruiz E, Alemany P, Alvarez S, Cano J (1997) J Am Chem Soc **119**: 1297
[25] Ruiz E, Alemany P, Alvarez S, Cano J (1997) Inorg Chem **36**: 3683
[26] Ruiz E, Cano J, Alvarez S, Alemany P (1999) J Comput Chem **20**: 1391
[27] Ruiz E, Alvarez S, Rodríguez-Fortea A, Alemany P, Pouillon Y, Massobrio C (2001) Magnetism: Molecules to Materials. In: Miller JS, Drillon M (2001) Magnetism: Molecules to Materials, vol 1, Wiley-VCH, Weinheim
[28] Cano J, Rodríguez-Fortea A, Alemany P, Alvarez S, Ruiz E (2000) Chem Eur J **6**: 327
[29] Becke AD (1993) J Chem Phys **98**: 5648
[30] GAUSSIAN98 (rev A7) Frisch MJ, Trucks GW, Schlegel HB, Scuseria GE, Robb MA, Cheeseman JR, Zakrzewski VG, Montgomery JA, Stratmann RE, Burant JC, Dapprich S, Millam JM, Daniels AD, Kudin KN, Strain MC, Farkas O, Tomasi J, Barone V, Cossi M, Cammi R, Menucci B, Pomelli C, Adamo C, Clifford S, Ochterski J, Petersson GA, Ayala PY, Cui Q,

Morokuma K, Malick DK, Rabuck AD, Raghavachari K, Foresman JB, Ciolowski J, Ortiz JV, Stefanov BB, Liu G, Liashenko A, Piskorz P, Komaromi I, Gomperts R, Martin RL, Fox DJ, Keith TA, Al-Laham MA, Peng CY, Nanayakkara A, Gonzalez C, Challacombe M, Gill PMW, Johnson BG, Chen W, Wong MW, Andres JL, Head-Gordon M, Replogle ES, Pople JA (1998) Gaussian, Pittsburgh, PA

[31] Schaefer A, Huber C, Ahlrichs R (1994) J Chem Phys **100**: 5829

[32] Schaefer A, Horn H, Ahlrichs R (1992) J Chem Phys **97**: 2571

Localized Orbitals *vs.* Pseudopotential-Plane Waves Basis Sets: Performances and Accuracy for Molecular Magnetic Systems

Carlo Massobrio[1,*] and **Eliseo Ruiz**[2]

[1] Institut de Physique et de Chimie des Matériaux de Strasbourg (IPCMS), F-67037 Strasbourg, France

[2] Departement de Química Inorgànica and Centre Especial de Recerca en Química Teorica (CeRQT), Universitat de Barcelona, E-08028 Barcelona, Spain

Received August 5, 2002; accepted August 9, 2002
Published online November 21, 2002

Summary. Density functional theory, in combination with *a*) a careful choice of the exchange-correlation part of the total energy and *b*) localized basis sets for the electronic orbitals, has become the method of choice for calculating the exchange-couplings in magnetic molecular complexes. Orbital expansion on plane waves can be seen as an alternative basis set especially suited to allow optimization of newly synthesized materials of unknown geometries. However, little is known on the predictive power of this scheme to yield quantitative values for exchange coupling constants J as small as a few hundredths of eV (50–300 cm^{-1}). We have used density functional theory and a plane waves basis set to calculate the exchange couplings J of three homodinuclear Cu-based molecular complexes with experimental values ranging from $+40\,cm^{-1}$ to $-300\,cm^{-1}$. The plane waves basis set proves as accurate as the localized basis set, thereby suggesting that this approach can be reliably employed to predict and rationalize the magnetic properties of molecular-based materials.

Keywords. Molecular modeling; Density functional calculations; Molecular magnetism; Magnetic properties.

Introduction

Molecular magnetism owes a large share of its impact on inorganic chemistry and material science to the growing capabilities of theoretical tools in predicting magneto-structural correlations for compounds of increasing complexity [1]. We refer here in particular to open-shell molecular structures containing paramagnetic centers coordinated through closed-shell ligands. The direct calculation of exchange

* Corresponding author. E-mail: Carlo.Massobrio@ipcms.u-strasbg.fr

couplings is nowadays possible via electronic structure-based techniques, by providing accurate values for quantities that are decisive in the design of molecular-scale magnets. Although in principle less accurate than *ab initio* multiconfiguration approaches as the configuration-interaction (CI), density functional (DFT) methods allow to investigate a wider range of systems sizes [1]. This includes fairly small homo- and heterodinuclear molecules as well as much larger polynuclear compounds and periodic solids. For these reasons, DFT has to be preferred when the primary goal is to obtain reliable quantitative estimate of exchange couplings as small as 50–100 cm^{-1} (0.005–0.01 eV) at an affordable computational cost, within an accessible and yet rigorous theoretical framework.

When adopting a DFT scheme among those made available in computational packages, two kinds of considerations are in order. On the one hand, the performances of DFT are crucially related to the quality of the exchange-correlation part of the energy functional. On the other hand, the desired level of manageability and computational convenience determines the choice of a basis set. For isolated magnetic molecules or magnetic crystals made of weakly interacting molecules, the two above criteria are largely fulfilled by the use of the hybrid B3LYP exchange correlation functional [2] in conjunction with localized orbitals as a basis set. These ingredients are implemented within the Gaussian code [3]. The evaluation of the singlet state $S = 0$ energy is carried out by using the so-called broken-symmetry (*i.e.* spin polarized with total spin $S = 0$) solution (BS) [4]. Indeed, a large amount of calculations has proved that the BS-DFT approach is best suited to provide quantitative estimates of J that are as reliable as those obtained from highly sophisticated (and much more expensive computationally) multireference *ab initio* calculations [1].

Experimentally, the magnetic susceptibility measurements are obtained from solid samples in which the packing forces can produce small deviations from the structure of the isolated molecule. Hence, in the calculations, neither the model nor the complete structures are optimized since very small variations in the geometry such as those induced by the packing forces can produce large changes in the calculated magnetic constant. Given the availability of structural data, calculations are typically performed for a fixed set of experimental coordinates. Along the same lines and by using the CRYSTAL [5] package, which is the periodic extension of the Gaussian scheme, exchange couplings can be obtained for periodic solids. In this case, several spin configurations corresponding to the same total spin (as in antiferromagnetic or ferrimagnetic situations) need to be taken into account. Then, the collection of total energies data resulting from the different spin configurations have to be mapped to an appropriate spin Hamiltonian.

The DFT schemes implemented in Gaussian and CRYSTAL are well suited to find minima in the energy if the starting configurations are close to local configurational minima. However, the methods based on plane waves are usually implemented with more efficient strategies to handle complex structural searches. This is for instance the case of a new family of solid-state molecular-based magnetic materials, the transition metal hydroxides $M_2(OH)_3X$, where M is a divalent transition metal ion and X an exchangeable anion [6, 7]. As in the case of copper hydroxide acetate $Cu_3(OH)_3CH_3COO$, knowledge of the atomic structure of these solid-state compounds is often incomplete [8]. Accordingly, it appears appropriate to tackle the atomic-scale study of molecular magnetic materials characterized by unknown or poor structural determination by resorting to plane waves (PW), more adapted to a flexible and efficient search of structural minima. The PW basis set is implemented for a periodic cell, by leading to the following expression for the single-particle orbitals entering the *Kohn-Sham* DFT Hamiltonian:

$$\phi_i^{\mathbf{k}} = e^{i\mathbf{kr}} \sum_{\mathbf{g}} c_i^{\mathbf{k}}(\mathbf{g}) e^{i\mathbf{gr}} \tag{1}$$

where $\mathbf{g}$ is a reciprocal lattice vector of the period cell, the wave vector $\mathbf{k}$ lies in the *Brillouin* zone of the reciprocal lattice of the periodic cell and the sum runs over the *Fourier* components $c(\mathbf{g})$ of the single-particle orbital. The advantage of PW over a localized basis set stems from the use of fast *Fourier* transform techniques for the calculation of the energy and the forces acting on the ion, operations which are simplified in this scheme since the PW do not depend on ionic coordinates [9]. PW are taken to represent valence orbitals, while the interaction of the valence electrons with the atomic cores is described by pseudopotentials. It should be stressed that pseudopotentials are devised to modify the potential felt by the valence electrons in the core region. The original potential would require an exceedingly high number of plane waves to account for the rapid variation of the valence states at very small distances. Such strong oscillations are induced by the orthogonalization condition between valence and core states. Because of the use of pseudopotentials, this effect is eliminated and the sum of Eq. (1) can be truncated to include an affordable number of plane waves, *i.e.* those with a kinetic energy $E_{\text{kin}} = 0.5(\mathbf{k} + \mathbf{g})^2$ less than a given energy E_{cut}. The value of E_{cut} depends on the nature of the system and on the quality of the pseudopotential [10]. PW are appealing and "friendly-user" since the convergence and the resulting accuracy of the calculations simply depend on the number of *Fourier* components included in the summation of Eq. (1).

By using PW and pseudopotentials, electronic and magnetic properties of copper hydroxonitrate $Cu_2(OH)_3NO_3$ have recently been investigated, providing an instructive picture of frustration effects in a prototypical layered magnetic material [11]. Turning to the quantitative estimate of the exchange coupling constants, the question arises on the predictive power of the PW scheme to evaluate quantities that can be as small as 0.05 eV. Typically, for solid-state and cluster science applications, energy values within this range are not significantly different, as it is the case when different isomers of a same cluster are obtained through structural optimization procedures. A source of error commonly invoked is related to the pseudopotential construction, characterized by a level of tolerance in reproducing the all-electron results close to a few hundredths of eV. We conclude that no clear-cut information exists on the accuracy of DFT-PW approaches for the calculation of singlet-triplet energy splittings in magnetic molecular systems.

In this work we assess the accuracy of DFT-PW calculations for magnetic molecular materials by calculating the exchange coupling of three isolated Cu_2-based complexes. For these molecules, both experimental and DFT results with localized basis have been obtained, yielding J values as low as $50\,\text{cm}^{-1}$. We are able to conclude that the DFT-PW scheme is well suited to provide quantitative information on the exchange couplings and that it can be reliably employed with the same purpose for larger systems and solid-state compounds.

Method

Our calculations are performed in the *Kohn-Sham* DFT framework with the BLYP generalized gradient approximation due to *Becke* (B) [12] for the exchange part and to *Lee*, *Yang* and *Parr* (LYP) [13] for the correlation part. The account of spin polarization for $S=0$ calculations corresponds to the broken-symmetry approach referred to in DFT applications of quantum chemistry [1]. All valence electrons are treated explicitly and on the same footing, *i.e. via Fourier* expansion of the associated orbitals in plane waves at the $\mathbf{k}=0$ point of the supercell. Norm conserving pseudopotentials generated following the scheme by *Trouiller* and *Martins* are used to account for core-valence interaction [14]. An energy cutoff E_{cut} equal to 90 Ry yields converged binding energies, equilibrium distances d_e and vibrational frequency ω for the Cu_2 and CuO dimers that were used in test calculations. Small differences are found when comparing our calculated d_e and ω of Cu_2 and CuO with the corresponding quantities obtained *via* the Gaussian-localized basis scheme. For Cu_2 the values are $d_e=2.27$ Å (PW) and $d_e=2.26$ Å (Gaussian), $\omega=270\,\text{cm}^{-1}$ (PW) and $\omega=256\,\text{cm}^{-1}$ (Gaussian). For CuO the values are $d_e=1.73$ Å (PW) and $d_e=1.75$ Å (Gaussian), $\omega=660\,\text{cm}^{-1}$ (PW) and $\omega=641\,\text{cm}^{-1}$ (Gaussian). When calculating the properties of an isolated system within a PW scheme, the simulation box has to be chosen large enough so as to minimize the interaction with the periodic images. This can be checked by recalculating the properties of interest for systems of increasing size. In our case, a converged value for the exchange coupling J is attained for a box length $L=18.5$ Å. We recall that J is defined as

$$J = E(S = 0) - E(S = 1) \tag{2}$$

where $E(S=0)$ and $E(S=1)$ are the total energies for the singlet and triplet state respectively. The above definition corresponds to the non spin-projected energy [15]. As discussed by several authors, the best results from DFT methods when calculating the singlet-triplet gap are obtained when the broken symmetry state is considered without spin projection, *i.e.* by using Eq. (2) [16–18]. This fact is due the inclusion of the non-dynamical correlation effects in the commonly used exchange functionals. The use of the spin projection simultaneously produces a suppression of such effects, as discussed recently by *Polo et al.* [19].

For each molecule, atomic positions are taken equal to the experimental values. Total energy minimization is carried out with respect to the coefficient of the PW expansion and stops when the largest gradient is smaller than 10^{-6} a.u. Our calculations are carried out by using the CPMD plane waves code [20]. For one of the studied systems, we have performed calculations with the Gaussian code, in order to complete previous results of Ref. [1]. We used analytical small core pseudopotentials proposed by *Hay* and *Wadt* [21] combined with B3LYP [2] and BLYP [12, 13] functional (using the LANL2DZ keyword in Gaussian code [3]) and all electron calculations using the PBE [22] and PBE0 [23] functionals.

Results and Discussion

The first molecule we considered is copper(II) acetate (Fig. 1). Copper(II) acetate has been adopted in extensive benchmark calculations comparing different electronic structure approaches for magnetic molecular materials [1]. The results reported in Ref. [1] are shown in Table 1, together with the value calculated in this work. The optimal performances of the hybrid B3LYP functional ($J = -299\,\mathrm{cm}^{-1}$ against $J = -297\,\mathrm{cm}^{-1}$, experimental value [24]) are clearly highlighted by the data. The present PW-BLYP calculations yield a J value ($J = -518\,\mathrm{cm}^{-1}$) which is closer to the experiment than the one obtained within Gaussian with the same functional using all electron basis sets or using *Hay-Wadt* pseudopotentials ($J = -779$ and $-754\,\mathrm{cm}^{-1}$, respectively). This change in the localized basis sets does not produce the same effect on the BLYP and B3LYP functionals, since for the BLYP the all electron value is more negative, while the opposite is found for the B3LYP one. The use of other GGA schemes with the localized basis sets, such as the PBE functional, also provides values close to the BLYP result with the same basis set. Due to the larger exact exchange contribution than within the B3LYP functional, the free parameter hybrid functional PBE0 gives less negative J values. We note that the inclusion of the exact exchange improves considerably the results in comparison with the other GGA functionals when using localized basis sets.

In a second test case, which belongs to the family of the extensively studied hydroxo-bridged Cu(II) dinuclear complexes [25], we considered a Cu-based dinuclear transition metal molecular complex, $[Cu_2(\mu\text{-}OH)_2(bipym)_2](NO_3)_2 4H_2O$ (*bipym* = bipyrimidine) (Fig. 2). In Ref. [15] the exchange coupling for this molecule was calculated *via* several energy functionals to elucidate the differences between the BLYP results and those obtained with hybrid methods. These latter were found to be dependent on the amount of exact exchange included.

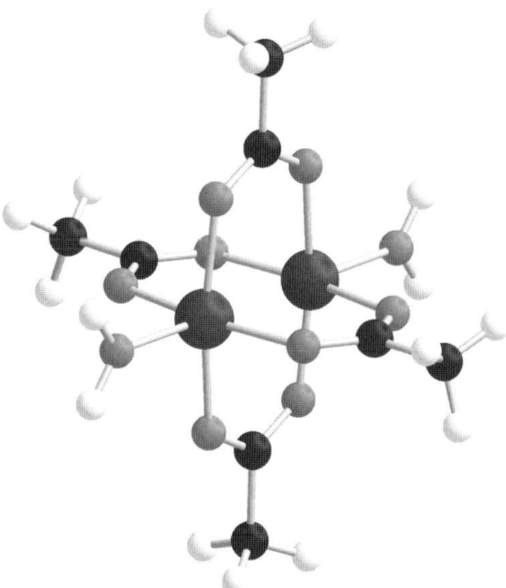

Fig. 1. Representation of the structure of the copper acetate molecule

Table 1. Value of exchange coupling constant, J, for copper(II) acetate calculated using several *ab initio* and density functional methods. Most of the results are taken from Ref. [1]. The calculation performed with the *Hay-Wadt* pseudopotentials combined with localized basis set are labeled as "pseudo". The BLYP-PW result corresponds to that obtained with the plane wave approach. All DFT and HF calcalations make use of the broken symmetry approach

Method	J (cm^{-1})
Experimental (Ref. [24])	−297
DDCI-2 (Ref. [1])	−77
DDCI-3 (Ref. [1])	−224
CASSCF (Ref. [1])	−24
CASPT2 (Ref. [1])	−117
UHF (Ref. [1])[a]	−54
Xα (Ref. [1])	−848
SVWN (Ref. [1])	−1057
B3LYP (Ref. [1])	−299
BLYP (Ref. [1])	−779
B3LYP pseudo	−346
BLYP pseudo	−754
PBE	−776
PBE0	−237
BLYP-PW	−518

[a] Using spin projection

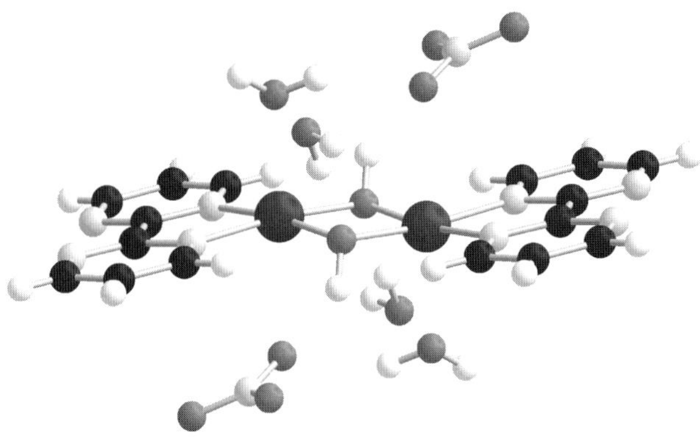

Fig. 2. Representation of the structure of $Cu_2(\mu\text{-}OH)_2(bipym)_2](NO_3)_2 4H_2O$

Again, the best agreement with experiments was obtained *via* the B3LYP functional in the spin non-projected version, which performs much better than spin-projected unrestricted *Hartree-Fock* calculations [15]. As shown in Table 2, the PW approach improves upon the localized basis one, yielding a value for J closer to experiments (BLYP-PW, $J = +95\,cm^{-1}$, experiment, $J = +114\,cm^{-1}$ [26]). A non-negligible difference is found between the two BLYP results ($+95\,cm^{-1}$ against $+221\,cm^{-1}$ with Gaussian).

In a final test, we look for a third molecule containing a pair of Cu centers and a smaller coupling constant, $[(dpt)Cu(\mu\text{-}Cl)_2Cu(dpt)]Cl_2$, ($dpt$ = dipropylenetriamine) (Fig. 3). This complex is an interesting example of dinuclear copper(II)

Table 2. Value of the exchange coupling constant, J, for $[Cu_2(\mu\text{-}OH)_2(bipym)_2](NO_3)_2$, calculated *via* DFT (broken symmetry) methods within Gaussian (Ref. [15]) and by using a PW basis set within DFT-BLYP (present calculations)

Method	J (cm^{-1})
Experimental (Ref. [26])	−114
B3LYP	−113
BLYP	−221
BLYP-PW	−95

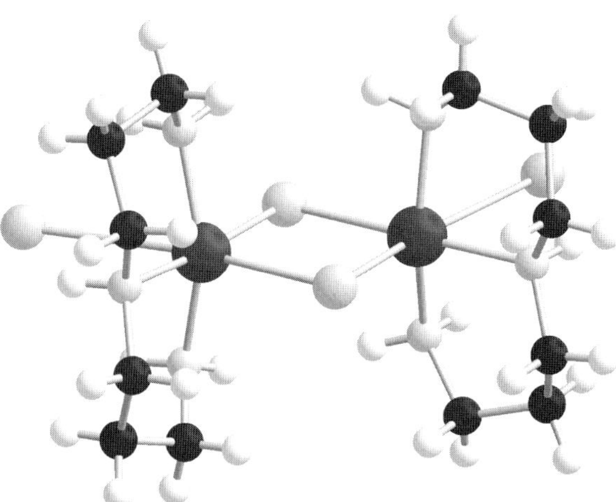

Fig. 3. Representation of the structure of $[(dpt)Cu(\mu\text{-}Cl)_2Cu(dpt)]Cl_2$, ($dpt$ = dipropylenetriamine)

system containing chloro bridging ligands with a $Cu(\mu\text{-}Cl)_2Cu$ motif. In Ref. [27] magnetic susceptibility data showed a ferromagnetic coupling between the two Cu centers, as corroborated by extended *Hückel* calculations. The data collected in Table 3 confirms that the combination of PW and BLYP is capable of reproducing fairly accurately the experimental J (BLYP-PW, $J = +61\,cm^{-1}$, experiment, $J = +42.9\,cm^{-1}$). As in the previous case, the value obtained *via* PW is about a factor of 2 lower than the one resulting from the localized orbital calculations with the same energy functional ($J = +100\,cm^{-1}$, Ref. [28]). While

Table 3. Value of the exchange coupling constant, J, for $[(dpt)Cu(\mu\text{-}Cl)_2Cu(dpt)]Cl_2$, calculated *via* DFT (broken symmetry) methods within Gaussian (Ref. [28]) and by using a PW basis set within DFT-BLYP (present calculations)

Method	J (cm^{-1})
Experimental (Ref. [27])	+42.9
B3LYP	+56
BLYP	+100
BLYP-PW	+61

further investigation are in order to elucidate the origins of this discrepancy, we recall that the use of pseudopotentials corresponds to the neglect of an explicit description of the atomic cores. On a quantitative basis, this may inhibit comparison with Gaussian basis results, as recently pointed out [29].

Overall, these results demonstrate that the DFT-PW is a reliable and accurate tool to calculate exchange coupling of transition metal complexes as small as $\sim 50\,\mathrm{cm}^{-1}$. A general recipe to investigate theoretically magnetic molecular-based materials becomes available. Provided the atomic structure is experimentally known, the DFT-PW method can be used directly to extract the exchange couplings from the mapping of a suitable spin Hamiltonian. For this, a collection of microscopic spin configurations compatible with the total magnetic character of the system is required. For each microscopic configuration, the ground state electronic structure can be efficiently attained within the DFT-PW scheme. Whenever a knowledge of the atomic structure is not available, structural optimization becomes a necessary prerequisite. The DFT-PW scheme is well suited to this task, allowing for a simultaneous optimization of both electronic and ionic degrees of freedom. This step can be followed by the determination of the coupling constant along the lines sketched above.

Despite the encouraging indications of this study the DFT-PW approach is also characterized by a few weak points that do not have to be overlooked. First and foremost the above mentioned difficulty to reproduce at the same level of accuracy all-electron results by using PW and pseudopotentials. In addition, we recall that using plane waves as a basis set corresponds to filling all portions of space with the same density of *Fourier* components, irrespective of the presence of atoms. The number of plane waves N_{pw} grows linearly with the volume and follows a power law $(E_{cut})^{3/2}$ as a function of the energy cutoff. For the third molecule we considered, $[(dpt)Cu(\mu\text{-}Cl)_2Cu(dpt)]Cl_2$, this cutoff amounts to as much as $2.4 \cdot 10^6$ plane waves for the expansion of the electronic density. In this case, the calculation of the electronic ground state required 5 hours of CPU time on 48 processors of an IBM-SP4 computer. These data point out the relatively high cost of these runs, together with the need of powerful parallel computers.

It was already mentioned that the hybrid B3LYP is the functional of choice for calculations of exchange couplings within the DFT formalism, at least in most transition metal complexes. However, this functional has not yet become operational in a PW code, as it would be desirable to exploit the flexibility of the plane waves basis set at a quantitative level. As a further point, we note that the non-localized character of this basis set prevents from a straightforward assignment of the initial spin density distribution. This is necessary to build up the catalog of all microscopic spin topologies corresponding to a given value for the total spin. Finally, by focusing on isolated molecules, it has to be reminded that the periodicity of the calculation cell is not well suited to calculations for charged systems, because of the long-range *Coulomb* interaction between the periodic images. Given the current pace of developments and improvements in this area, none of these shortcomings are stumbling blocks [29, 30]. Therefore, we are confident on the appearance of an increased number of DFT-PW applications for the theoretical study of molecular-based magnetic materials.

Acknowledgments

One of the authors (*CM*) is indebted to Mauro Boero (RICS-AIST, Tsukuba, Japan) for enlightening exchanges and help with the use of pseudopotentials. Calculations have been carried out on the facilities of the IDRIS and CINES French national computer centers. Financial help within the PICASSO France-Spain bilateral agreement is gratefully acknowledged.

References

[1] Ruiz E, Alvarez S, Rodríguez-Fortea A, Alemany P, Pouillon Y, Massobrio C (2001) Electronic Structure and Magnetic Behavior in Polynuclear Transition-metal Compounds. In: Miller JS, Drillon M, Magnetism: Molecules to Materials II, Molecule-Based Materials, Wiley-VCH, p 227

[2] Becke AD (1993) J Chem Phys **98**: 5648

[3] Gaussian98 (1998): Frisch MJ, Trucks GW, Schlegel HB, Scuseria GE, Robb MA, Cheeseman JR, Zakrzewski VG, Montgomery JA, Stratmann RE, Burant JC, Dapprich S, Millam JM, Daniels AD, Kudin KN, Strain MC, Farkas O, Tomasi J, Barone V, Cossi M, Cammi R, Mennucci B, Pomelli C, Adamo C, Clifford S, Ochterski J, Petersson GA, Ayala PY, Cui Q, Morukuma K, Malick DK, Rabuck AD, Raghavachari K, Foresman JB, Ciolowski J, Ortiz JV, Stefanov BB, Liu G, Liashenko A, Piskorz P, Komaroni I, Gomperts R, Martin RL, Fox DJ, Keith T, Al-Laham MA, Peng CY, Nanayakkara A, Gonzalez C, Challacombe M, Gill PMW, Johnson BG, Chen W, Wong MW, Andres JL, Head-Gordon M, Replogle ES, Pople JA, Gaussian, Pittsburgh, PA (USA)

[4] Noodelmann L (1981) J Chem Phys **74**: 5737

[5] CRYSTAL98 (1999) Saunders VR, Dovesi R, Roetti C, Causà M, Harrison NM, Orlando R, Zicovich-Wilson CM, Università di Torino, CCLRC Daresbury Laboratory

[6] Rabu P, Drillon M, Awaga K, Fujita W, Sekine T (2001) Hybrid organic-Inorganic Multilayer Compounds: Towards Controllable and/or Switchable Magnets. In: Miller JS, Drillon M, Magnetism: Molecules to Materials II, Molecule-Based Materials, Wiley-VCH, p 357

[7] Rabu P, Rouba S, Laget V, Hornick C, Drillon M (1996) J Chem Soc Chem Comm 1107

[8] Masciocchi N, Corradi E, Sironi A, Moretti G, Minelli G, Porta P (1997) J Solid State Chemistry **131**: 252

[9] Galli G, Parrinello M (1992) Ab-initio Molecular Dynamics: Principles and Practical Implementation. In: Meyer M, Pontikis V, Computer Simulation in Material Science, Interatomic Potentials, Simulations Techniques and Applications, NATO-ASI Series, series E: Applied Science, vol 205, Kluwer Academic Publishers, p 283

[10] Gillan MJ (1992) Calculating the Properties of Materials from Scratch. In: Meyer M, Pontikis V, Computer Simulation in Material Science, Interatomic Potentials, Simulations Techniques and Applications, NATO-ASI Series, series E: Applied Science, vol 205, Kluwer Academic Publishers, p 257

[11] Massobrio C, Rabu P, Drillon M, Rovira C (1999) J Phys Chem B **103**: 9387

[12] Becke AD (1988) Phys Rev A **38**: 3098

[13] Lee C, Yang W, Parr RG (1988) Phys Rev B **37**: 785

[14] Trouiller N, Martins JL (1991) Phys Rev B **43**: 1993

[15] Ruiz E, Cano J, Alvarez S, Alemany P (1999) J Comp Chem **20**: 1391

[16] Wittbrodt JM, Schlegel HB (1996) J Chem Phys **105**: 6574

[17] Perdew JP, Savin A, Burke K (1995) Phys Rev A **51**: 4531

[18] Perdew JP, Erzerhof M, Burke K, Savin A (1997) Int J Quantum Chem **61**: 197

[19] Polo V, Kraka E, Cremer E (2002) Theor Chem Acc **107**: 291

[20] CPMD 3.0f (1998) Hutter J, Ballone P, Bernasconi M, Focher P, Fois E, Goedecker S, Marx D, Parrinello M, Tuckerman M

[21] Wadt WR, Hay PJ (1985) J Chem Phys **82**: 370

[22] Perdew JP, Burke K, Ernzerhof M (1996) Phys Rev Lett **77**: 3865

[23] Perdew JP, Burke K, Ernzerhof M (1997) Phys Rev Lett **78**: 13965
[24] Güdel HU, Stebler A, Fürrer A (1979) Inorg Chem **18**: 21
[25] Ruiz E, Alemany P, Alvarez S, Cano J (1997) J Am Chem Soc **119**: 1297
[26] de Munno G, Julve M, Lloret F, Faus J, Verdaguer M, Caneschi A (1995) Inorg Chem **34**: 157
[27] Rodríguez M, Llobet A, Corbella M, Martell AE, Reibenspies J (1999) Inorg Chem **38**: 2328
[28] Rodríguez-Fortea A, Alemany P, Alvarez S, Ruiz E (2002) Inorg Chem **41**: 3769
[29] Martyna GJ, Tuckermann ME (1999) J Chem Phys **110**: 2810
[30] Füsti-Molnar L, Pulay P (2002) J Chem Phys **116**: 7795

SpringerChemistry

Wolfgang Linert (ed.)

Highlights in Solute-Solvent Interactions

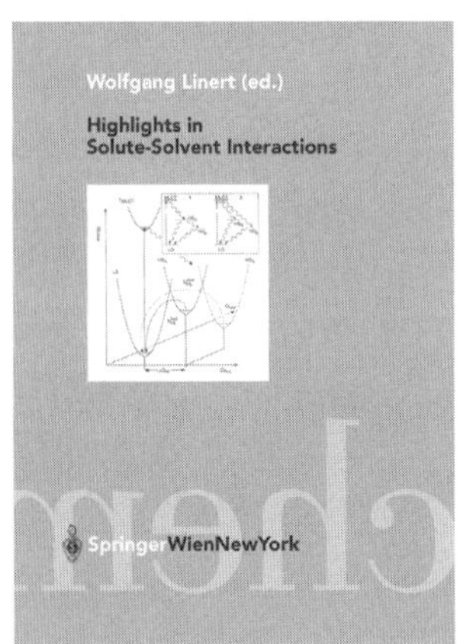

With a Foreword by Henry Taube.
2002. IX, 214 pages. 87 figures.
Hardcover EUR 116,–
(Recommended retail price)
Net-price subject to local VAT.
(Special Edition of "Monatshefte für Chemie/
Chemical Monthly", Vol.132, No. 11, 2001)
ISBN 3-211-83731-0

This book emphasizes a broad spectrum of features in solution chemistry, reaching from recent developments in the empirical characterization of solvent-solute interactions up to modern theoretical descriptions of liquid-state systems, from solid-liquid interfaces to preferential solvation in mixed solvents. Accordingly, this collection presents the most important, though strongly different approaches to the understanding of the mutual influences between solute and solvent. Descriptions of actual and practically useful applications of these concepts are included.

Please visit our website: **www.springer.at**

A-1201 Wien, Sachsenplatz 4–6, P.O. Box 89, Fax +43.1.330 24 26, e-mail: books@springer.at, Internet: **www.springer.at**
D-69126 Heidelberg, Haberstraße 7, Fax +49.6221.345-229, e-mail: orders@springer.de
USA, Secaucus, NJ 07096-2485, P.O. Box 2485, Fax +1.201.348-4505, e-mail: orders@springer-ny.com
Eastern Book Service, Japan, Tokyo 113, 3–13, Hongo 3-chome, Bunkyo-ku, Fax +81.3.38 18 08 64, e-mail: orders@svt-ebs.co.jp

SpringerChemistry

Microchimica Acta

An International Journal on Micro and Trace Analysis

Microchimica Acta is an international journal that publishes articles covering all aspects of modern microanalytical sciences, including fundamental studies, practical applications, and new instrumental approaches. It has a strong focus on analytical methods that apply to microchemistry, such as new spectroscopic, electroanalytical, or sensory schemes. It also reports on methods with extremely low limits of detection, on microtechniques and on nanotechniques. Articles cover modern applications of such techniques to areas like environmental, clinical, or process analysis, to geochemistry, material sciences, or in studies on objects of art.

Subscription Information
ISSN 0026-3672 (print), ISSN 1436-5073 (electronic)
2003. Vols. 143–145 (4 issues each). Title-No. 604
EUR 1544,– plus carriage charges

View table of contents and abstracts online at: **www.springer.at/mca**

SpringerWienNewYork

A-1201 Wien, Sachsenplatz 4–6, P.O. Box 89, Fax +43.1.330 24 26, e-mail: journals@springer.at, Internet: **www.springer.at**
USA, Secaucus, NJ 07096-2485, P.O. Box 2485, Fax +1.201.348-4505, e-mail: orders@springer-ny.com
Eastern Book Service, Japan, Tokyo 113, 3–13, Hongo 3-chome, Bunkyo-ku, Fax +81.3.38 18 08 64, e-mail: orders@svt-ebs.co.jp